AF879941

FORD FALCON
289 - 302 V8
OWNERS WORKSHOP MANUAL
1965-1971

Models covered:
Ford Falcon 289 V8 1965-1971
Ford Falcon 302 V8 1965-1971

ISBN 9780851472638

Brooklands Books

BROOKLANDS BOOKS LTD.
P.O. BOX 904, AMERSHAM,
BUCKS, HP6 9JA. UK

FO71WH www.brooklandsbooks.com 12W5/2893

INTRODUCTION

This do-it-yourself Workshop Manual has been specially written for the owner who wishes to maintain his car in first class condition and to carry out his own servicing and repairs. Considerable savings on garage charges can be made, and one can drive in safety and confidence knowing the work has been done properly.

Comprehensive step-by-step instructions and illustrations are given on all dismantling, overhauling and assembling operations. Certain assemblies require the use of expensive special tools, the purchase of which would be unjustified. In these cases information is included but the reader is recommended to hand the unit to the agent for attention.

Throughout the Manual hints and tips are included which will be found to be invaluable, and there is an easy to follow fault diagnosis at the end of each chapter.

Whilst every care has been taken to ensure correctness of information it is obviously not possible to guarantee complete freedom from errors or to accept liability arising from such errors or omissions.

Instructions may refer to the right-hand or left-hand sides of the vehicle or the components. These are the same as the right-hand or left-hand of an observer standing behind the car and looking forward.

By the Autobooks Team of Technical Writers

Brooklands Books Ltd., PO Box 904, Amersham, Bucks, England. HP6 9JA

brooklands-books.com

www.brooklandsbooks.com

CONTENTS

INDEX

HINTS ON MAINTENANCE AND OVERHAUL

There are few things more rewarding than the restoration of a vehicle's original peak of efficiency and smooth performance.

The following notes are intended to help the owner to reach that state of perfection. Providing that he possesses the basic manual skills he should have no difficulty in performing most of the operations detailed in this manual. It must be stressed, however, that where recommended in the manual, highly-skilled operations ought to be entrusted to experts, who have the necessary equipment, to carry out the work satisfactorily.

Quality of workmanship:
The hazardous driving conditions on the roads today demand that vehicles should be as nearly perfect, mechanically, as possible. It is therefore most important that amateur work be carried out with care, bearing in mind the often inadequate working conditions, and also the inferior tools which may have to be used. It is easy to counsel perfection in all things, and we recognize that it may be setting an impossibly high standard. We do, however, suggest that every care should be taken to ensure that a vehicle is as safe to take on the road as it is humanly possible to make it.

Safe working conditions:
Even though a vehicle may be stationary, it is still potentially dangerous if certain sensible precautions are not taken when working on it while it is supported on jacks or blocks. It is indeed preferable not to use jacks alone, but to supplement them with carefully placed blocks, so that there will be plenty of support if the car rolls off the jacks during a strenuous manoeuvres. Axle stands are an excellent way of providing a rigid base which is not readily disturbed. Piles of bricks are a dangerous substitute. Be careful not to get under heavy loads on lifting tackle, the load could fall. It is preferable not to work alone when lifting an engine, or when working underneath a vehicle which is supported well off the ground. To be trapped, particularly under the vehicle, may have unpleasant results if help is not quickly forthcoming. Make some provision, however humble, to deal with fires. Always disconnect a battery if there is a likelihood of electrical shorts. These may start a fire if there is leaking fuel about. This applies particularly to leads which can carry a heavy current, like those in the starter circuit. While on the subject of electricity, we must also stress the danger of using equipment which is run off the mains and which has no earth or has faulty wiring or connections. So many workshops have damp floors, and electrical shocks are of such a nature that it is sometimes impossible to let go of a live lead or piece of equipment due to the muscular spasms which take place.

Work demanding special care:
This involves the servicing of braking, steering and suspension systems. On the road, failure of the braking system may be disastrous. Make quite sure that there can be no possibility of failure through the bursting of rusty brake pipes or rotten hoses, nor to a sudden loss of pressure due to defective seals or valves.

Problems:
The chief problems which may face an operator are:

1 External dirt.
2 Difficulty in undoing tight fixings
3 Dismantling unfamiliar mechanisms.
4 Deciding in what respect parts are defective.
5 Confusion about the correct order for reassembly.
6 Adjusting running clearances.
7 Road testing.
8 Final tuning.

Practical suggestions to solve the problems:

1. Preliminary cleaning of large parts - engines, transmissions, steering, suspensions, etc., should be carried out before removal from the car. Where road dirt and mud alone are present, wash clean with a high pressure water jet, brushing to remove stubborn adhesions, and allow to drain and dry. Where oil or grease is also present, wash down with a proprietary compound (Gunk, Teepol etc.,) applying with a stiff brush - an old paint brush is suitable, into all crevices. Cover the distributor and ignition coils with a polythene bag and then apply a strong water jet to clear the loosened deposits. Allow to drain and dry. The assemblies will then be sufficiently clean to remove and transfer to the bench for the next stage.

   On the bench, further cleaning can be carried out, first wiping the parts as free as possible from grease with old newspaper. Avoid using rag or cotton waste which can leave clogging fibres behind. Any remaining grease can be removed with a brush dipped in paraffin. Avoid using paraffin or petrol in large quantities for cleaning in enclosed areas, such as garages, on account of the high fire risk.

When all exteriors have been cleaned, and not before, dismantling can be commenced. This ensures that dirt will not enter into interiors and orifices revealed by dismantling. In the next phases, where components have to be cleaned and keep the containers covered except when in use. After the components have been cleaned, plug small holes with tapered hard wood plugs cut to size and blank off larger orifices with greaseproof paper and masking tape. Do not use soft wood plugs or matchsticks as they may break.

2. It is not advisable to hammer on the end of a screw thread, but if it must be done, first screw on a nut to protect the thread, and use a lead hammer. This applies particularly to the removal of tapered cotters. Nuts and bolts seem to 'grow' together, especially in exhaust systems. If penetrating oil does not work, try the judicious application of heat, but be careful of starting a fire.

   Tight bushes or pieces of tail-pipe rusted into a silencer can be removed by splitting them with an open-ended hacksaw. Tight screws can sometimes be started by a tap from a hammer on the end of a suitable screwdriver. Many tight fittings will yield to the judicious use of a hammer, but it must be a soft-faced hammer if damage is to be avoided, use a heavy block on the opposite side to absorb shock. Any parts of the steering system which have been damaged should be renewed, as attempts to repair them may lead to cracking and subsequent failure, and steering ball joints should be disconnected using a recommended tool to prevent damage.

3. It often happens that an owner is baffled when trying to dismantle an unfamiliar piece of equipment. So many modern devices are pressed together or assembled by spinning-over flanges, that they must be sawn apart. The intention is that the whole assembly must be renewed. However, parts which appear to be in one piece to the naked eye, may reveal close-fitting joint lines when inspected with a magnifying glass, and, this may provide the necessary clue to dismantling. Left-handed screw threads are used where rotational forces would tend to unscrew a right handed screw thread.

   Be very careful when dismantling mechanisms which may come apart suddenly. Work in an enclosed space where the parts will be contained, and drape a piece of cloth over the device if springs are likely to fly in all directions. Mark everything which might be reassembled in the wrong position, scratched symbols may be used on unstressed parts, or a sequence of tiny dots from a centre punch can be useful. Stressed parts should never be scratched or centre-popped as this may lead to cracking under working conditions. Store parts which look alike in the correct order for reassembly. Never rely upon memory to assist in the assembly of complicated mechanisms, especially when they will be dismantled for a long time, but make notes, and drawings to supplement the diagrams in the manual, and put labels on detached wires. Rust stains may indicate unlubricated wear. This can sometimes be seen round the outside edge of a bearing cup in a universal joint. Look for bright rubbing marks on parts which normally should not make heavy contact. These might prove that something is bent or running out of truth. For example, there might be bright marks on one side of a piston, at the top near the ring grooves, and others at the bottom of the skirt on the other side. This could well be the clue to a bent connecting rod. Suspected cracks can be proved by heating the component in a light oil to approximately 100°C, removing, drying off, and dusting with French chalk, if a crack is present the oil retained in the crack will stain the French chalk.

4. In determining wear, and the degree, against the permissible limits set in the manual, accurate measurement can only be achieved by the use of a micrometer. In many cases the wear is given to the fourth place of decimals; that is in ten-thousandths of an inch. This can be read by the vernier scale on the barrel of a good micrometer. Bore diameters are more difficult to determine. If, however, the matching shaft is accurately measured, the degree of play in the bore can be felt as a guide to its suitability. In other cases, the shank of a twist drill of known diameter is a handy check.

   Many methods have been devised for determining the clearance between bearing surfaces. Today the best and simplest is by the use of Plastigage, obtainable from most garages. A thin plastic thread is laid between the two surfaces and the bearing is tightened, flattening the thread. On removal, the width of the thread is compared with a scale supplied with the thread and the clearance is read off directly. Sometimes joint faces leak persistently, even after gasket renewal. The fault will then be traceable to distortion, dirt or burrs. Studs which are screwed into soft metal frequently raise burrs at the point of entry. A quick cure for this is to chamfer the edge of the hole in the part which fits over the stud.

5. Always check a replacement part with the original one before it is fitted.

   If parts are not marked, and the order for reassembly is not known, a little detective work will help. Look for marks which are due to wear to see if they can be mated. Joint faces may not be identical due to manufacturing errors, and parts which overlap may be stained, giving a clue to the correct position. Most fixings leave identifying marks especially if they were painted over on assembly. It is then easier to decide whether a nut, for instance, has a plain, a spring, or a shake-proof washer under it. All running surfaces become 'bedded' together after long spells of work and tiny imperfections on one part will be found to have left corresponding marks on the other. This is particularly true of shafts and bearings and even a score on a cylinder wall will show on the piston.

6. Checking end float or rocker clearances by feeler gauge may not always give accurate results because of wear. For instance, the rocker tip which bears on a valve stem may be deeply pitted, in which case the feeler will simply be bridging a depression. Thrust washers may also wear depressions in opposing faces to make accurate measurement difficult. End float is then easier to check by using a dial gauge. It is common practice to adjust end play in bearing assemblies, like front hubs with taper rollers, by doing up the axle nut until the hub becomes stiff to turn and then backing it off a little. Do not use this method with ball-bearing hubs as the assembly is often pre-loaded by tightening the axle nut to its fullest extent. If the split-pin hole will not line up, file the base of the nut a little.

   Steering assemblies often wear in the straight-ahead position. If any part is adjusted, make sure that it remains free when moved from lock to lock. Do not be surprised if an assembly like a steering gearbox, which is known to be carefully adjusted outside the car, becomes stiff when it is bolted in place. This will be due to distortion of the case by the pull of the mounting bolts, particularly if the mounting points are not all touching together. This problem may be met in other equipment and is cured by careful attention to the alignment of mounting points.

   When a spanner is stamped with a size and A/F it means that the dimension is the width between the jaws and has no connection with ANF, which is the designation for the American National Fine thread. Coarse threads like Whitworth are rarely used on cars to-day except for studs which screw into soft aluminium or cast iron. For this reason it might be found that the top end of a cylinder head stud has a fine thread and the lower end a coarse thread to screw into the cylinder block. If the car has mainly UNF threads then it is likely that any coarse threads will be UNC, which are not the same as Whitworth. Small sizes have the same number of threads in Whitworth and UNC, but in the 1/4 inch size for example, there are twelve threads to the inch in the former and thirteen in the latter.

7. After a major overhaul, particularly if a great deal of work has been done on the braking, steering and suspension systems, it is advisable to approach the problem of testing with care. If the braking system has been overhauled, apply heavy pressure to the brake pedal and get a second operator to check every possible source of leakage. The brakes may work extremely well, but a leak could cause complete failure after a few miles.

   Do not fit the hub caps until every wheel nut has been checked for tightness, and make sure the tyre pressures are correct. Check the levels of coolant, lubricants and hydraulic fluids. Being satisfied that all is well, take the car on the road and test the brakes at once. Check the steering and the action of the handbrake. Do all this at moderate speeds on quiet roads, and make sure there is no other vehicle behind you when you try a rapid stop.

   Finally, remember that many parts settle down after a time, so check for tightness of all fixings after the car has been on the road for a hundred miles or so.

8. It is useless to tune an engine which has not reached its normal running temperature. In the same way, the tune of an engine which is stiff after a re-bore will be different when the engine is again running free. Remember too, that rocker clearances on push-rod operated valve gear will change when the cylinder head nuts are tightened after an initial period of running with a new head gasket. Trouble may not always be due to what seems the obvious cause. Ignition, carburation and mechanical condition are interdependent and spitting back through the carburetter, which might be attributed to a weak mixture, can be caused by a sticking inlet valve. For one final hint on tuning, never adjust more than one thing at a time or it will be impossible to tell which adjustment produced the desired result.

CHAPTER 1

THE ENGINE

1:1 Description

All the models covered by this manual are fitted with V8 engines of conventional design. Earlier models are fitted with an engine of 289 cu in displacement. The general design of both sizes of engine is similar in each case. Full details of bore, stroke and other dimensions are given in **Technical Data** at the end of this manual.

A sectioned view of the engine is shown in **FIG 1:1**. If a high-performance engine is fitted, it is similar in design but has solid valve lifters. The crankshaft rotates in five main bearings attached to the crankcase. Four sets of throws and crankpins are integral with the crankshaft, and each crankpin has a pair of opposed connecting rods attached to it. The camshaft is driven, at the front of the engine, by a timing chain and two sprockets, so that it revolves at half crankshaft speed. Renewable bearings are fitted for camshaft, crankshaft and big-ends.

Hydraulic valve lifters, which give zero lash in the valve train, operate from the cams on the camshaft. The motion is transmitted using hollow pushrods with ball ends to the rocker arms on the cylinder heads, and the rocker arms open the valves against the pressure of their return springs.

Firing order and gas flow:

The cylinders are numbered counting from the front of each bank. Those on the right bank are numbered 1, 2, 3, 4 while those on the left bank are numbered 5, 6, 7, 8. The firing order is 1-5-4-2-6-3-7-8.

Each barrel of the carburetor feeds a set of cylinders through the manifold passages, as shown in **FIG 1:2**. Initial heating of the inlet manifold is by exhaust gases passing through the gas crossover passage in the inlet manifold, shown in **FIG 1:3**. Coolant flows through the manifold to ensure that it is heated when the engine is hot. Warming of the inlet manifold ensures full vaporization of the fuel droplets before they enter the combustion chamber.

Inlet valves can be identified by their larger heads, and the porting arrangement is shown in **FIG 1:4**. For models fitted with an automatic choke, the heat chamber is shown in **FIG 1:5**.

Lubrication:

A sectioned view of the engine showing a typical lubrication system and oil flow is shown in **FIG 1:6**. The oil is drawn from the sump of the oil pan, through a

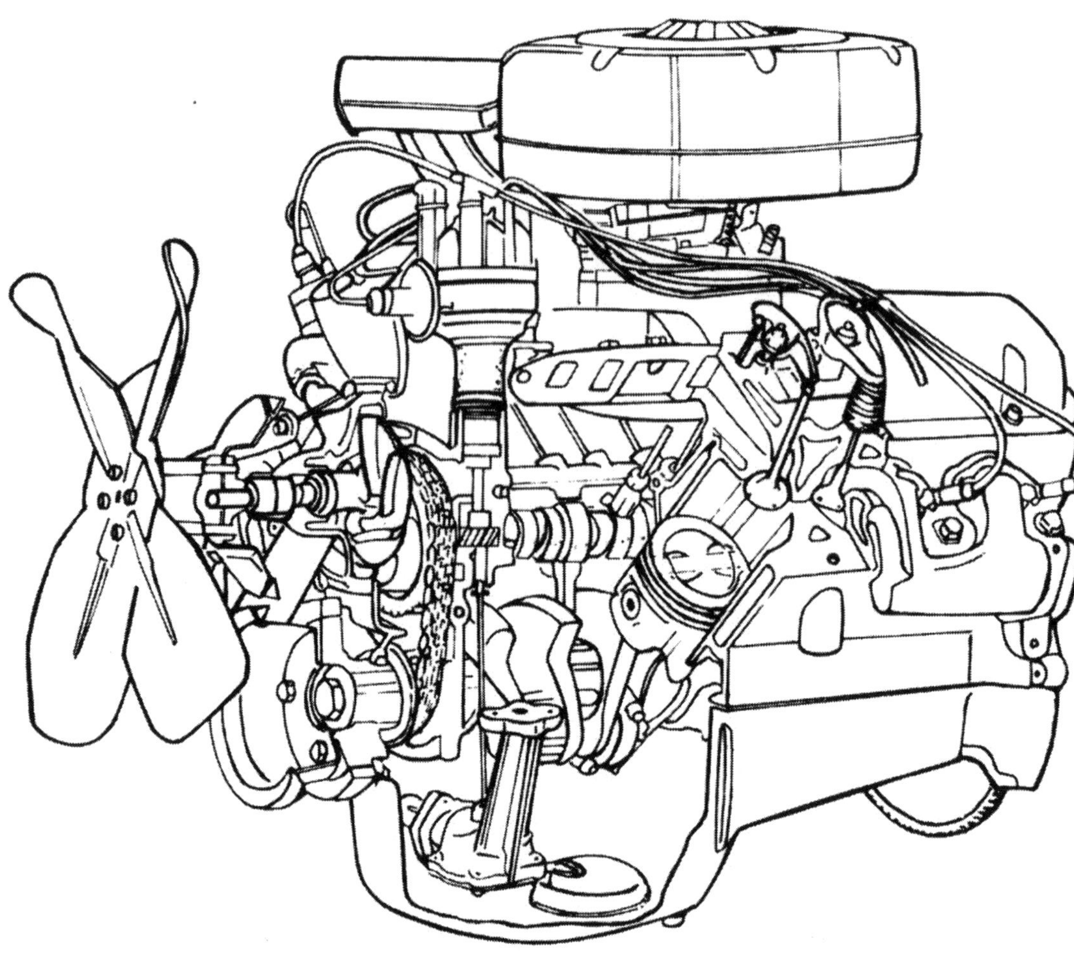

FIG 1:1 Sectioned view of typical engine

gauze strainer, by the oil pump. The pump is driven by gears from the camshaft. The pressurized oil from the pump first passes a relief valve, which opens when the maximum oil pressure is reached and returns surplus oil to the inlet of the pump. The relief valve ensures that the maximum oil pressure is limited when the engine is cold. The oil then passes through a full flow oil filter before entering the main oil gallery, which is located on the right side of the camshaft. Internally drilled oil passages lead the oil from the main gallery to the crankshaft main bearings, and internal passages lead the oil from the main bearings to the camshaft bearings, each main bearing feeding its adjacent camshaft bearing. Oil from the main bearings is also led to the connecting rod bearings by passages through the crankshaft itself.

The oil filter is fitted with a bypass valve. If the element becomes choked there will be a difference of oil pressure across it and when this pressure difference is sufficiently high, it will open the bypass valve and allow unfiltered oil to pass through to the engine. **This is a safety precaution against oil starvation but the filter should have been changed before this occurs.**

The hydraulic valve lifters are also connected directly to the oil gallery by ports. Each valve lifter is fitted with a metering valve which allows oil to pass up the hollow pushrod to lubricate the rocker and valve gear. There are slight differences in the mode of lubricating the timing chain and cylinder bores but generally the remaining parts are lubricated by oil mist and oil thrown from the bearings.

Crankcase ventilation and exhaust emission control:

On all models the system is designed for crankcase fumes to be drawn into the inlet manifold and burnt in the combustion chambers. On the very early models the air for ventilating the crankcase is drawn in through the filter in the oil filler cap.

The closed crankcase ventilation system is shown in **FIG 1:7.** Air for the system is drawn in through the carburetter air cleaner and then passed through to the oil filler cap by a tube. The oil filler cap is otherwise sealed. The air is drawn into the rocker arm cover and drawn through the pushrod guide holes and around the timing cover. The rotation of the crankshaft causes the air to flow up into the righthand rocker cover and it passes through a regulator valve before being drawn into the inlet manifold. Sectioned views of the valve are shown in **FIG 1:8.** At idle speed, when there is a high vacuum in the manifold, the suction overcomes the pressure of the spring and the valve moves to the low speed position where the flow through it is restricted. At high speeds the spring pressure opens the valve to provide a less restricted flow for the ventilating air. If the regulator valve was not present the flow through the engine would be excessive at low speeds, and would be sufficient to alter the mixture strength.

On models fitted with manually-operated transmission, a Thermactor system is fitted. This injects extra air into the exhaust ports so that hydrocarbon emissions are

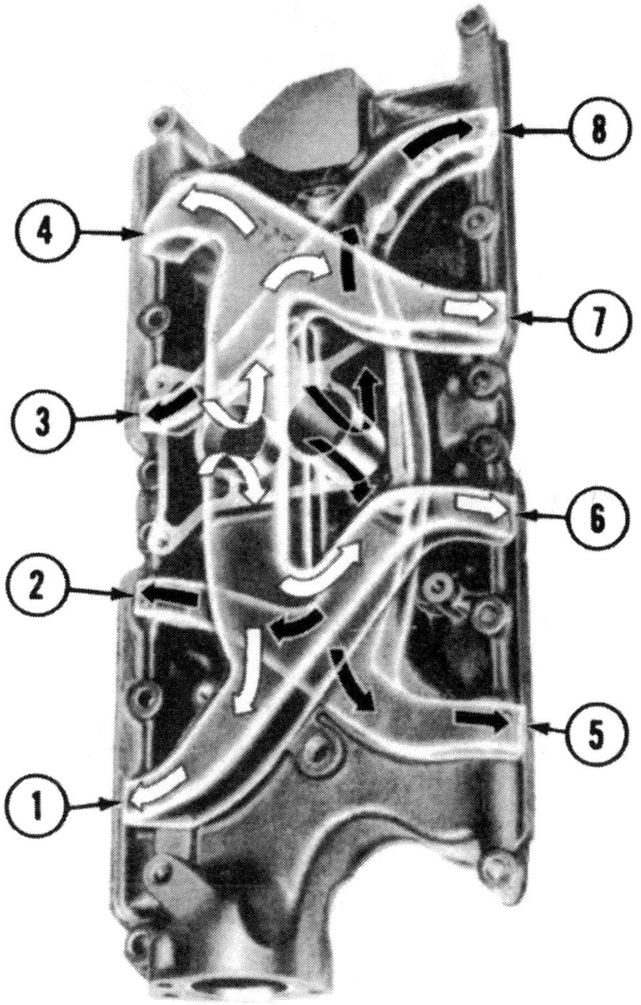

1, 4, 6, AND 7 ⇨

2, 3, 5, AND 8 ➡

FIG 1:2 The flow of the air/fuel mixture through a typical intake manifold

burnt before they leave the exhaust system. Further details of the system are given in **Section 1:13.**

On models fitted with automatic transmission, an IMCO system is used and the reduction in exhaust contaminants is obtained by using specially calibrated carburetors and distributors.

1:2 Engine removal and installation

It is essential to remove the engine from the vehicle in order to remove the crankshaft or camshaft bearings. All other operations can be carried out with the engine fitted to the vehicle, though for major overhauls it will be more convenient to have the engine on the bench.

If the owner is not a skilled automobile mechanic it is suggested that he reads 'Hints on Maintenance and Overhaul' at the start of this manual, as it contains useful hints and tips.

Removal:

The engine is removed leaving the transmission fitted to the vehicle.

1 Drain the crankcase and remove the oil filter. Remove the hood (see **Chapter 13**). Drain the cooling system,

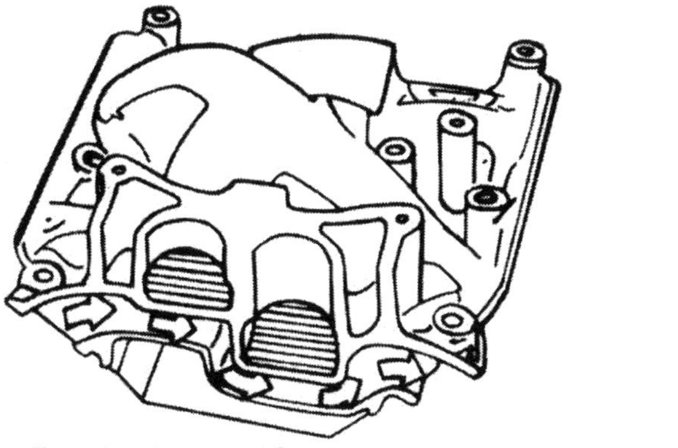

FIG 1:3 The intake manifold exhaust gas crossover passage

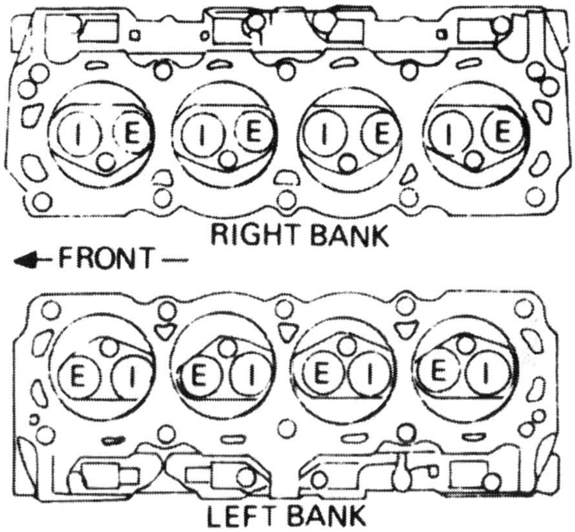

FIG 1:4 The valve port arrangement in the cylinder heads

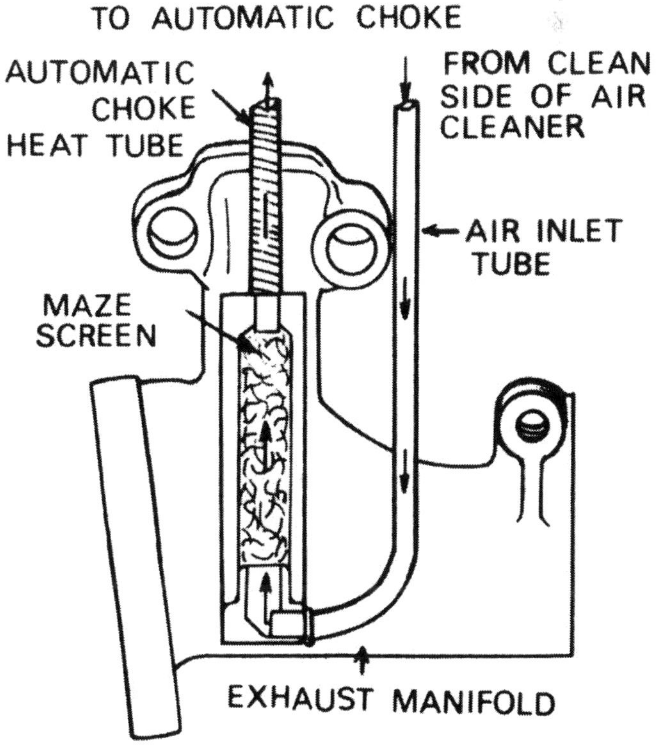

FIG 1:5 The heat chamber for the automatic choke

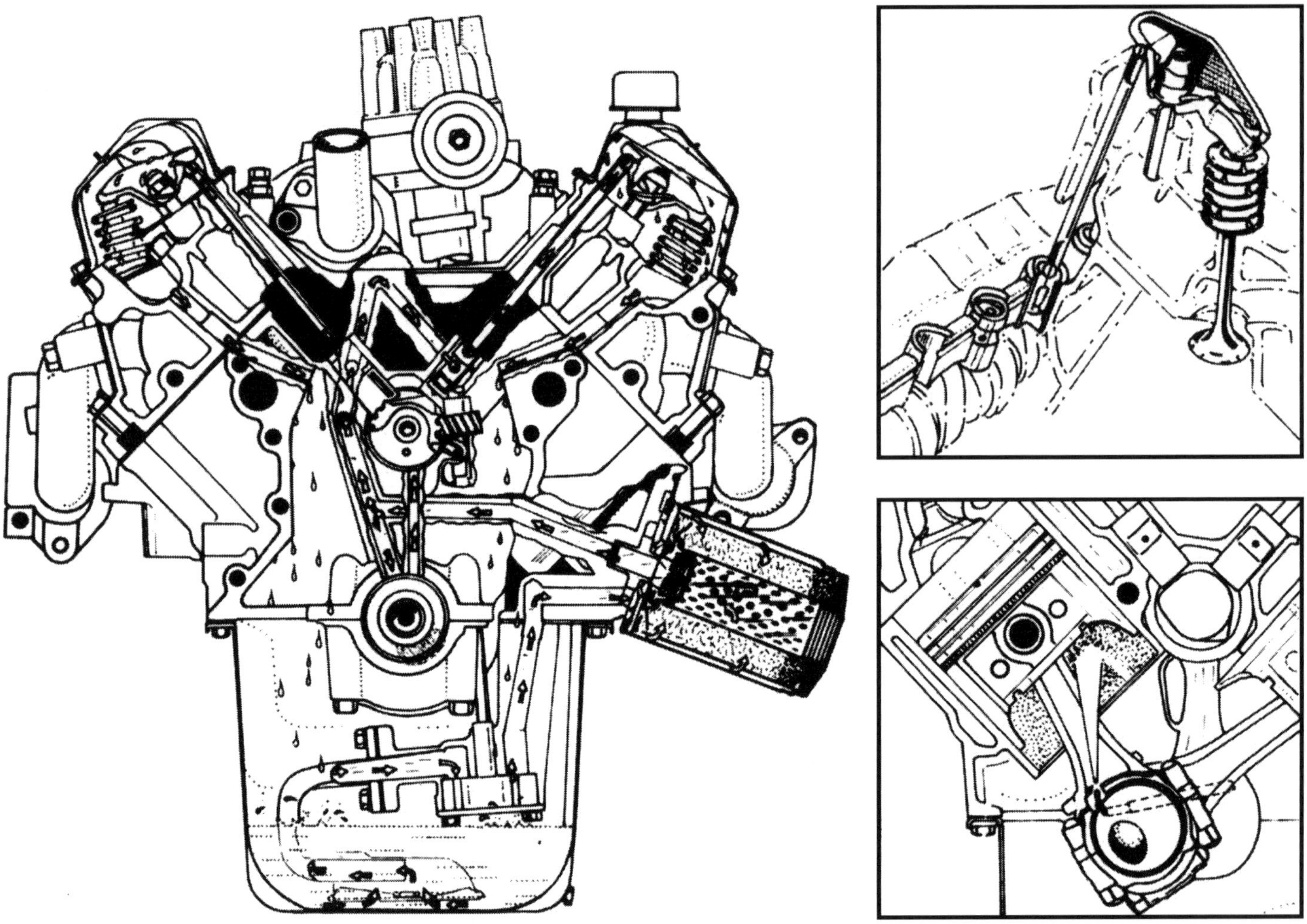

FIG 1 : 6 Typical engine lubrication system

disconnect the radiator hoses and remove the radiator. Remove the fan, pulley and belt (see **Chapter 4**). On a vehicle fitted with automatic transmission it will be necessary to disconnect the oil cooler lines at the radiator. Disconnect the battery ground cable at the cylinder block.

2 Disconnect the automatic choke heat chamber air inlet hose at the inlet tube near the right rocker cover. Disconnect the closed crankcase ventilation system and remove the air cleaner as a complete assembly with its inlet duct.

3 Loosen the alternator bolts and allow it to swing down out of the way, after disconnecting its wires. Disconnect the oil pressure sending unit wire at the sending unit. Remove the heater hose from the automatic choke housing. Disconnect the heater hoses at the water pump and intake manifold. On a vehicle with power brakes, or other vacuum-assisted units, disconnect the vacuum hose from the inlet manifold.

4 Disconnect the accelerator rod from the linkage. On models fitted with air-conditioning and power-steering, remove the speed-up control assembly from the bellcrank. Disconnect the flexible line from the fuel tank line and plug the fuel tank line to prevent petrol syphoning out.

5 **Models fitted with automatic transmission.** Disconnect the throttle valve vacuum line at the inlet manifold. Disconnect the manual shift rod and remove the retracting spring. Disconnect the transmission filler tube from its bracket on the block (see **Chapter 7**).

6 **Models fitted with power-steering.** Disconnect the power steering pump bracket from the cylinder head. Remove the drive belt and tie the pump safely out of the way so that oil does not drain from it (see **Chapter 10**).

7 **Models fitted with air conditioning.** Isolate the compressor and remove it as described in **Chapter 13.** Take care during this operation.

8 **Models fitted with Thermactor emission control.** Remove the air pump filter if it is not connected to the engine.

9 Remove the upper bolts connecting the flywheel or converter housing to the engine. Raise the front of the car. Disconnect the starter cables and remove the starter with its dust shield. Disconnect the muffler inlet pipes from the exhaust manifolds. Disconnect the engine support insulators, shown in **FIG 1 : 9,** from the brackets on the frame underbody.

10 On a vehicle with manual shift transmission, remove the remaining bolts that secure the flywheel housing to the engine, and lower the car back to the ground with the transmission supported on a jack and pad of wood.

11 On a vehicle with automatic transmission, remove the converter housing inspection cover. Disconnect the

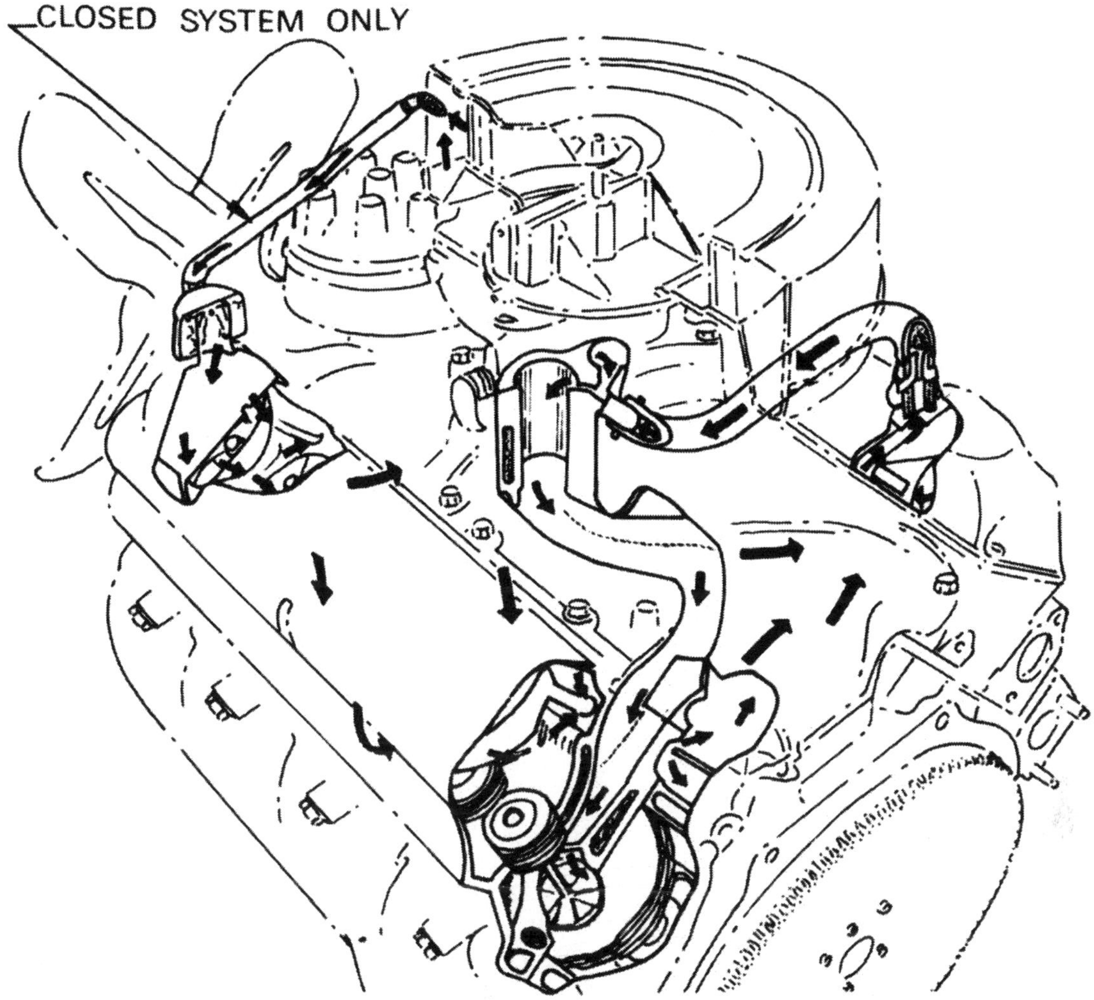

FIG 1:7 The positive crankcase ventilation system. On early models the air is drawn in through the oil filler cap direct

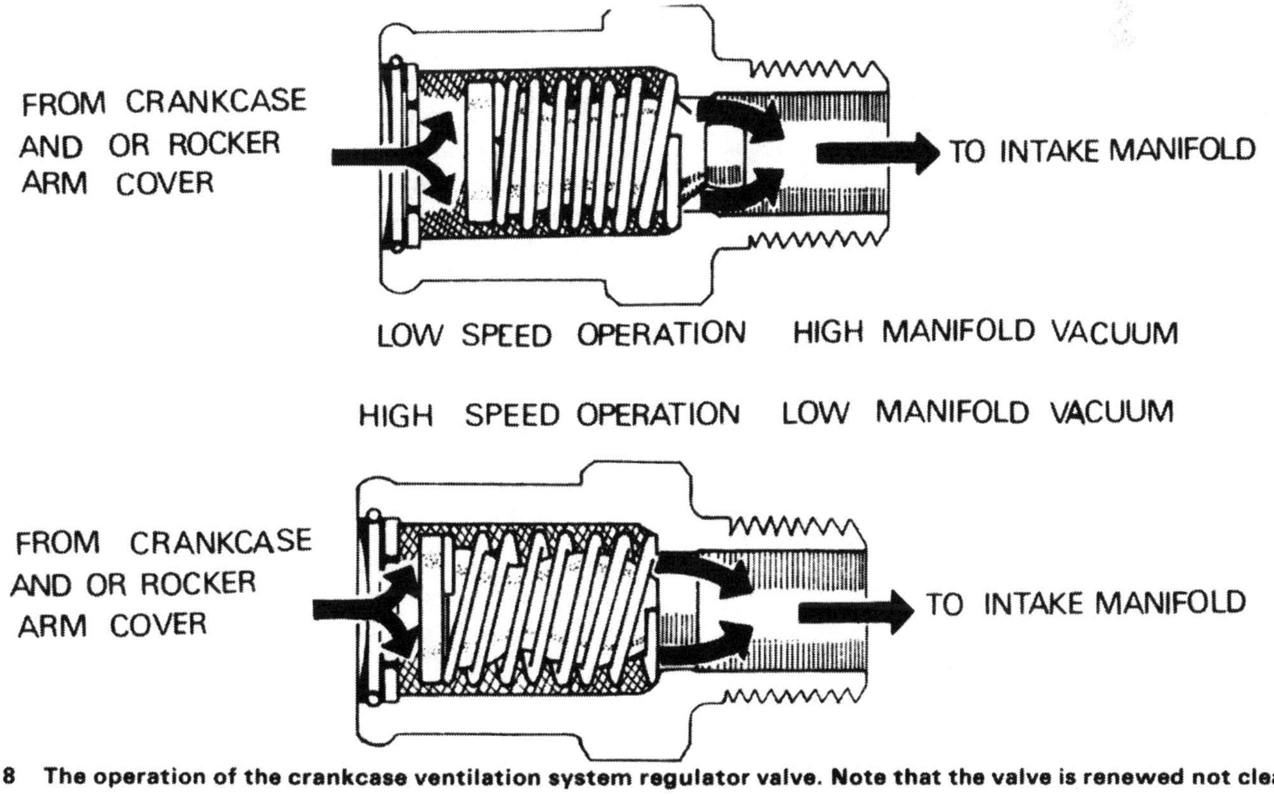

FIG 1:8 The operation of the crankcase ventilation system regulator valve. Note that the valve is renewed not cleaned

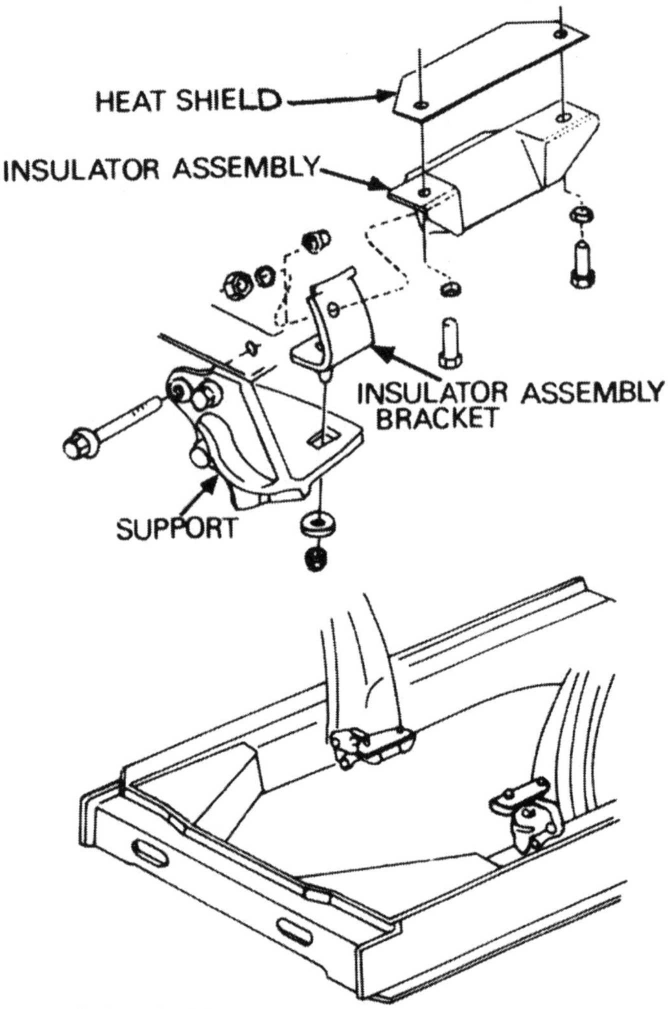

FIG 1:9 Typical engine support mountings

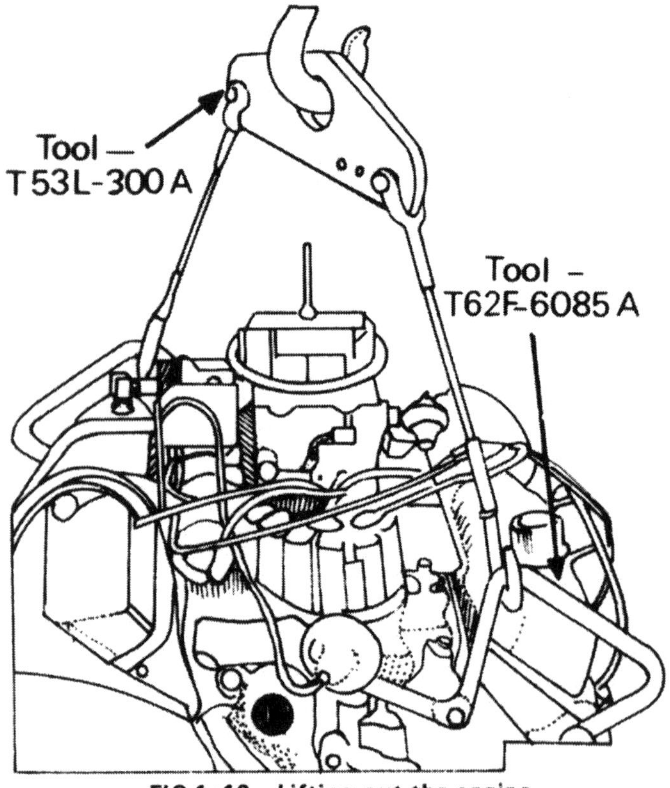

FIG 1:10 Lifting out the engine

flywheel from the converter. Secure the converter in the housing (see **Chapter 7**) and remove the remaining converter housing to engine bolts.

Lower the car back to the ground, and then support the transmission from underneath using a jack and pad of wood.

12 Install the lifting tools to the engine, as shown in **FIG 1:10.** Remove the air cleaner duct stud from the exhaust manifold. Remove the battery. Raise the engine slightly to clear it from its mountings and then pull it forward until the input shaft of the transmission (converter pilot on automatic transmissions) is well clear. The engine can then be lifted out of the car, taking care not to let it catch or damage components. **Do not allow the weight of the engine to hang on the input shaft.**

Installation:

The engine is installed in the reverse order of removal. The main point to watch is when mating the engine to the transmission. Do not force the engine back into place and if the shaft hangs up, lock the transmission and rotate the engine until the splines are in line. Before lowering the engine back into place, fit new gaskets on the muffler inlet pipes.

Make sure that all connections are tight and correctly made. Fill and bleed the cooling system (see **Chapter 4**). Install the oil filter and fill the crankcase to the correct level with fresh oil. Adjust all drive belt tensions correctly (see **Chapter 4**). Adjust transmission linkages as required (see **Chapter 6** for manual shift transmission and **Chapter 7** for automatic transmissions).

Check for leaks before starting the engine. Start the engine and run it at a fast idle and again check for leaks at hose connections and all gaskets.

1:3 The manifolds

Exhaust manifolds:

On the right exhaust manifold, disconnect the automatic choke heat chamber hose at the inlet tube and remove the heat tube.

Remove the air cleaner and intake duct assembly. Disconnect the manifold at the muffler inlet pipe. Remove the retaining bolts and their tabwashers so that the manifold can be removed from the cylinder head.

Install the manifold in the reverse order of removal. Clean the mating surfaces of the cylinder head and manifold as well as those of the muffler inlet pipe and manifold. Apply **graphite grease** to the mating surfaces and position the manifold on the cylinder head. Secure the manifold in place with the retaining bolts and new tabwashers. Working from the centre to the ends, progressively tighten the bolts to the correct torque load and lock the bolts by bending one tab of the washer over a flat on the bolt.

Fit new gaskets to the muffler inlet pipe, correctly position the inlet pipe and torque the retaining nuts to the correct load.

Install the pipes for the automatic choke. Start the engine and check for exhaust leaks.

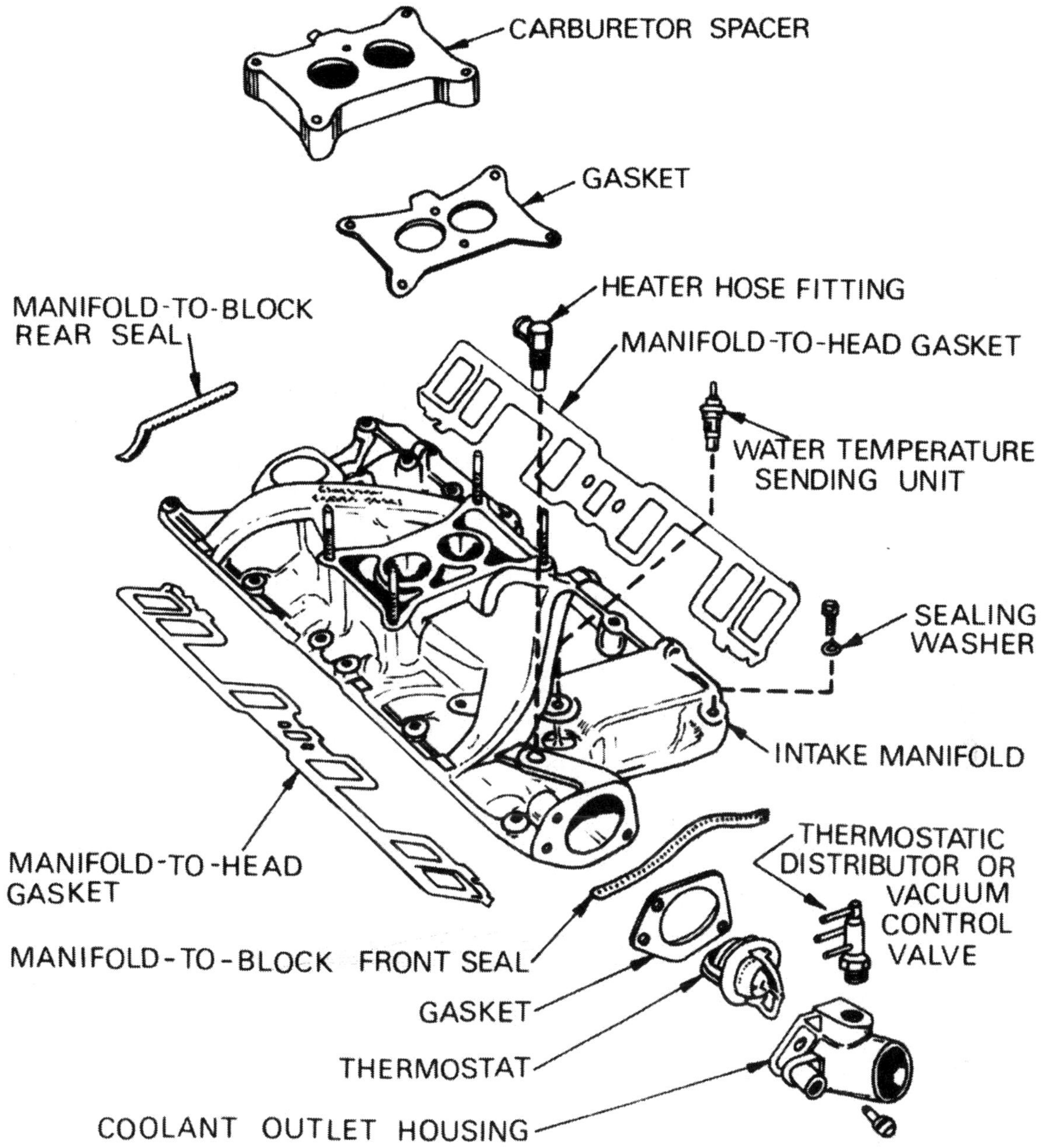

FIG 1:11 Typical intake manifold components

Inlet manifold:

A typical inlet manifold is shown in **FIG 1:11.** Drain the cooling system (see **Chapter 4**). Remove the air cleaner and duct assembly after disconnecting the automatic choke tubes from the manifold and the hoses for the crankcase ventilation system. Disconnect any vacuum lines from the manifold. Disconnect the accelerator rod at the carburetor and remove the return spring. Disconnect the high tension lead and wires from the ignition coil. Disconnect the high tension leads from the sparking plugs, after identifying them with labels, **by pulling and twisting on the moulded caps only.** Remove the distributor cap complete with the high tension leads. Take out the distributor hold-down bolt and remove the distributor. Disconnect the fuel feed pipe from the carburetor. Loosen the clamps and disconnect the heater hoses and the outlet hose. Disconnect the wire from the temperature sending unit. On models fitted with

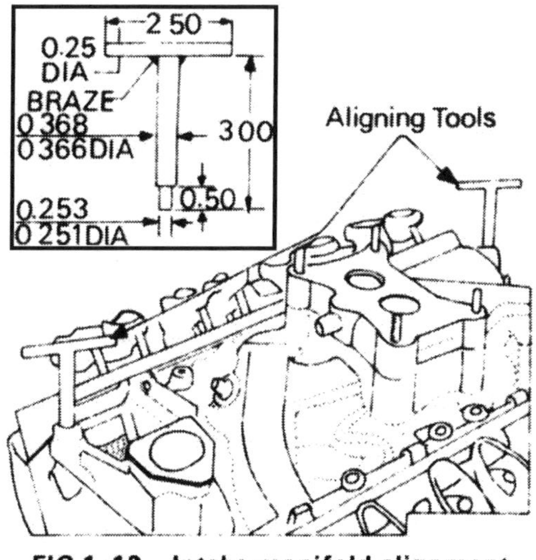

FIG 1:12 Intake manifold alignment

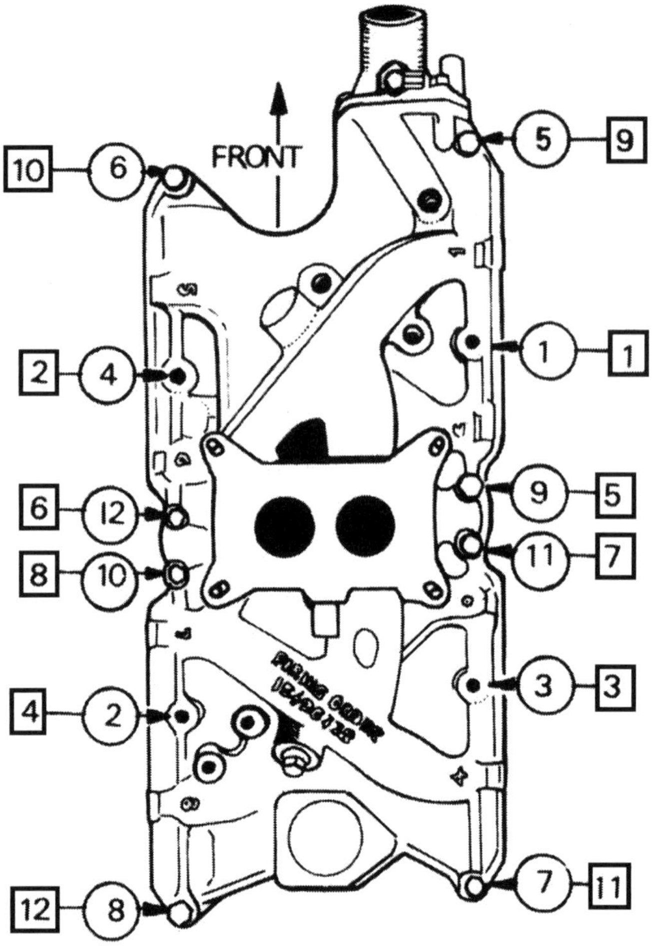

FIG 1:13 Intake manifold attachment bolt torque sequence. The sequence in circles represents the sequence for 1968 and onward. The sequence in squares represents the sequence for 1967 and prior

Thermactor, remove the air manifold hoses at the bypass valve and position them out of the way. On models fitted with air conditioning remove the brackets that secure the compressor to the manifold. Remove the manifold retaining bolts and their sealing washers. Lift off the manifold assembly, prying it off carefully if it sticks and taking great care not to damage the mating faces.

If the manifold parts are dismantled, make sure that the mating surfaces are clean before reassembling the parts. The temperature sending unit should be screwed back into place **using an electrically-conducting sealer.** The coolant outlet housing should be secured using a new gasket and waterproof sealer.

Thoroughly clean the mating surfaces of the manifold and cylinder heads, using suitable solvent to remove all traces of oil or grease. Coat the cylinder head surfaces lightly with contact adhesive. Position the new seals on the cylinder block and the new gaskets on the cylinder heads so that the gaskets are interlocked with the seal tabs. Apply non-hardening sealing compound at the four junctions of the seals and gaskets, making sure that the gaskets align with the ports in the cylinder heads. Carefully lower the manifold assembly back into place, using the tools shown in **FIG 1:12** to align it as shown. **Run a finger around the seals and if the seals have shifted the manifold assembly must be removed and the seals correctly positioned.** Refit the attachment bolts, **using new sealing washers,** and torque-load them in the sequence shown in **FIG 1:13.** Install and connect the remainder of the parts in the reverse order of removal. Run the engine, after filling and bleeding the cooling system, until it has reached its normal operating temperature—while checking for leaks or defects.

When the engine temperature has stabilized, re-torque the manifold attachment bolts and adjust the carburetor and ignition settings.

1:4 The cylinder heads

Removal:

Remove the intake manifold as described in the previous section. To save confusion between parts it is advisable to remove and service one cylinder head at a time.

Remove the rocker cover. On the left cylinder head, isolate and remove the air conditioning pump (if fitted), disconnect the power steering pump bracket from the cylinder head and move the pump safely out of the way (if fitted), disconnect the hose from the check valve if a Thermactor system is fitted. On the right cylinder head remove the alternator mounting bracket bolt and spacer, ignition coil and air cleaner inlet duct assembly.

Loosen the rocker arm securing nuts, shown in **FIG 1:14,** so that the rocker arms can be rotated to one side and the pushrods withdrawn. **Store the pushrods in order as they must be returned to their original positions.** Install the engine lifting tools T62F-6085-A on either end of the cylinder head. Remove the cylinder head attaching bolts and use the special tools to lift the cylinder head off from the engine. The special tools are not essential but they do give a good handle to lift with and they can be used as stands to support the cylinder head.

Pushrods:

These are removed after the rocker arm nuts, shown in **FIG 1:14,** have been slackened off sufficiently to allow the rocker arms to be rotated off the pushrods. **Keep the pushrods in order.**

Wash the pushrods in gasoline or a suitable solvent and blow out the oil passage with compressed air (or a

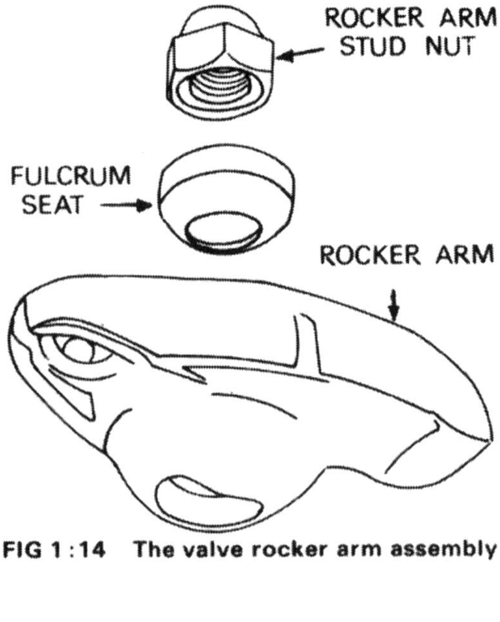

FIG 1:14 The valve rocker arm assembly

tire pump). Check the pushrod ends for nicks, wear or damage. Check the pushrods for straightness. This can be done visually or by rolling them along a flat surface, though a better method is to lay each pushrod in turn onto V-blocks and rotate it using a dial indicator to measure the runout. If the pushrods are bent by more than .015 inch, **do not attempt to straighten them** but fit new ones in their place, as well as renewing pushrods that have damaged ends.

Rocker arms:

The parts are shown in **FIG 1:14.** The parts can be removed after unscrewing the self-locking nuts from the studs on the cylinder head. If the pad at the valve end of the rocker arm is lightly scored it can be trued up by careful use of an oil stone but badly worn pads require renewal of the rocker arm.

Renew the nut if the breakaway torque is outside the limits of 4.5 to 15 lb ft for earlier models (undoing the nut) and on later models 18 to 22 lb ft after the nut contacts the shoulder. If renewing the nut does not bring the breakaway torque within limits, or the stud is damaged, the stud must be renewed. A service kit is supplied for renewing defective studs, including reamers and oversize studs.

Valve spring and seal removal:

If the cylinder head has been removed from the engine, this includes removing the valve itself. The valve springs and seals can be removed, with the cylinder head still fitted to the engine. If the cylinder head is still fitted, turn the engine until both valves on the cylinder are fully closed. Remove the sparking plug and fit an adaptor so that compressed air can be injected into the combustion chamber and keep the valves closed when the springs are removed. Typical valve spring, valve and seal parts are shown in **FIG 1:15.** Use a valve spring compressor to compress the spring, with the valve head supported by air pressure, or by the tool itself, and remove the locks. Gently release the tool and pressure on the spring so that the retainer, spring and seal can be removed. If the cylinder head is free from the car then the valve can be slid out of the combustion chamber. **Store the valves in the correct order as they must be installed into their original positions.** Discard the old oil seals.

The valves:

Check the stems for wear, scoring or bending and renew any valves that show these defects. The end of the valve stem may be ground down by a maximum of .010 inch if the tip is worn or scored. There must be $\frac{1}{32}$ inch minimum of head left above the seat area, and if the valve has been reground so much that this depth is not present, or the valve has a sharp edge in the combustion chamber, a new valve must be fitted in its place.

Valve springs:

These should be checked on a special rig, so that the length under load can be measured. As a rough test mount a new spring end onto the old one and compress the pair between the jaws of a vice. If the old spring is appreciably shorter than the new one, while under pressure, then the spring is weak and the set should be renewed.

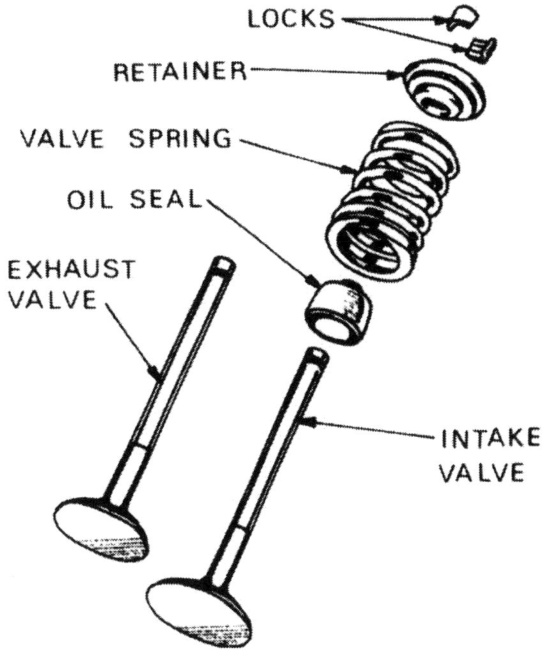

FIG 1:15 The valve assembly

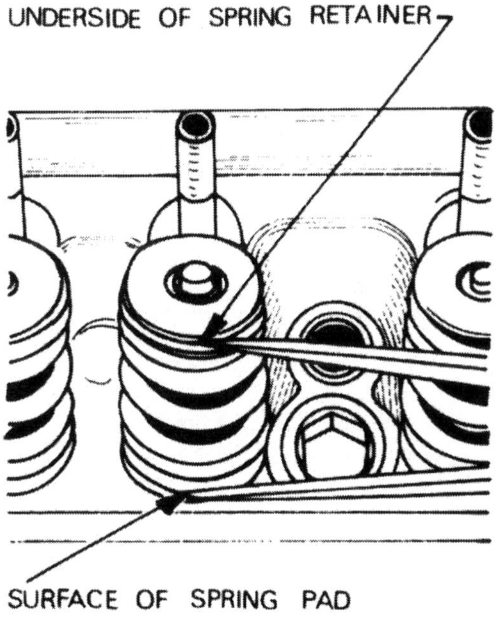

FIG 1:16 Checking the valve spring assembled height

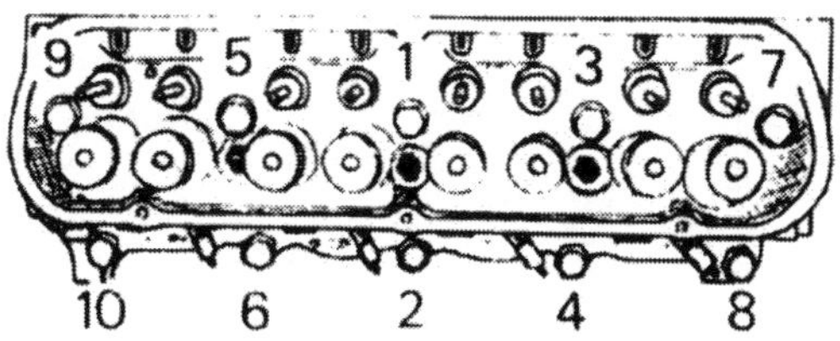

FIG 1:17 The cylinder head bolt torque sequence

Place the valve spring, closed coils downwards, onto a perfectly flat surface and use a square to check that the spring is vertical. Rotate the spring and if the gap between the top of the spring and square exceeds $\frac{5}{64}$ inch, the spring should be rejected.

Valve guides:

Special dial gauges are made for checking the wear of the stems and guides. If in doubt, try a new valve in the guide and check for excessive rock.

Valves with oversize stems are available to take up wear, but the guide must first be reamed to the correct size. Service kits of reamers are available—.003 oversize with standard pilot, .015 oversize with .003 pilot and .030 oversize with .015 pilot. If the wear is such that the next oversize valve stem cannot be used, ream out the valve guide in stages.

Once the valve guides have been reamed, the seats in the cylinder head must be recut to ensure concentricity with the new valve.

Valve seats:

Check the efficiency of the seats by inverting the cylinder head and making a dam around the valve with plasticine or suitable material. With the valve held closed by its springs, fill the dam with gasoline and any leaks through the valve will then be obvious.

If the seats are pitted or leak, they should be recut using special grinding tools at a service garage. If the seats in the cylinder head are then too wide, they should be reduced using cutters. **Lapping-in of valves is not recommended.** The cutters should be used so that the finished valve seat in the head contacts the seat on the valve approximately centrally (use Prussian blue to check the area of contact).

Decarbonizing:

Remove the deposits from the combustion chambers before taking out the valves, as the valves will protect the seats in the head. Scrape out the worst deposits using a sharpened piece of perspex or hardwood. Final cleaning can then be carried out using a piece of worn emerycloth dipped in kerosene. On cast iron components, a rotary wire brush can be used to remove deposits but under no circumstances may this treatment be used on light-alloy components.

When the combustion chambers are clean, remove the valves and remove deposits from the valves, **taking great care not to damage the ground portion of the stem or the valve seat.** Ports and manifolds can then be cleaned out, making sure that all dirt is blown out using compressed air.

Turn the engine until a pair of pistons is at TDC, and remove deposits (using a sharpened piece of perspex or hardwood) from the piston crowns. **Do not use any form of abrasive and plug oilways and water passages with rag to prevent dirt from falling down them.** Clean the other pair of pistons in a similar manner. Smear a little grease around the tops of the bores. Blow away all loose dust and dirt from the top of the engine. Turn the engine over so that the piston rings scrape the grease and any dirt to the top of bores where it can be wiped away.

Reassembly and installation:

Make sure that all the parts are thoroughly cleaned and washed with suitable solvent. Pull thin pieces of rag soaked in petrol through the valve guides to clean them. Make sure that there are no specks of dirt left on the mating surfaces of the cylinder head, block or manifold. Blow through the manifolds to remove any dirt or swarf that may have accumulated in them.

Check the cylinder head for bow or distortion using a steel straightedge and feeler gauges. Small nicks and high spots can be smoothed down with an oil stone. If the surface is not flat the cylinder head must be sent away for planing or grinding, but not more than .010 inch material may be removed and if the flatness cannot be restored a new cylinder head must be fitted.

Install the valves into their original positions, lubricating their stems liberally with engine oil. Use the compressor to hold down the springs, closed coils downwards, while installing the locks. When the valve and spring has been refitted, measure the height of the valve spring, as shown in **FIG 1:16.** If the height is not within the correct limits (see **Technical Data**) install sufficient .030 inch shims between the cylinder head and valve spring pad to bring the height within limits. **Do not fit excess shims as the extra strain can lead to failure of parts in the valve train.** Do not forget to install new valve stem seals before installing the valve springs.

Lay a new cylinder head gasket into position on the cylinder block, locating it on the dowels. Sealing compound is not required on the composition gaskets used. Lower the cylinder head into place and secure it with the cylinder head bolts finger tight. The cylinder head bolts are tightened in three progressive steps, in the sequence shown in **FIG 1:17.** Tighten the bolts to 50 lb ft, in sequence, then to 60 lb ft and finally tighten them to 65 to 72 lb ft still in sequence.

The remainder of the parts are then installed in the reverse order of removal. The pushrods must be installed into their original positions and the setting of the rocker arms is dealt with in the next section.

1:5 The hydraulic valve lifters

This section deals with the valve lifters, and the rocker arm adjustment after the parts have been installed.

Normally the hydraulic valve lifters are removed after the cylinder head has been taken off. They can be lifted out using a magnetic rod but if they stick, an expanding finger tool will be required.

The valve lifters can be removed without having to take off the cylinder head. Remove the rocker cover, slacken the rocker arm nuts until the rocker arms can be rotated and the pushrods taken out—remembering to store the pushrods in order. Fit a probe lamp into a pushrod aperture so that the parts can be seen, and use an expanding-finger tool to draw the lifter out of its bore. In some cases it will be necessary to lift out the unit, move it along and then remove through a larger aperture. Store the valve lifters in a suitable rack in their correct order, as **they must be installed in their original positions.**

Earlier models are fitted with Type 1 hydraulic lifters, shown in **FIG 1:18,** while later models may be fitted with either Type 1 lifters or the later Type 2 lifter, shown in **FIG 1:19.** The lifters themselves are interchangeable, and both types may be used on the same engine, but the parts are not interchangeable between the two types and dismantling procedures vary. **Each lifter has its parts matched in manufacture so if any part of a lifter is found to be defective, the complete unit must be renewed.**

Noisy valve lifters:

The operation of a type 2 hydraulic valve lifter is shown in **FIG 1:20.** When the valve is closed, the lifter is on the base circle of the cam and the spring moves the plunger upwards until the rocker arm contacts the valve stem and zero lash is established. While this occurs there is a reduced pressure in the compression chamber allowing the check valve to open and oil to pass into the chamber from the engine oil gallery. Once the compression chamber is full, there is no pressure differential and the check valve closes. As the cam rotates and raises the lifter, the oil in the compression chamber, being non-compressible, ensures that full movement is transferred to the plunger and from there to the valve train, with the lifter acting as a hydraulic ram. A slight leakage past the plunger is allowed, as this prevents the lifter from over-compensating and holding the valve open and leakage on each stroke is replenished on the next closed position. A metering valve allows oil to pass through for valve train lubrication.

When the engine has been left overnight or for a lengthy period the oil will drain down from the lifters, particularly those which are holding valves open, and excess lash will be set up. The valve train will therefore sound noisy for the first few minutes of operation and then quieten as the compression chambers of the lifters fill. This is perfectly normal and no action need be taken.

Intermittent clickings are mainly caused by incorrect initial adjustment or a piece of dirt momentarily holding the check valve open, though they may be caused by a defective check valve.

General noise throughout the valve train is caused by lack of oil or incorrect adjustment. It should be noted that if the crankcase is overfilled the connecting rods will thrash in the oil, causing foam which will make the lifters operate noisily, as the air in the compression chambers will compress preventing the lifter acting solidly.

A moderate rapping noise can be caused by a worn valve lifter or incorrect initial adjustment, while a heavy rapping noise is caused by the plunger sticking in the body.

If the operation is noisy, first check the oil level in the crankcase and then make sure that the initial adjustment is correct. If the fault still is not cured, remove the lifters for cleaning and checking.

Testing:

The hydraulic valve lifters are tested after cleaning by using a special test rig, which applies a constant load of 50 lb, the time being noted for the plunger to move through its full range of travel. **The lifter must be free from engine oil and completely filled with special test fluid, otherwise the results will be inaccurate.** If it is suspected that the lifters are at fault, they should be dismantled and cleaned and then reassembled without oil and taken to a service garage for checks.

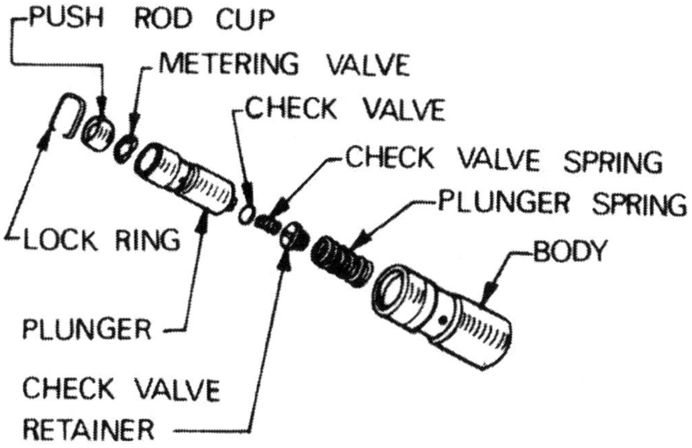

FIG 1:18 Type 1 hydraulic valve lifter

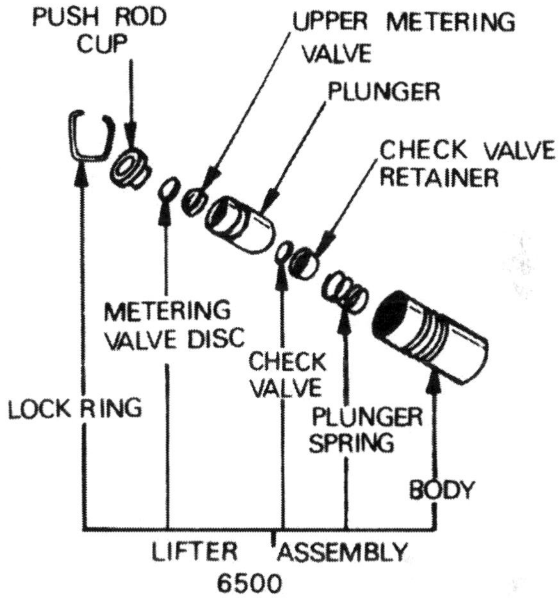

FIG 1:19 Type 2 hydraulic valve lifter

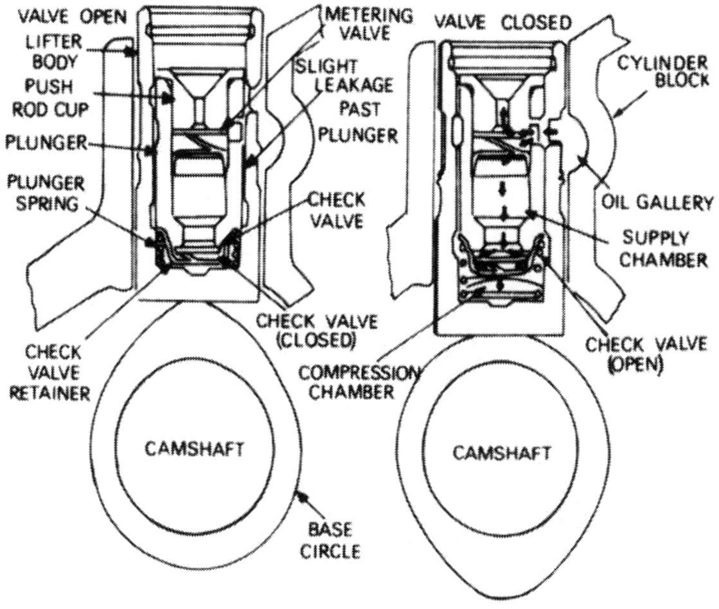

FIG 1:20 Operation of a type 2 hydraulic valve lifter

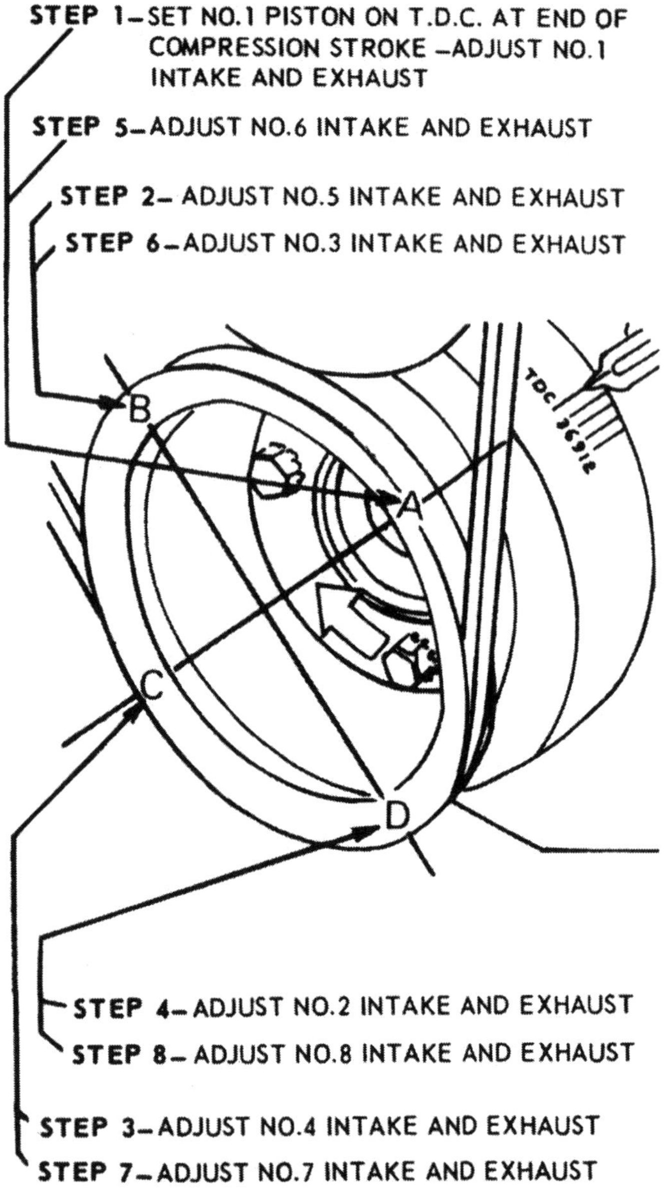

FIG 1:21 The sequence for adjusting the valve clearance

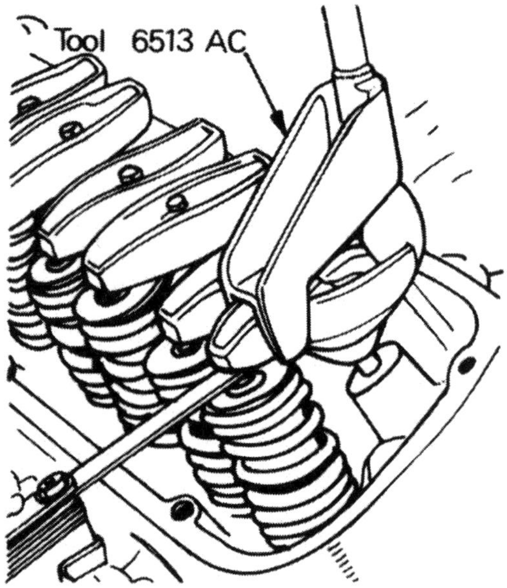

FIG 1:22 The best method of adjusting the valve clearance

When the unit has been dismantled, wash the parts in petrol, denatured alcohol or trichlorethylene to remove all dirt, gum or varnish. Check the parts for wear or damaged, paying particular attention to the valves and the fit of the plunger in the bore. Renew the unit if parts are defective or the bore shows signs of fretting or picking up. The bench must be perfectly clean when reassembling the units, and each valve lifter should be serviced in turn to ensure that parts are not accidentally interchanged.

Type 1 valve lifter:

The parts are shown in **FIG 1:18.** Remove the lock ring using a pair of needle-nosed pliers. It may be necessary to press down the plunger while removing the lock ring. Remove the pushrod cup, metering valve disc, plunger and spring assembly. Remove the plunger spring from the body of the plunger then carefully remove the check valve retainer and check valve spring, followed by the check valve itself. On some models balls will be used instead of discs for the check valve.

The parts are cleaned and checked as described for testing. Before installing the lifter in the engine, fill it with test fluid or engine oil

Stand the plunger inverted on a bench and fit the check valve back into place. Install the check valve spring followed by the retainer and push the retainer down into place. Place the plunger spring back into the body and install the plunger assembly so that the spring is correctly located. Filling the unit with oil before reassembly will make it difficult to press down the plunger. Install the metering valve disc, followed by the pushrod cup, and lock the parts in place with the lock ring.

If the lifter is not to be tested at a service garage, stand the unit upright in a container of fresh engine oil and press the plunger down into the body. Release the plunger to allow oil to be drawn in and then repeat the operation. When the unit is filled with oil it will be extremely difficult to press down the plunger.

Type 2 valve lifter:

The parts are shown in **FIG 1:19** and the unit is identified by the three rings around the body. The unit is dismantled using the same method as for the Type 1 lifter but **great care must be taken not to bend, damage or distort the metering valve or its tensioning finger. The same care must be taken with the check valve parts.**

Clean and check the parts as described for testing.

Stand the plunger inverted onto a clean bench. Place the check valve disc centrally into it. Install the check valve retainer, using a slight twisting motion and taking great care not to distort it or bend the preformed fingers, until it is fully in place. Slide in the plunger spring and, still leaving the plunger inverted, slide down the body onto it. Use the fingers to hold the parts together and turn them over. Install the upper metering valve into the plunger, **taking great care not to damage or distort it and making sure that it is central in the plunger without being tilted.** Place the metering valve disc into position followed by the pushrod cup. Secure the parts with the lock ring.

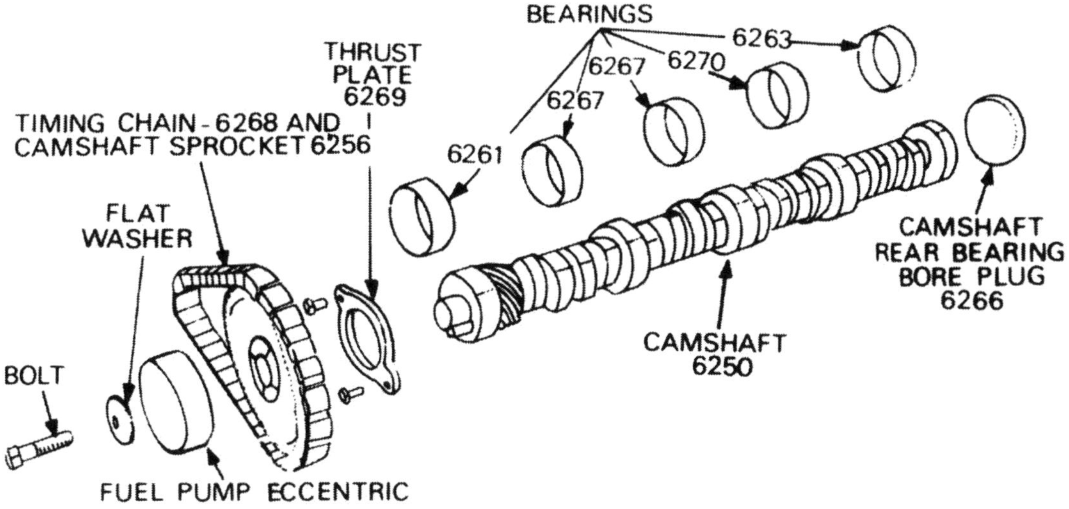

FIG 1 : 23 The camshaft and its related parts

Fill the unit with oil by pumping it in a container of fresh engine oil. If the unit is being tested at a service garage, it will be filled with test fluid on the test rig.

Installation and adjustment:

Install the valve lifters into their original positions, making sure that they slide freely in their bores. Place the pushrods back into position, making sure that the pushrod ends seat correctly into the cups in the valve lifters. Smear the pivots for the rocker arms, pushrod ends and tops of the valve stems with Lubriplate or similar lubricant, and rotate the rocker arms so that they are in correct position on the valve stems and pushrods. Remove the sparking plugs so that the engine will be easier to turn over. Make four chalk marks, at approximate right angles to each other, on the crankshaft damper (as shown in **FIG 1 : 21**) so that the mark A aligns with the TDC mark on the pulley. Turn the engine until the mark A aligns with the timing pointer and No. 1 (front on right bank) piston is at the end of its compression stroke. The compression stroke can be found by blocking the plug hole with a finger or thumb and feeling for the rise in pressure as the piston moves upwards on the compression stroke. If the engine cannot be turned by pulling on the drive belt and rotating the damper, the starter motor must be used instead. Disconnect the HT lead from between coil and distributor to prevent the engine from firing and preferably use an auxiliary switch across the terminals of the solenoid.

Once the engine is correctly set, adjust both valves of No. 1 cylinder. Press firmly down on the rocker arm until the hydraulic valve lifter is fully collapsed and hold it down in this position. Use narrow feeler gauges to check the clearance between the pad on the rocker arm and top of the valve stem, as shown in **FIG 1 : 22.** On earlier models the limits for the gap are .082 to .152 inch but on later models the limits are extended to .067 to .167 inch. On all models the adjustment should be set as close as possible to .117 inch for optimum performance. To close the gap, screw the rocker arm nut downwards and upwards to increase. **The gap must be measured with the hydraulic valve lifter collapsed and pressure applied to keep it closed.**

An alternate method is to slacken off the rocker arm nut until the pushrod can just be rotated freely, indicating zero lash in the valve train, and then tighten down the nut by $\frac{3}{4}$ turn, so as to set the lifter in the middle of its range. This method is not so accurate and the previous method should always be used for preference.

Once the valves on No. 1 cylinder have been adjusted, turn the engine through 90 deg. until the mark B aligns with the timing pointer. Set the valves on No. 5 (front on left bank) in a similar manner. Rotate the engine through a further 90 deg. and adjust the next set of valves in the firing order and carry on until all cylinders have their valves correctly set.

Install the rocker covers. If the cover gaskets are damaged, peel out the old ones and stick new ones back into place, using a little non-setting jointing compound.

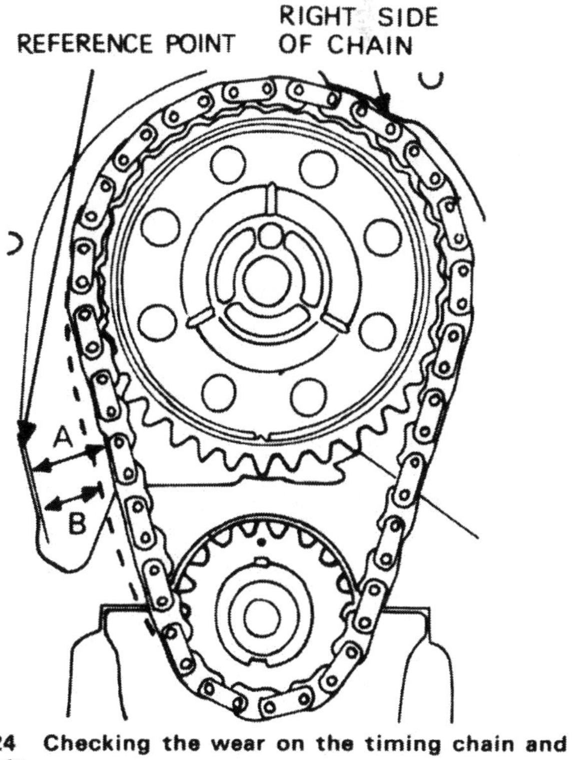

FIG 1 : 24 Checking the wear on the timing chain and sprockets

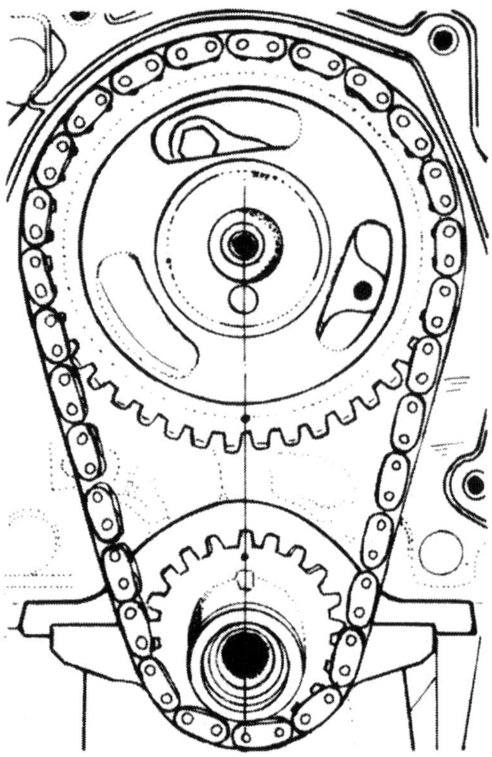

FIG 1 : 25 The valve timing marks correctly aligned

Restore the starter circuit to normal if an auxiliary switch was used. Start the engine. For the first few minutes the valve train will sound noisy as the lifters settle but the noise should rapidly fall away. Check for oil leaks around the rocker covers after the engine has been running for a while.

1 : 6 The valve timing gear and camshaft

All the parts can be removed and installed while the engine is fitted in the vehicle, except the camshaft bearings. The camshaft parts are shown in **FIG 1 : 23**. In order to renew the bearings the engine must be removed from the vehicle and then dismantled.

Special tools are required to withdraw the old bearings and draw the new bearings back into place, and the bearings will be damaged if the incorrect procedures are used. For this reason it is advisable to dismantle the engine and take the crankcase to a service agent for renewal of the camshaft bearings.

Front cover:

1 Drain the cooling system and crankcase. Disconnect the battery. Disconnect the hoses from the water pump. Slacken the alternator attachments so that the drive belt can be removed, disconnect the alternator bracket from the water pump and swing the alternator down out of the way. On models with Thermactor emission control, free the pump and its brackets. On models fitted with power steering and/or air conditioning, remove the drive belts. Remove the cooling fan, spacer and water pump pulley.

2 Remove the crankshaft pulley from the crankshaft damper. Take out the damper retaining bolt and draw off the damper using a suitable extractor (preferably T58P-6316-A).

3 Disconnect the fuel outlet line from the fuel pump. Take out the pump attachment bolts, withdraw the pump from the engine and lay it out of the way, still attached to its flexible feed line.

4 Remove the crankcase dipstick. Take out the bolts that connect the oil pan to the cylinder front cover. Take out the remainder of the attachment bolts and remove the cylinder front cover with the dipstick tube and water pump attached to it. If a new cover is to be fitted, remove the dipstick tube and water pump from the old cover, fitting them onto the new cover before installation.

The parts are refitted in the reverse order of removal. A new oil seal should be fitted into the cover and a new gasket used between cover and crankcase.

The old oil seal should be driven out using a suitable pin punch, but taking care not to damage the cover itself. Coat the new oil seal with grease and drive it fully back into the cover so that it seats in its recess. The lips of the seal must face into the engine and the spring must be correctly in place.

Thoroughly clean the mating surfaces of the cover and crankcase, and coat them lightly with sealing compound. Position a new gasket on the crankcase. When refitting the cover, it is best to use the centralizing tool T61P-6019-B or 6059-F slid onto the crankshaft, as this will centralize the oil seal around the crankshaft. Tighten the attachment bolts before removing the tool. Do not forget to refit the bolts that secure the oil pan to the cover.

Install any parts removed in the reverse order of removal and correctly set the drive belt tensions. Fill and bleed the cooling system (see **Chapter 4**). Fill the crankcase to the correct level on the dipstick with fresh oil of the correct grade. Operate the engine at a fast idle and check for oil or coolant leaks.

Timing chain and sprockets:

Remove the cylinder front cover as just described. Before removing the parts, check them for wear as shown in **FIG 1 : 24.** Pull the chain tight to the left and measure the dimension A from a convenient reference point. Pull the chain tight to the right and measure the dimension B from the same reference point. The difference between the two dimensions represents the deflection of the chain. If the maximum deflection exceeds .5 inch, the chain and sprockets are worn and new parts must be installed on reassembly. **Do not renew the timing chain alone as the sprocket teeth are also worn and will quickly tear up a new chain.**

When the deflection has been checked, turn the engine until the valve timing marks align, as shown in **FIG 1 : 25. Once the sprockets and chain have been removed, do not turn either the camshaft or the crankshaft otherwise a valve may hit the piston crown causing internal damage.** Remove the bolt and washers that secure the fuel pump eccentric to the camshaft sprocket, as well as securing the sprocket itself. Remove the oil flinger from the crankshaft. Grip the camshaft sprocket and chain with one hand and grip the crankshaft sprocket and chain with the other hand. Slide both sprockets off the shafts, keeping them roughly in alignment. Once the sprockets are off the engine, the timing chain can be removed from around them.

The parts are refitted in the reverse order of removal. Encircle the sprockets with the chain so that their timing marks are in alignment. Press the sprockets back into position, keeping them in rough alignment so as not to strain the chain, and making sure that the timing marks still align. Refit the fuel pump eccentric so that it is held by its dowel and secure the parts with the bolt and washer. Install the oil slinger onto the crankshaft.

Refit the front cover in the reverse order of removal. When installing the damper to the crankshaft, apply Lubriplate to the rubbing surface to prevent damage to the oil seal in the cover and apply a mixture of white lead and oil to the front of the damper. Pull the damper back into place on its key using a suitable tool.

Camshaft:

Before removing the camshaft, check the valve lift and the camshaft end float. Mount a dial indicator onto the crankcase so that its stylus rests vertically on the end of the camshaft. Slacken off all the rocker arm nuts so that no load is being applied to the lobes of the camshaft. **This is essential otherwise the load will prevent the camshaft from moving and prying on the camshaft sprocket will damage the sprocket.** Pull and push the camshaft in and out of the crankcase, noting the total end float on the dial indicator. Up to the end of 1966 the limits are .001 to .007 inch with a maximum wear limit of .012 inch but on all later models the limits are .0005 to .0055 inch with a wear limit of .007 inch. The end float is controlled by the thrust plate shown in **FIG 1 : 23.** If the end float is below limits the cause is either dirt behind the thrust plate or nicks on the surface of the thrust plate. If the end float is beyond limits, a new thrust plate must be fitted.

Remove the rocker arms or swing them to one side to expose the ends of the pushrods. Mount a Dial indicator on the cylinder head so that its stylus rests vertically on the end of the pushrod. Crank the engine over slowly and measure the lobe lift (see **Technical Data**). If the lift is more than .005 inch below the given dimension there is wear on the cams or the bearings.

Before the camshaft can be removed, the valve lifters must be taken out (see **Section 1 : 5**) and the distributor removed (see **Chapter 3**). Remove the front cover, sprockets and timing chain. Remove the radiator (see **Chapter 4**). Take out the two bolts that secure the thrust plate and carefully withdraw the camshaft from the engine, taking great care not to damage or score the bearings.

Refit the camshaft in the reverse order of removal. Install the front cover and all relevant parts, setting the correct tensions on the drive belts. Connect up all hoses. Fill and bleed the cooling system. Fill the crankcase to the correct level with fresh oil. Install the valve lifters and valve train in the reverse order of removal, making sure that pushrods and lifters are installed in their original positions. Adjust the valve clearances as described in **Section 1 : 5** before refitting the rocker covers and air cleaner assembly.

1 : 7 The oil filter

The oil filter and engine oil should be changed at regular intervals of 6000 miles maximum. The oil does not lose its lubricating properties but it does collect combustion products and sludge from the engine. The oil filter element traps particles and would eventually choke if not renewed. Most of the sludge and acids, which would cause corrosion in the engine, are drained out with the old oil.

Before draining out the oil, take the car for a short run to warm and circulate the oil so that dirt is held in suspension. **If the oil is drained when the engine has stabilized at its normal operating temperature, the oil will be hot enough to cause serious scalding if it runs over the hand.** Drain the old oil into a suitable container by removing the drain plug from the oil pan. Place a drip tray under the oil filter. Two types of filter may be fitted; the cartridge type shown in **FIG 1 : 26,** and the renewable element type shown in **FIG 1 : 27.** If the

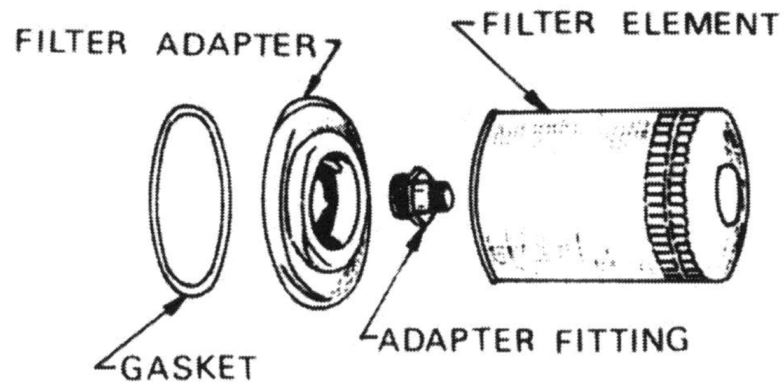

FIG 1 : 26 Cartridge type oil filter

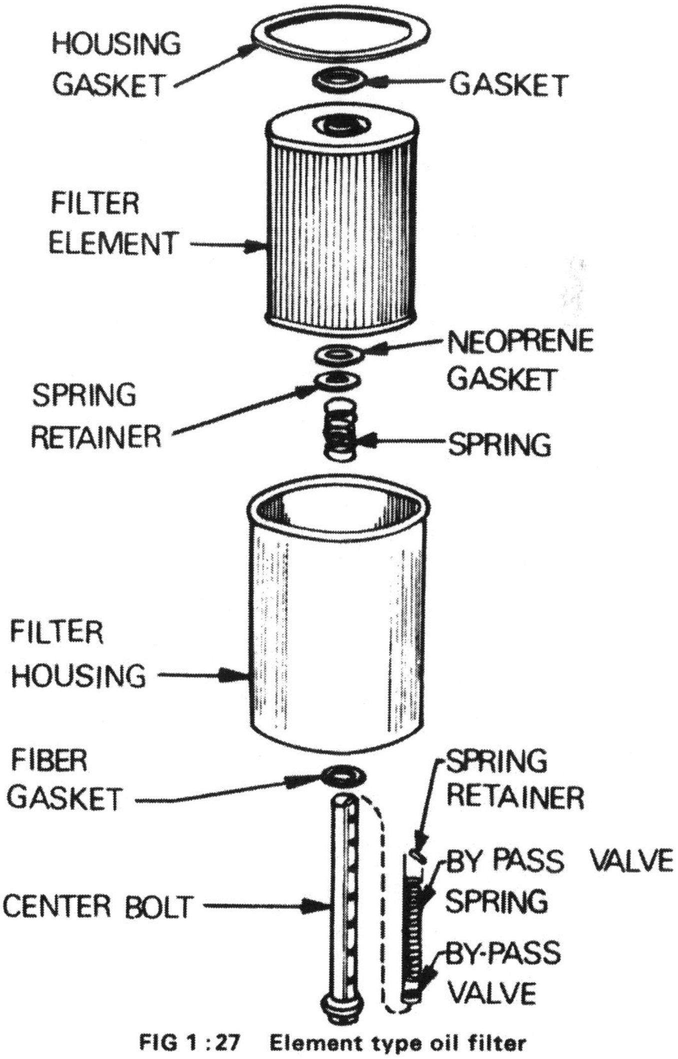

FIG 1 : 27 Element type oil filter

filter is of the cartridge type, unscrew the old cartridge from the adaptor and discard it complete. Check the gasket on the new cartridge and smear it lightly with oil. Screw the cartridge back into place by hand until the gasket touches the adaptor face, and then tighten the canister by a further ½ turn.

On the element type filter, remove the assembly by unscrewing the centre bolt. Remove the old element and its neoprene gaskets from the housing and discard them. Withdraw the centre bolt from the housing and discard the fibre washer. Pry out the housing gasket from the adaptor, using a small screwdriver. In some cases it will be necessary to use two small screwdrivers, one to pull the gasket partly out of the adaptor and the other to pass between gasket and adaptor so that the gasket can be pulled out. Wash the housing, centre bolt assembly, spring and spring retainer in petrol or suitable solvent. Fit a new fibre washer to the centre bolt and slip it back into the housing. Install the spring, spring retainer and new neoprene gasket, followed by a new element and upper gasket. Install the new housing gasket squarely and fully into the adaptor housing. Install the filter assembly in the housing and tighten the centre bolt finger tight. Rotate the housing slightly in each direction to ensure that it seats fully into the gasket. Tighten the centre bolt sufficiently to ensure that the housing is well seated and will not leak. **Do not overtighten the centre bolt,** as leaks are caused by incorrect installation of the housing gasket or by the housing not being square to the adaptor, and overtightening will only cause further distortion.

Refit the oil pan drain plug and tighten it to a torque of 15 to 20 lb ft. Fill the crankcase with approximately 5 quarts of oil, which includes the quart required for the oil filter change. The level will be high. Start the engine and check for oil leaks immediately. Allow the engine to run at a fast idle for a few minutes to ensure that the oil filter is filled. Stop the engine and allow it to stand for five minutes before checking the level on the dipstick. Top up as required to bring the level to the full mark but **do not overfill the crankcase.**

A further oil strainer is fitted to the inlet tube assembly of the oil pump. This strainer is only accessible after the oil pan has been removed. Whenever the oil pan is removed, for whatever reason, the inlet tube assembly should be removed from the oil pump, washed in gasoline and checked for tears in the gauze. Install the assembly using a new gasket.

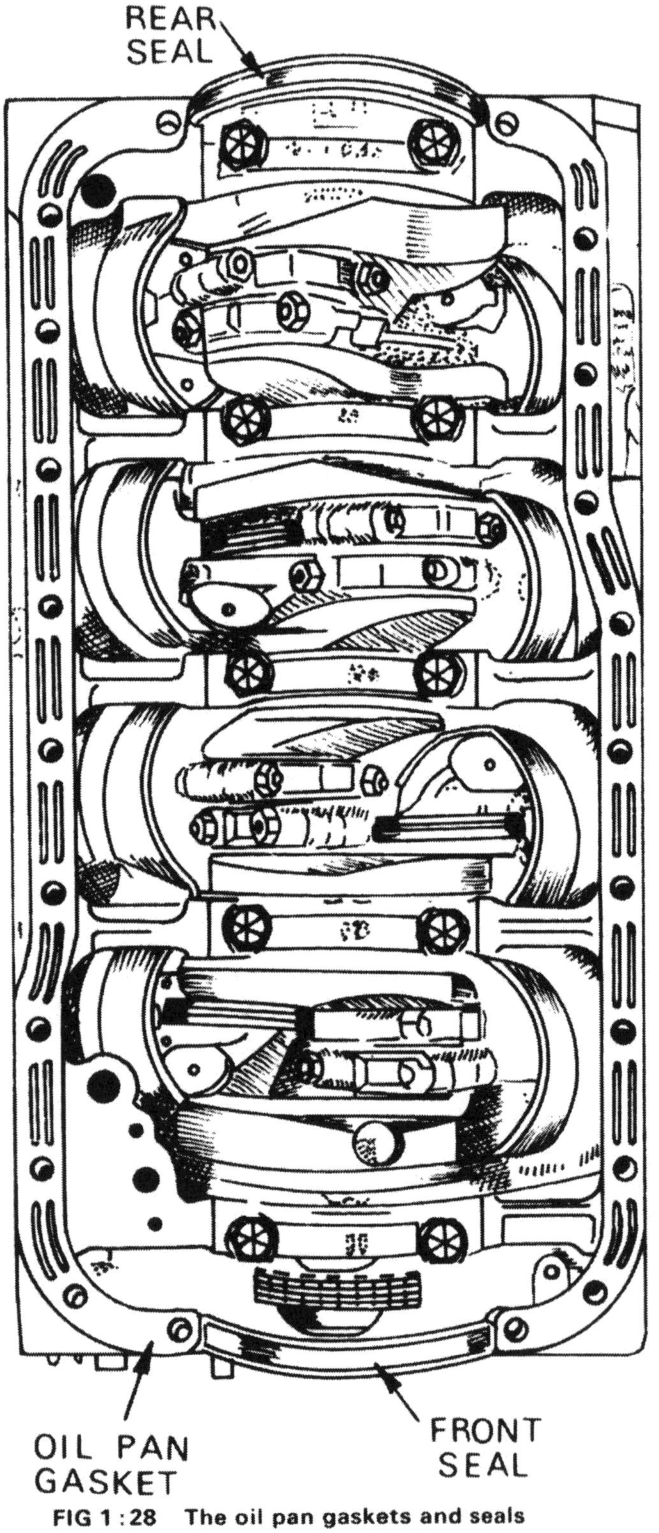

FIG 1:28 The oil pan gaskets and seals

1:8 The oil pan and oil pump

Oil pan removal:

Drain the crankcase, by removing the oil pan drain plug, and remove the oil level dipstick. Disconnect the stabilizer bar from the frame and lower it (see **Chapter 9**). Take out all the oil pan retaining bolts, except one on either side. Support the oil pan by hand and take out the two retaining bolts. Lower the oil pan and crank the engine until the crankshaft throws will give the maximum clearance and then remove the oil pan from under the car.

Take out the two bolts that secure the oil pump inlet tube assembly to the oil pump and remove the inlet tube assembly. Discard all the old seals and gaskets.

Wash the components in casoline, using rags to wipe out the worst of the sludge from the bottom of the oil pan. Remove all parts of old gasket or seal from the sump and crankcase mating surfaces. Make sure that the baffle is secure and that there is no dirt under it.

Install the inlet tube and screen assembly onto the oil pump, using a new gasket. Lightly coat the faces of the sump and crankcase with sealing compound. Position the oil pan gaskets on the crankcase. Position the front seal on the cylinder front cover and the rear seal on the rear main bearing cap, as shown in **FIG 1:28**, making sure that the tabs on the seals are over the oil pan gaskets. Raise the oil pan back into position, taking care not to disturb the gaskets or seals, and secure it with one bolt fingertight on either side. Fit the remainder of the securing bolts, guiding them through the holes in the gaskets. Tighten the bolts, working from the centre outwards, to a torque of 9 to 11 lb ft (the $\frac{5}{16}$ inch bolts to 18 lb ft). Make sure that the oil pan drain plug is in place and fill the crankcase with fresh oil. **If the oil filter has not been renewed, the level must not be allowed to go above the full mark on the dipstick.**

Oil pump removal:

Remove the oil pan and inlet tube assembly. The parts of the oil pump are shown in **FIG 1 : 29.** Remove the retaining bolts and withdraw the oil pump as well as its intermediate drive shaft. Discard the old gasket fitted above the pump.

Install the oil pump in the reverse order of removal. Position the intermediate shaft firmly and fully up into the distributor socket. The stop on the shaft should touch the roof of the crankcase, and if it does not do so remove the shaft and position the stop so that it will. Fit a new gasket

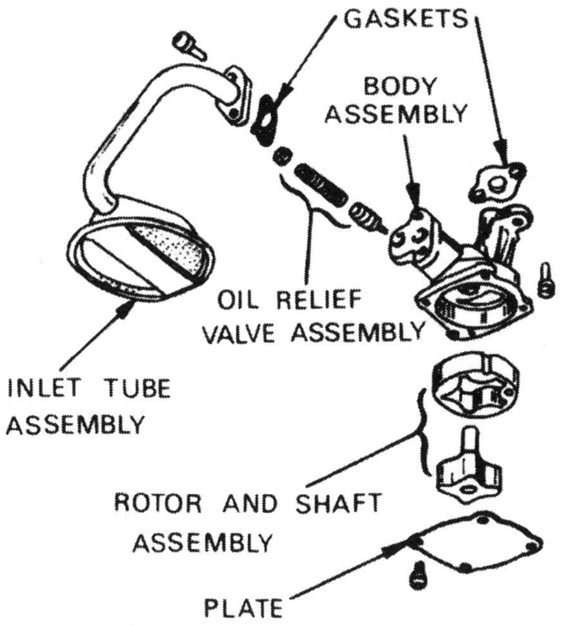

FIG 1 : 29 The oil pump components

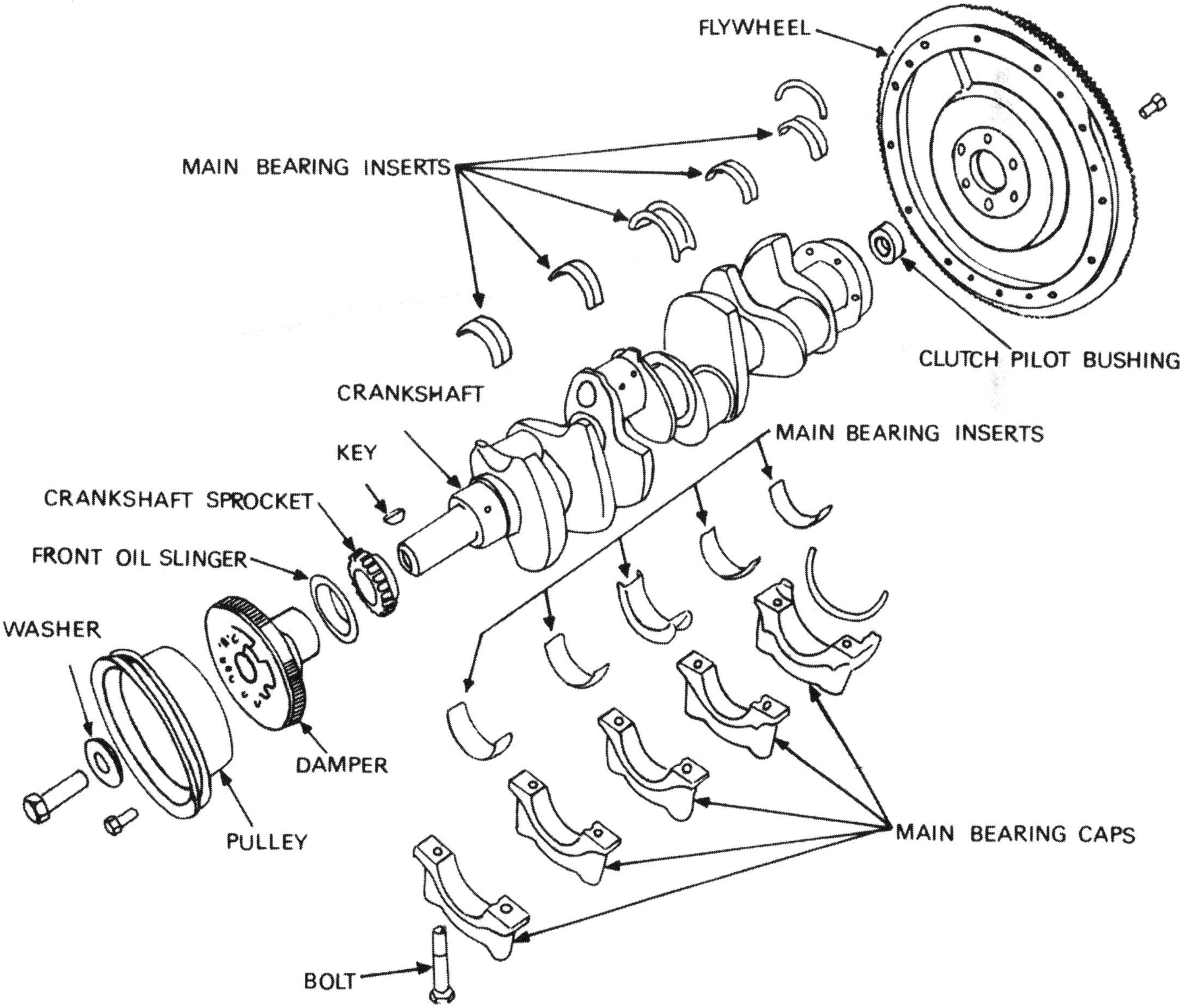

FIG 1 : 30 The crankshaft and its related parts

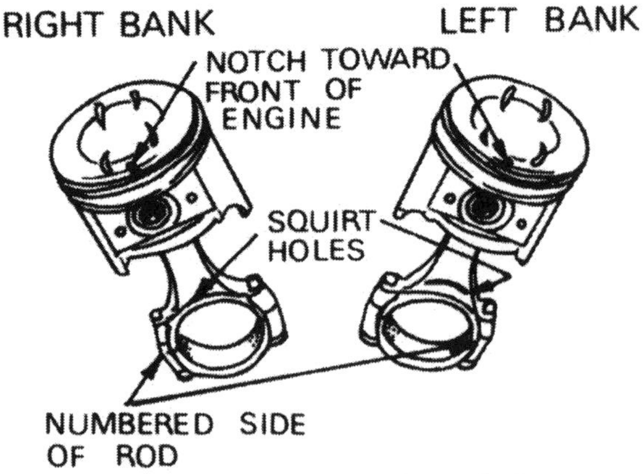

FIG 1 :31 The correct positioning of the connecting rods and pistons. Later models may not have the oil squirt holes

to the oil pump and gently press the pump back into position. If the pump will not go back, turn the drive shaft to a new position to allow the hex. to align. **Do not attempt to force the pump back into position.** Secure the pump with its bolts, refit the inlet tube assembly followed by the oil pan.

Oil pump servicing:

The pump takes the full flow of oil through the engine and should therefore give long satisfactory service with minimum wear. Failure to change the oil at regular intervals, leaving the oil full of dirt, will be the main cause of wear in the oil pump.

Remove the cover retaining screw and the cover. Withdraw the inner rotor and shaft assembly followed by the outer rotor. Screw a self-threading screw into the cap for the relief valve and use the screw to draw out the cap. Shake out the spring and plunger.

Wash all the parts in gasoline or suitable solvent. Check the lobes of the rotors for obvious wear or scoring. If the rotors are defective, they must be renewed as a set as they are matched on manufacture. Refit both rotors back into the body assembly. Make sure that the identification dimple on the outer rotor is facing outwards on the same side as the dimple on the inner rotor. Insert feeler gauges between the outer race and housing to measure the clearance between them. If the clearance exceeds .006 to .012 inch, try renewing the rotors but if the clearance is still excessive a new pump must be fitted. Lay a steel straightedge across the end of the housing and use feeler gauges to measure the end float of the rotors. If the end float exceeds .001 to .004 inch, try renewing the rotors and if this fails to cure the fault renew the complete pump.

Check the end plate and renew it if it shows scoring or wear on its surface. Make sure that the rotor marks are correctly positioned before securing the end plate back into position.

Check the relief valve spring for weakening (dimensions given in **Technical Data**). Examine the plunger for scoring or fret marks. Install the plunger and spring back into the body assembly and secure them with a new end cap. Pour fresh engine oil in through the inlet or outlet and rotate the pump to spread the oil throughout the pump, as this will help in priming the pump when the engine starts. Check that the pump rotates freely.

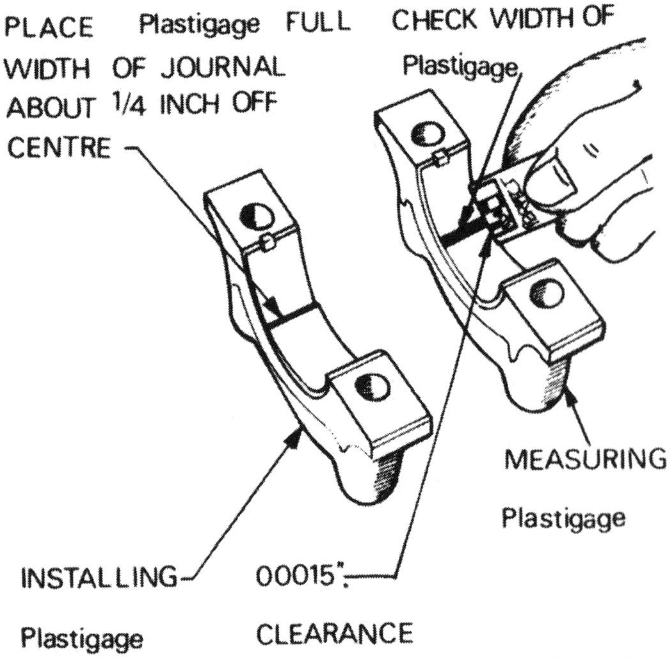

FIG 1 :32 Checking the clearance on a bearing using Plastigage

1 :9 The crankshaft bearings

All the crankshaft bearings, including both the sets of main and connecting rod bearings, can be renewed with the engine still fitted in the vehicle. **If the crankshaft journals or crankpins are found to be worn or scored, the engine must be removed from the vehicle and dismantled in order to remove the crankshaft.**

The crankshaft and its associated parts are shown in **FIG 1 :30.** The correct position of the notches on the pistons and numbers on the connecting rods are shown in **FIG 1 :31.**

It should be noted that if the rear crankshaft seal is found to be leaking, the engine must be removed from the vehicle and the crankshaft removed from the engine in order to renew the upper portion of the seal.

Drain the crankcase, remove the oil pan and oil pump as described in the previous section. Check that the connecting rods and all bearing caps are identified by numerals. If these numerals are not present, identify the parts with a code of light centre punch marks, so that they will be installed not only in their original positions but also facing in their original directions.

Standard bearing shells are supplied in two wall thicknesses, to allow the diametrical clearance of the bearing to be set with precision. The shells marked with a blue daub of colour give less clearance, and should be tried first, while those marked with a daub of red colour have a greater clearance. It is permissible to intermix both red and blue shells throughout the engine, even to the extent of having one red and one blue shell on one bearing. Undersize bearings do not have any grades and the journal or crankpin must be accurately ground to undersize. The diametrical clearances for the models are given in **Technical Data.**

To check the diametrical clearance of a bearing, use a Plastigage as shown in **FIG 1 :32.** With the engine fitted to the vehicle, renew one main bearing set at a time and when checking the clearance of a bearing, support the adjacent counterweight of the crankshaft on a jack to prevent the weight of the crankshaft from altering the clearance. Make sure that the journal, or crankpin, as well as the bearing cap and shells are scrupulously clean. Lay a strip of gauge into place as shown and refit the cap, tightening the bolts or nuts to the correct torque load. **Do not crank the engine.** Remove the cap and measure the spread of the strip with the gauge provided. Check at the widest point to get the minimum clearance and at the narrowest point to get the maximum clearance, the difference representing the taper.

If the correct clearance cannot be obtained using a combination of a blue and red shell, try one —.002 undersize shell and a red or blue shell. **If more than one —.002 shell is fitted to the journals, make sure that all the undersize shells are on the same side to ensure that the crankshaft is kept in alignment. Never file the caps or connecting rods in an attempt to reduce the bearing clearance.** This will make the bores non-circular, causing pinch loads on the bearing shells followed by rapid failure.

When the bearing has been disconnected, clean the crankpin thoroughly and measure it at several points with a micrometer gauge. Ovality must not exceed .0004 inch and taper must not exceed .0003 inch per inch. These figures also apply to the journals. If the ovality, taper or wear exceed the limits, or if there are scores, then the crankshaft must be removed (after the engine has been taken out of the vehicle) and sent away for specialist regrinding to the next undersize.

Main bearings:

If the engine is out of the vehicle, all the bearing caps can be removed. When changing the shells with the engine in the vehicle, service one bearing at a time so that the crankshaft is always supported by the remainder of the bearings. Remove the bolts securing the cap and pull off the cap. Slide out the shell insert from the cap. Fit the tool 6331 to the oil hole, shown in **FIG 1:33** in the crankshaft and slowly crank the engine so that the tool catches on the edge of the upper shell and pushes it out

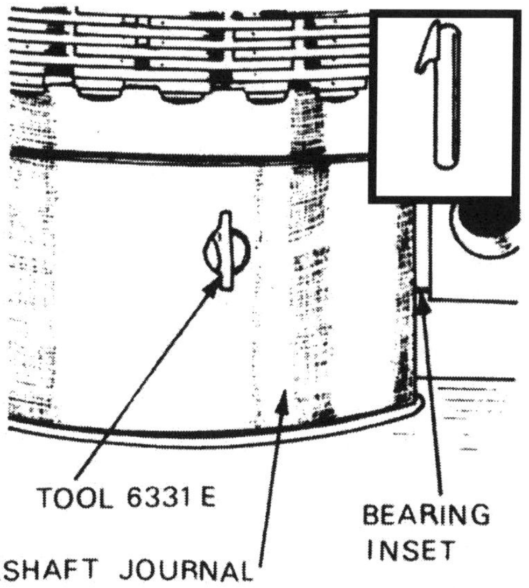

FIG 1:33 The special tool for removing the upper main bearing shell inserts, with the engine installed in the vehicle

in the direction of the locating tag. Flush out between the crankshaft and bore in the crankcase with a syringe to remove all particles of dirt. Lubricate the new shell with fresh oil and fit it back into position, using the tool in the crankshaft to push it back and making sure that the locating tag seats into its recess. Check the bearing diametrical clearance as described earlier, using a Plastigage, after tightening the bolts to a torque of 60 to 70 lb ft. When the clearance has been satisfactorily set, by selective fitting of the shell inserts, liberally oil the journal and refit the cap. Tighten the cap bolts to a torque of 60 to 70 lb ft and check that the crankshaft rotates freely before servicing the next bearing.

No. 3 (centre) main bearing is fitted with special shell inserts which control the crankshaft end float. Fit the cap with the bolts fingertight. Pry the crankshaft forward against the thrust surface of the upper half of the bearing and hold it in this position. Pry the cap to the rear so as to align the thrust surfaces. Tighten the cap bolts to the correct torque while still holding the crankshaft forward. The method is shown in **FIG 1:34.**

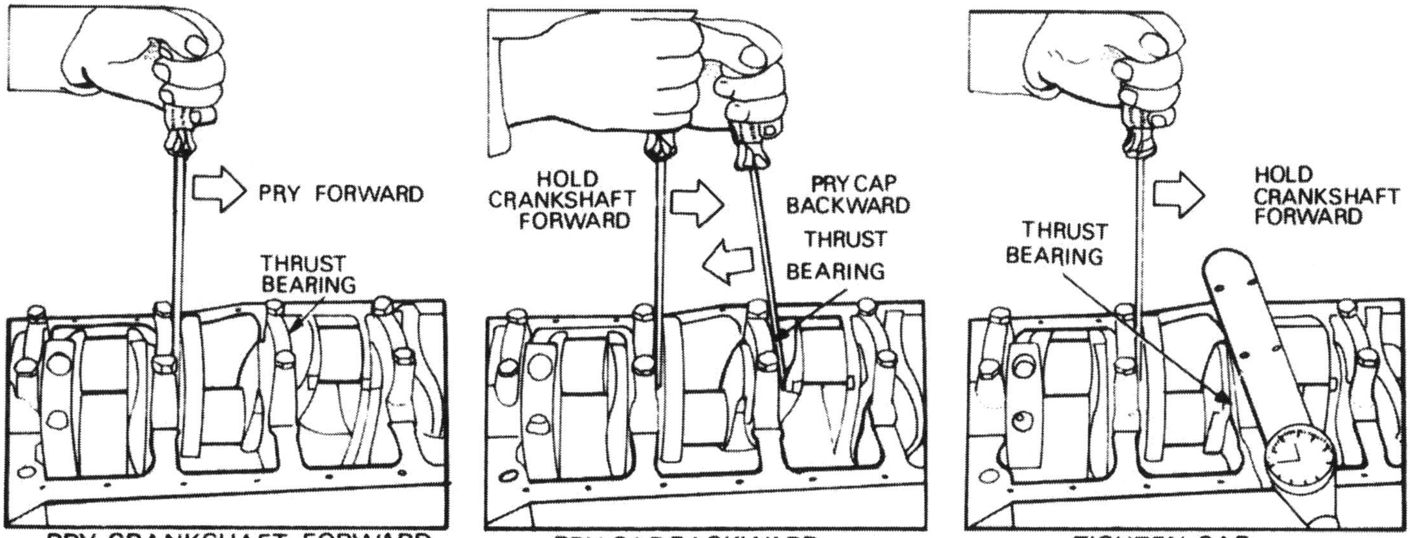

FIG 1:34 The correct method of fitting and securing the centre main bearing cap to ensure that the thrust faces align

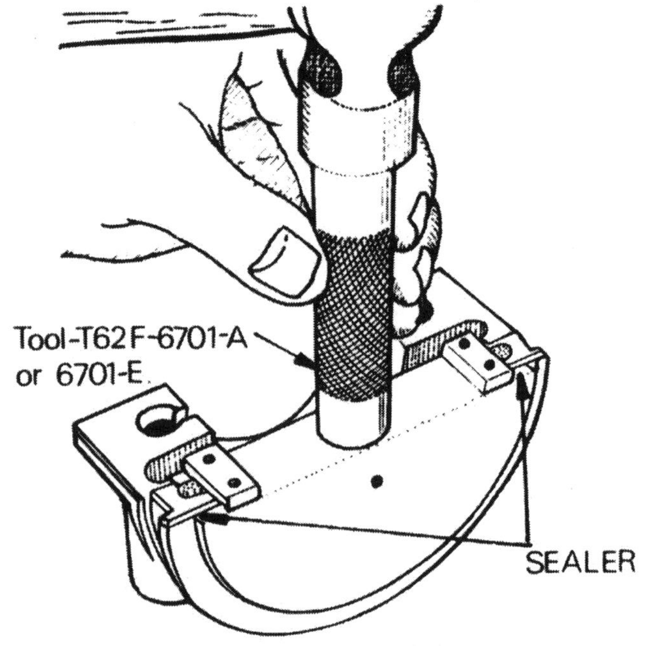

FIG 1:35 The forming tool for fitting the rear crankshaft seal halves

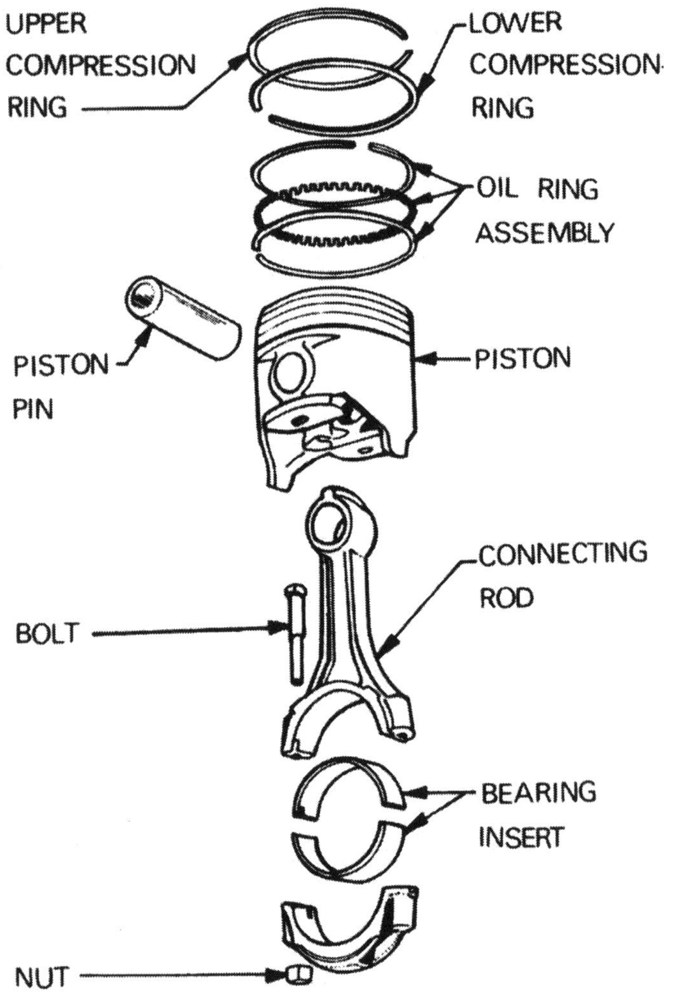

FIG 1:36 Typical piston and connecting rod assembly parts

The old bottom half of the oil seal in the rear main bearing cap should be removed and a new piece installed in its place. Preform the new seal by gently bending it with the fingers. Make sure that the cap is perfectly clean. Insert the seal into its recess, pressing in the centre first and then working outwards so that the seal extends equally on both sides. The seal forming tool should then be used as shown in **FIG 1:35** to ensure that the seal is fully in position. Use a sharp knife to cut off the protruding ends of the seal and lightly apply sealer to the areas shown. **Do not apply sealer forwards of the oil slinger groove.** Refit the rear main bearing cap.

Connecting rod bearings:

These are treated in a similar manner to the main bearings, noting that there are two connecting rods on each crankpin. Crank the engine until the pistons are at their lowest position and then undo the nuts that secure the caps. Pull off the caps and push the connecting rods and pistons up the bores so that the connecting rods clear the crankpins. The shell inserts can then be removed from both the cap and connecting rod. Renew the shells if they are worn, scored or pitted. Check the crankpins with a micrometer gauge.

Selectively fit new shell inserts, measuring the diametrical clearance using a Plastigage strip.

Install the new shell inserts, liberally lubricate the crankpins with clean oil and pull the connecting rods down into place on crankpins. Refit the bearing caps and secure them by tightening the nuts to a torque of 19 to 24 lb ft. Make sure that the identification numerals face the correct direction. **The connecting rod bolts must be fitted with their flats outwards otherwise the bolts will loosen and fail.**

Check the clearance between each pair of connecting rods, using feeler gauges, and ensure that the crankshaft rotates freely. The crankshaft should be rotated with great care if any connecting rods are disconnected as there is a danger of trapping the end of the rod between a throw on the crankshaft and the side of the crankcase.

When all the bearings and seals are in place, install the oil pump and oil pan. Fill the crankcase to the correct level with fresh oil. It is advisable to crank the engine on the starter, with the sparking plugs removed, not only to prime the oil pump and oil lines but to ensure that there are no snags with the bearings.

1:10 The pistons and connecting rods

These can be removed with the engine installed in the vehicle. Remove the appropriate cylinder head or both heads if all the pistons are to be removed (see **Section 1:4**). Remove the oil pan and oil pump (see **Section 1:8**) and then disconnect the appropriate connecting rods from the crankpins (see previous section). Before removing the pistons and connecting rod assembly, remove (with service garage cutters) the unworn ridge or ring of carbon around the top of the bore. Lay rags into the bores to catch the swarf and dirt from the cutter. Press the connecting rod up the bore of the cylinder, using a wooden rod when it is too far up to push by hand, until the piston extends beyond the top of the bore and the assembly can be withdrawn from the engine.

Cylinder bores:

These should be examined whenever the cylinder head is removed. For accurate checks a special dial indicator is required and the bore should be measured at several points both with and across the thrust axis so that ovality, taper and wear can be determined. A rough guide to the cylinder wear can be judged from the depth of the unworn ridge around the top of the bore. The taper should be within the limit of .001 inch and the wear limit is .010 inch. The out-of-round should be within the limit of .001 inch and the wear limit is .005 inch. If the wear limits are exceeded, the cylinder requires reboring and fitting with a new oversize piston. It should be noted that all pistons, standard and oversize, are of the same weight, and therefore if one bore only is accidentally scored or damaged it alone may be bored out and an oversize piston fitted.

The unworn ridge around the top of the bore should be removed (using special cutters) before extracting the pistons, as otherwise the rings may be damaged as they pass over the ridge. A slight risk can be taken and the ridge left in place if it is only shallow. If new piston rings only are to be fitted, to cure high oil consumption without the wear requiring reboring, then the unworn ridge must be removed and the whole surface of the bore lightly honed. This ensures that the top compression ring will not hit the ridge and break up, and allows the new rings and cylinder bore to run in together.

Piston pins:

The pins are an interference fit in the connecting rods and a floating fit in the pistons. Special tools, held by a service garage, are essential for removing or refitting the piston pin. The piston should not be removed unnecessarily from the connecting rod and all cleaning operations can be carried out with the piston pin fitted.

If the pistons are removed from the connecting rods, it is essential that they are installed so that the notches on the crown face forwards, the identification numbers face outwards and the oil squirt holes face inwards, as shown in **FIG 1:31**.

Piston rings:

The piston, rings, piston pin and connecting rod parts are shown in **FIG 1:36**.

Special ring expanding tools are made for removing and refitting piston rings. The tool grips the ends of the ring and then gently parts them so that the ring can be lifted up and out of the piston. If these tools are not available, three shims (such as discarded feeler gauges) can be used instead. Slide a shim under one end of the ring and work the shim around under the ring, at the same time pressing the ring as it comes free gently onto the land of the piston above it. When all the ring is on the land, slip the other two shims equidistantly under the ring. Gently part the ends of the ring with the thumb nails and slide the ring up and off the piston to remove it. **Take great care to prevent the back of the ring from scoring the piston.**

Use a broken piston ring, or special tool, to remove all the carbon and deposits from the ring grooves. **Take great care not to remove metal from the piston otherwise the oil consumption will be increased.**

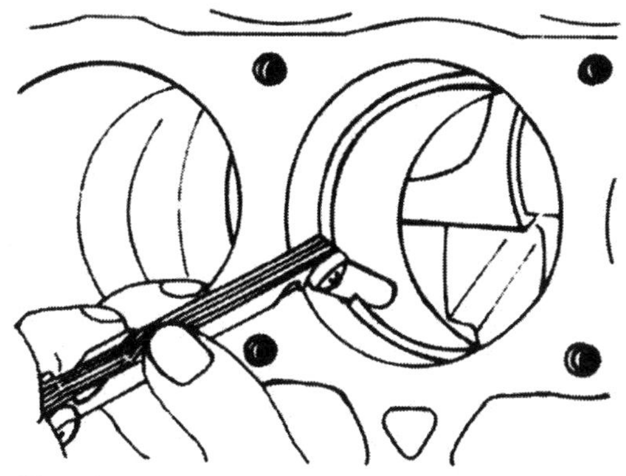

FIG 1:37 Checking the gap on the ends of the rings, when fitted

Clean out the oil return holes behind the oil control ring with a blunt piece of wire.

Spring the new rings, in turn, into the top of the bore and press them squarely down with an inverted piston until they are approximately 1 inch from the top of the bore. Measure the gap between the ends with feeler gauges, as shown in **FIG 1:37**. If the gaps are too small (see **Technical Data**) file the ends carefully with a smooth file or try another ring. If the gap is too large, check that the correct ring size has been obtained and try another ring.

Check the grooves in the pistons. If the lower lands have high steps the pistons should be renewed. Fit the rings back into place and use a feeler gauge to check the side clearance on the compression rings, as shown in **FIG 1:38**. If the clearance is excessive, new pistons must be fitted. Check that the ring slides freely around in its groove.

Pistons:

Pistons are graded to enable them to be selectively fitted to give the correct clearance between piston and bore. Accurate measuring equipment is required for measuring the piston OD and bore ID so that the actual clearance can be determined, and the selection of pistons is best left to a service garage. Once the pistons have been selected and matched to bores, they must be fitted to their correct bores.

The piston crown may be lightly polished with worn emerycloth wetted with kerosene as all traces of abrasive

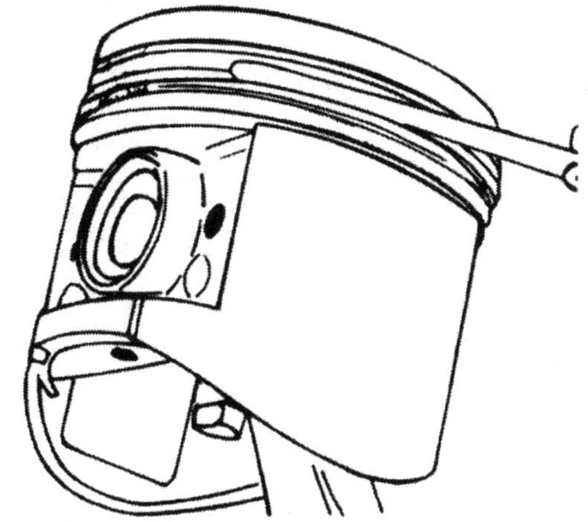

FIG 1:38 Checking the piston ring side clearance

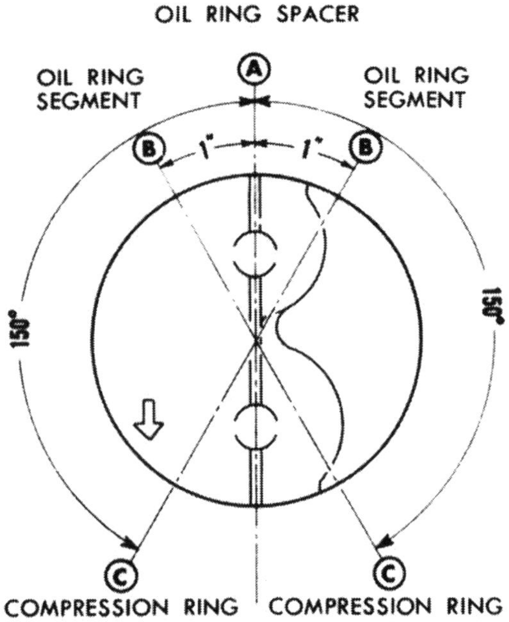

FIG 1 : 39 The correct spacking of the gaps in the piston rings

can be washed off afterwards. Lacquering or gum on the side of the piston, indicative of gas blow-by past the rings, must not be removed with any form of abrasive. Solvents such as denatured alcohol, carbon tetrachloride or trichlorethylene may be used. The latter two solvents must be used in a well ventilated space only.

Installation:

If the piston has been removed from the connecting rod, make sure that they are correctly assembled, as shown in **FIG 1 : 31**. Install the piston rings and turn them until their gaps are situated as shown in **FIG 1 : 39**. Compress the rings into the piston with a suitable ring clamp. If a proper ring clamp is unavailable, a large worm-driven hose clip may be used instead—though more care will be needed. Oil the rings and piston lightly before fitting the clamp. Lower the connecting rod down the bore of the correct cylinder, **making sure that its identifying number is outwards and the notch in the piston crown faces forwards.** Carefully enter the skirt of the piston into the bore and press down the piston, making sure that the appropriate throw on the crankshaft is at BDC, until the clamp touches the block. Keep gently pressing the piston down into the bore, allowing the clamp to slide off as each ring enters the bore. **Do not allow the ring to escape from the clamp before it is in the bore and do not attempt to force the pistons into place.** The piston rings are very brittle and any mishandling will snap them.

When the piston has fully entered the bore, install the shell inserts in the connecting rod and bearing cap. Pull the connecting rod down onto the crankpin and fit the cap as described in the previous section. Install the cylinder head(s), manifolds, oil pump and oil pan in the reverse order of removal.

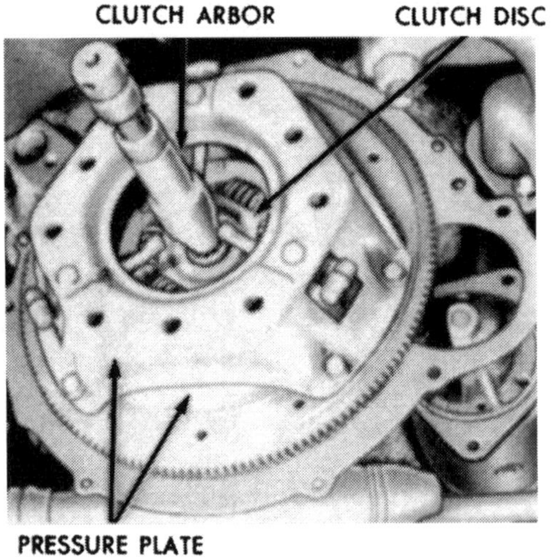

FIG 1 : 40 The clutch attachments, showing the locating arbor in place

1 : 11 The clutch and flywheel

These are only accessible after the transmission has been removed from the vehicle (see **Chapter 6**) or the engine has been removed. This section deals only with the removal and refitting of the clutch. All details of clutch maintenance and servicing are dealt with in **Chapter 5.**

Removing clutch:

With the transmission removed as described in **Chapter 6**, disconnect the return spring from the clutch release lever. Disconnect the starter cables and remove the starter motor. Remove the bolts that secure the lower part of the flywheel housing to the engine rear plate, as well as the bolts that secure the housing to the cylinder block, so that the flywheel housing can be removed.

The attachments of the clutch, with the centralizing arbor fitted, are shown in **FIG 1 : 40.** If the clutch is to be used again, make alignment marks on the cover and flywheel so that the clutch can be installed back into its original position. Evenly and diagonally slacken the six clutch attachment bolts, while holding the cover in place. Remove the bolts and lift off the clutch cover assembly, taking care not to allow the clutch disc to fall. **Avoid handling the parts with greasy hands.**

The clutch is installed in the reverse order of removal. Secure the clutch disc in place with a centralizing arbor and then install the cover assembly. **It is essential that the arbor is left in place while the attachment bolts are progressively tightened, otherwise the input shaft of the gearbox will not pass through the clutch disc.**

Flywheel:

Remove the clutch, or automatic transmission as described in **Chapter 7.** Remove the retaining bolts and withdraw the flywheel from the crankshaft.

The flywheel is installed in the reverse order of removal. Make sure that the mating surfaces of both crankshaft and clutch are scrupulously clean and free from any specks of dirt. Check both the flywheel face runout and starter ring gear runout using a dial indicator mounted on the block.

If the flywheel friction face is scored or damaged the damage may be machined off, provided that not more than .045 inch of stock is removed. If the damage cannot be cleaned off without exceeding this limit a new flywheel must be fitted.

Flywheel ring gear:

This is renewable, and a new ring should be fitted if the teeth are broken or worn. Drill partially through the root of a tooth and split the ring with a chisel so that it can be removed. **Take great care not to mark or damage the flywheel itself.**

Thoroughly clean the periphery of the flywheel, using a wire brush to remove dirt and emerycloth to smooth down high spots. Heat the new ring evenly until it has expanded sufficiently to be fitted to the flywheel. **Do not heat the ring beyond 500°F as this will ruin the temper of the metal and avoid localized hot spots.** Use tongs to lay the ring into position and drive it fully against the shoulder of the flywheel using drifts made of soft metal. Allow the ring to cool naturally before moving the flywheel.

Pilot bushing:

This is fitted into the end of the crankshaft to guide the front end of the transmission input shaft. Check that the bush is a tight fit in the crankshaft and that the ID is within limits and not bell-mouthed. Renew the bush if it is worn or shows fret marks.

The old bush is best removed using an internal two-legged extractor coupled with an impact hammer. The bushing can be removed using a length of snug fitting rod. Find or turn a suitable piece. Pack the bushing recess with grease, leaving no air pockets. Press the end of the rod into the bush and give it a few blows on the end with a hammer. The hydraulic action of the grease will force out the bushing.

Install the new bushing using a suitable stepped drift then lubricate the bushing lightly with wheel bearing grease.

1:12 The crankshaft

The crankshaft parts are shown in **FIG 1:30** and its bearings are dealt with in **Section 1:9**. This section deals with the removal of the crankshaft.

A stand will be required to support the engine in an inverted position. If a suitable stand is not available, remove all parts that can be easily damaged (alternator, distributor, carburetor, Thermactor pump) and stand the engine inverted on old tires after removal.

1 Drain the oil from the crankcase. Remove the engine from the vehicle (see **Section 1:2**). Remove the vulnerable parts and invert the engine. Remove the spark plugs from all cylinders.

2 Remove the oil pan and oil pump (see **Section 1:8**). Remove the cylinder front cover, timing chain and sprockets (see **Section 1:6**). Remove the clutch and flywheel (see previous section).

3 Check that all the connecting rods and bearing caps are correctly numbered in sequence. Disconnect each connecting rod in turn from the crankpins and push the connecting rod and piston assemblies as far as they will go into the bores. Loosely refit the connecting rod bearing caps to their connecting rods.

4 Evenly slacken all the bolts that secure the main bearing caps and remove the caps. Carefully lift out the crankshaft from the engine. **Handle the crankshaft with care to prevent damage to the ground surfaces or fractures.**

5 Remove the upper and lower rear seals from the crankcase and rear main bearing cap. Loosely refit the main bearing caps to their correct locations.

If the crankshaft is scored or worn, it should be sent away for specialist regrinding. The correct undersize bearing shell inserts will be returned with it. New shell inserts are fitted exactly as received, apart from cleaning off any protective, and do not require boring or scraping. It is advisable to check the bearing diametrical clearances, as described in **Section 1:9.**

Before installing the crankshaft, clean it and the bearing bores thoroughly. Blow through the oilways in the crankshaft with gasoline or kerosene and then compressed air. A syringe and tire pump can be used. If a degreasing solvent is used, the bright metal parts should be wiped over with a piece of clean lint-free cloth dipped in clean oil, so as to prevent corrosion. Cleaning out the oilways is particularly important if the journals and crankpins have been reground, otherwise particles of swarf may remain in the crankshaft and be forced into the bearings by the pressure of the oil.

The crankshaft is installed in the reverse order of removal. Remove the bearing caps and discard the old bearing shell inserts if new ones are being fitted. If the old bearing shell inserts are being used again, they must be refitted into their original positions. Locate the new bearing shells in place, ensuring that their tags are in their recesses.

Fit new upper and lower rear seals into the recesses in the rear main bearing cap and crankcase. Both seals are fitted as described in **Section 1:9** for fitting the seal to the rear main bearing cap.

Lubricate the crankshaft bearings liberally with fresh oil. Lay the crankshaft back into place, taking great care not to damage the bearing surfaces, and secure it in place with the main bearing caps. Mount a dial indicator on the crankcase so that the stylus rests vertically on the end of the crankshaft. Firmly lever the crankshaft backwards and forwards and use the indicator to measure the crankshaft end play. The limits should be .004 to .008 inch and the wear limit is .012 inch. If the wear limit is exceeded, fit new centre bearing shell inserts. If the end play is below limits, check the thrust faces of the bearing and crankshaft for chips, dirt or nicks. If the thrust faces are satisfactory, they are probably not correctly aligned. Slacken the bolts for the centre main bearing and set them finger tight. Carry out the procedure shown in **FIG 1:34** to bring the thrust faces into alignment and again check the end play.

When the end play is correctly set and the main bearings all secured, check that the crankshaft rotates freely. Install the connecting rod bearings and again check that the crankshaft rotates freely. The engine is then reassembled in the reverse order of dismantling and installed back into the vehicle.

1:13 Emission control

The crankcase breathing system has been briefly dealt with in **Section 1:1.** The regulator valve fitted into the system is not cleanable and it should be renewed at regular intervals. Clean the filter in the oil filler cap with kerosene but do not blow through it with compressed air. A defective crankcase breathing system can cause loping

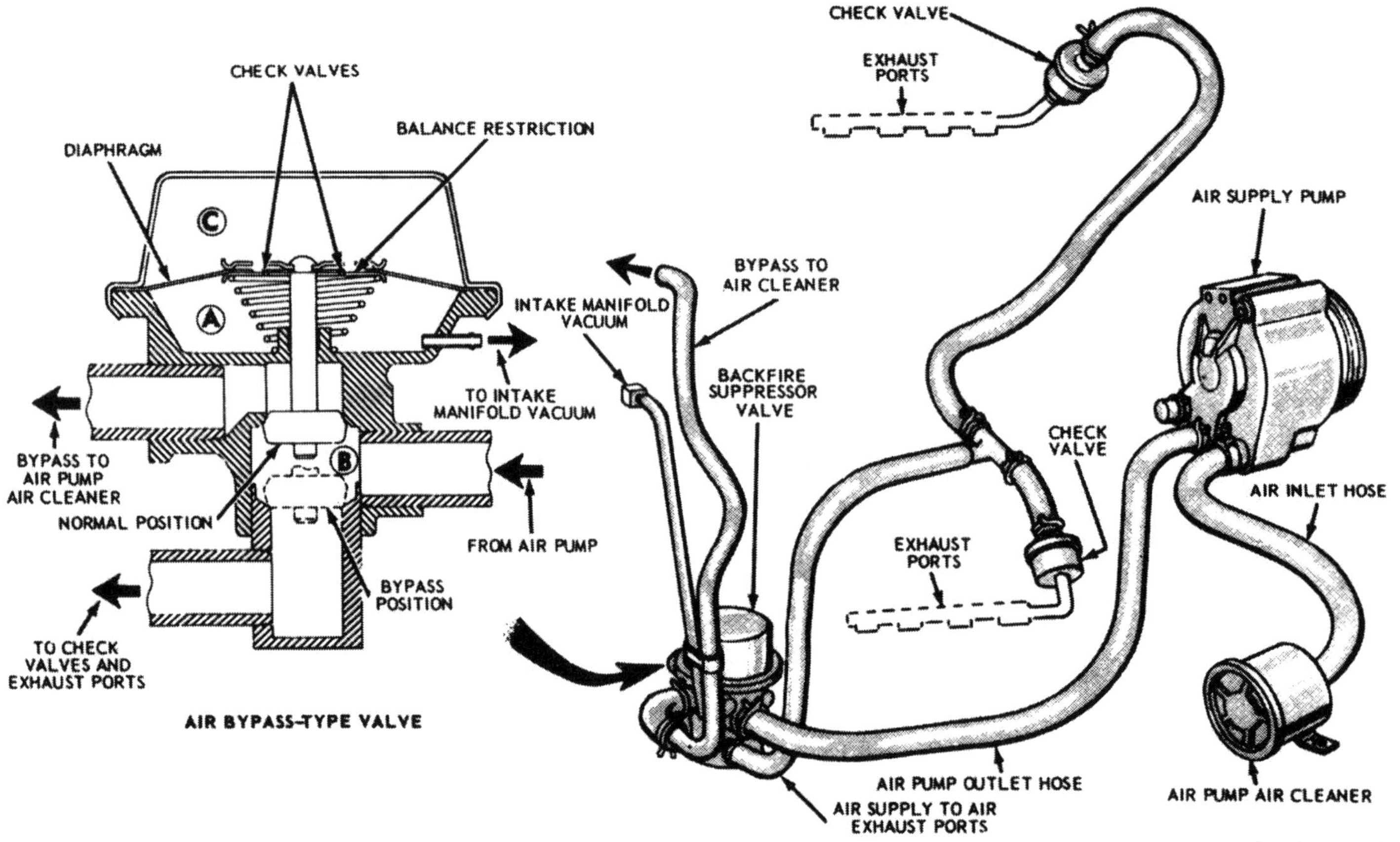

FIG 1:41 The schematic layout of the Thermactor emission control system

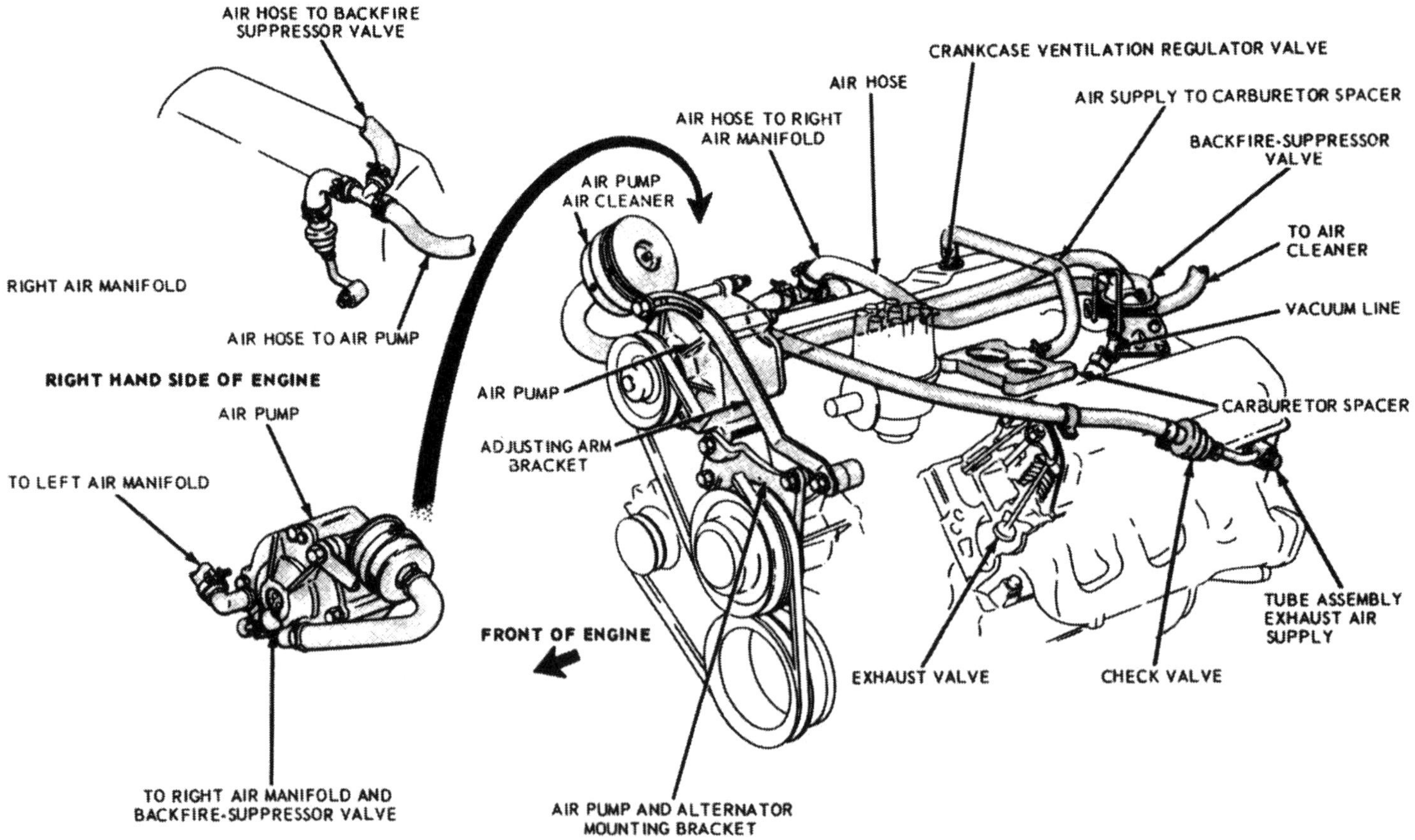

FIG 1:42 Typical installation of the Thermactor emission control system

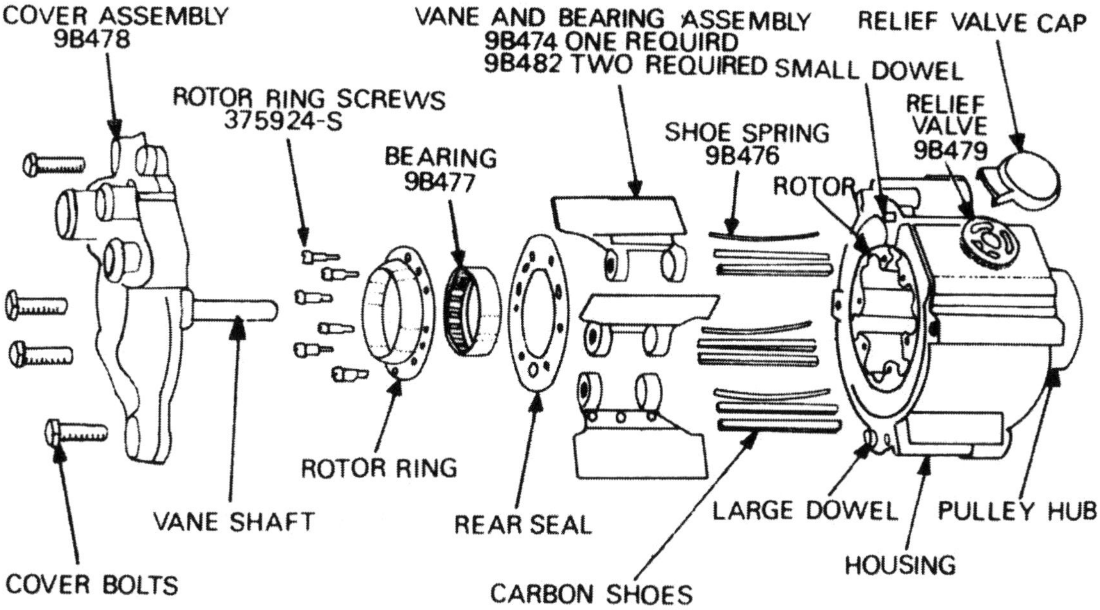

FIG 1:43 The components of the early air supply pump

or rough engine idle as well as increasing fuel consumption. If the idle is poor, check all the hoses in the system for loose connections, blockage, kinks or splits. If no fault is found, remove the old regulator valve and fit a new one in its place. If the idle is now satisfactory, leave the new valve in place and discard the old one. If the fault is still not cured it is advisable to have the system checked using a special tester at a service garage. If the system is found to be satisfactory other causes for the poor idling must be found.

The Imco system of emission control, used on vehicles fitted with automatic transmissions, relies on special carburetters and distributors and the parts will be dealt with in the relevant chapters (see **Chapters 2** and **3**).

Thermactor system:

The schematic layout of the parts in the system are shown in **FIG 1:41**. The air supply pump is driven by a belt from the engine and ensures a continuous supply of air under pressure. On earlier models the pump can be dismantled for servicing but on later models the pump is a sealed unit and must be exchanged if defective. The air for the pump is drawn in through a cleaner to remove any particles in the air. The air cleaner should be given regular attention. Hoses lead the air to the bypass and back fire valve. The valve prevents backfiring and also bypasses the air to the carburetor air intake on deceleration. From the bypass valve the air passes through hoses to the check valves for each bank of cylinders. The check valves close if the exhaust manifold pressure is greater than the pump pressure, to prevent a reverse flow of hot exhaust gases. The air is injected directly into the exhaust ports through nozzles, and mixes with the hot exhaust gases to ensure complete combustion of any remaining hydrocarbons.

Maintenance:

The installation of the parts on the engine is shown in **FIG 1:42**. There will be slight differences between models and on some engines the manifolds for the Thermactor air are integral in the cylinder heads while on others separate external manifolds are used.

Maintenance mainly consists of checking that the drive belt tension is correct and the hoses free without kinks or blockage, and renewing the air cleaner at regular intervals.

When the cylinder head is removed, check that the air passages are clean and free from deposits, cleaning them with a $\frac{5}{16}$ inch wire brush. Renew the nozzles if they are burnt or damaged. If there are apparent faults in the system, check that they are not caused by engine defects, incorrect carburetor adjustments or incorrect ignition settings.

Check valve test:

Start and run the engine until it has reached its normal operating temperature. Disconnect the air supply hoses at the check valves. The valve plate inside the valve should be lightly against its seat, away from the manifold. At idle some flutter may be expected because of the exhaust pressure pulsations, but the valve should still remain on its seat. Use a pencil to press the valve open and check that the valve returns fully and freely as soon as it is released. Slowly increase the idle to 1500 rev/min. There should be no exhaust gas leakage and the valve should stay firmly on its seat. If the valve does not return freely after moving by probe, or if it leaks at higher speeds, the valve is defective and must be renewed.

Air leakage:

Leaks in the system will make it impossible to set the idle speed correctly, making it either rough or impossible to reduce to the correct speed.

Disconnect the vacuum hose to the valve and plug the end of the hose. If the idle can now be correctly set the valve is defective and must be renewed.

Backfire suppressor valve functional test:

Disconnect the hose connecting the pump to the valve at the valve. Snap the throttle open and shut. A loud sucking nose should be heard from the valve if it is

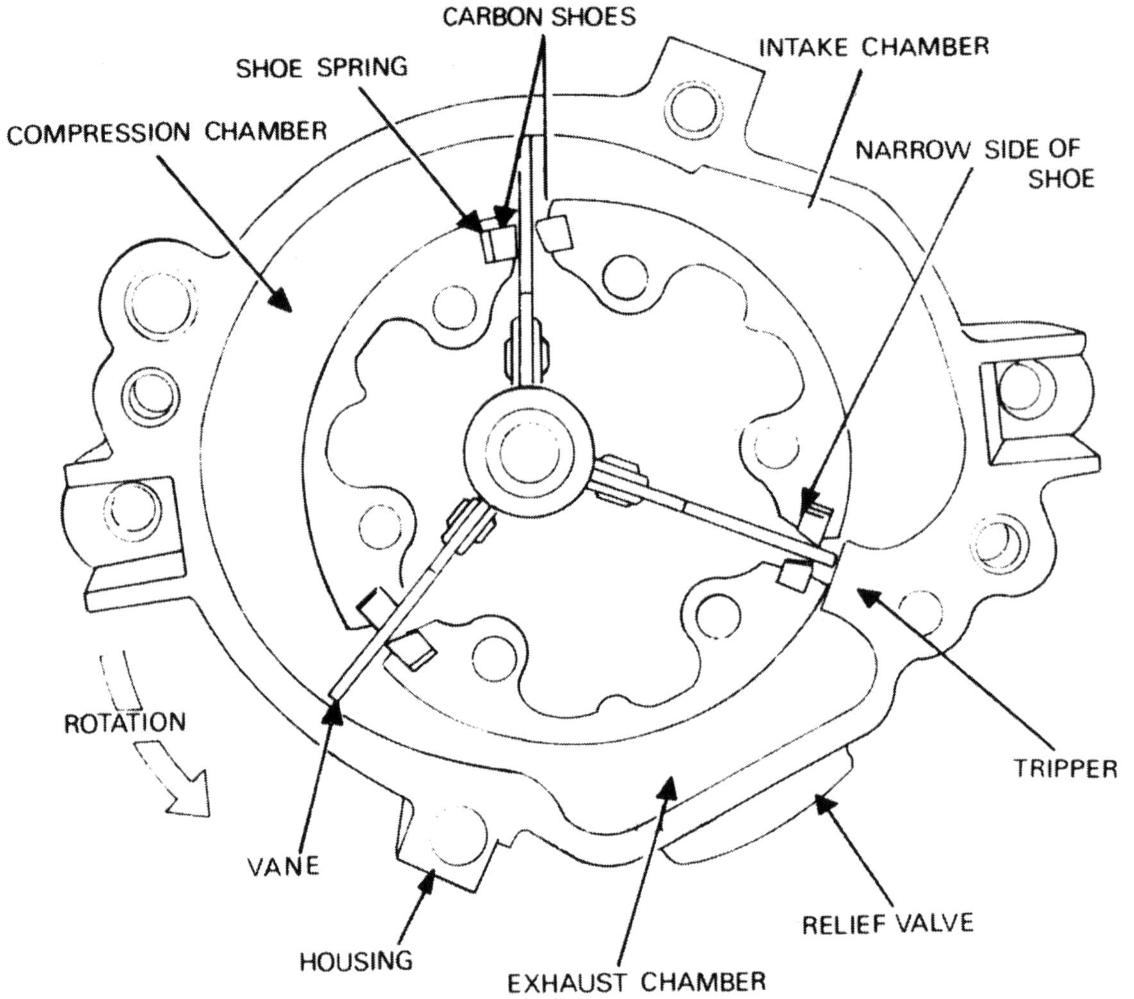

FIG 1 : 44 The correct assembly of the vanes, shoes and springs on the early air supply pump

functioning correctly. If no sucking noise is heard and the engine backfires during deceleration, the valve is defective and must be renewed.

With the hose between pump and valve connected, disconnect the hose between the air cleaner and valve at the valve. Again snap the throttle open and closed. Compressed air should be heard escaping from the valve, particularly during deceleration. If the air is not heard escaping, again the valve is defective and must be renewed.

Pump output pressure:

The output pressure should be tested if the pump is suspected of being defective. Take the car to a service garage which will have the necessary adaptors and gauges to check the pressure.

Early Thermactor pump:

The components are shown in **FIG 1 : 43** and a view of the vanes, carbon shoes and springs installed is shown in **FIG 1 : 44.**

To dismantle the pump, secure the hub of the pulley in the padded jaws of a vice. **Do not clamp the housing itself.** Remove the cover securing bolts, tap the cover off the dowels with a rawhide or plastic mallet and then remove the cover by hand. Take out the rotor ring securing screws and remove the rotor ring and bearing assembly. The bearing and rotor assembly are matched on production so if the bearing is defective a new rotor must be fitted. Removal of the rotor itself from the housing is not recommended for the owner, so if the bearing is found to be defective the complete pump should be taken to a service garage. Do not remove the bearing from the ring housing as special tools are required to install the bearing.

Remove and discard the carbon rear seal, noting that a new rear seal must be fitted on reassembly. Select the vane which has its end nearest to the end cover and gently pull it partially out. Alternately pull on each vane in turn so that they come out. If the carbon shoes start to come out with the vanes, press down the shoe with a blunt screwdriver and remove the spring. This will also make it easier to remove the vanes. Use a pair of tweezers to carefully withdraw the carbon shoes, noting that they are very fragile.

Lubricate the vane assembly bearings with special air pump grease. Hold a finger over one end of the vane bearing and partially press in the vane shaft so that the lubricant is forced into the needle rollers. Wipe off the surplus lubricant and repeat the procedure on the remaining vane bearings. Mesh the vanes so that their ends are level and install the vanes into the rotor, after blowing out carbon dust and wiping the parts clean. Slide the carbon shoes back into position, making sure that they are installed as shown in **FIG 1 : 44.** Install the springs so that their centres act against the shoe and the ends against the rotor. Lubricate the bearing and wipe away surplus grease. Refit the bearing and rotor ring, using a little Loctite on the threads of the securing screws, and a new carbon rear seal. A special tool T66L-9A486-C is made for aligning the vane bearings but if care is taken they can be worked into alignment with a piece of rod. Carefully align the cover assembly with the rotor bearings and gently work the vane shaft through the bearings, wiggling or turning as required. **Do not attempt to force the shaft through.** Align the cover with the dowels and press it back into position. Tighten and torque load the cover bolts in a diagonal sequence.

The pump relief valve can be renewed without having to remove the pump from the engine.

1:14 Trouble shooting guide

(a) Engine will not start

1 Defective ignition coil
2 Water on sparking plug leads, distributor cap or ignition coil
3 Loose wires or defective insulation in the ignition low tension circuit
4 Defective ignition contact points
5 Defective ignition condenser
6 Defective automatic choke
7 Faulty or jammed starter
8 Spark plug leads incorrectly connected
9 Blocked fuel filter or carburetor jets
10 Leaking valves
11 Sticking valves
12 Valve timing incorrect
13 Incorrect ignition timing

(b) Engine stalls

1 Check 1, 3, 4, 6, 9, 10 and 11 in (a)
2 Spark plugs defective
3 Retarded ignition
4 Mixture too weak
5 Water in fuel
6 Fuel tank vent system blocked
7 Faulty hydraulic valve lifters or incorrectly adjusted rocker arms

(c) Rough engine idle

1 Check 4, 6, 8, 10, 11, 12 and 13 in (a) and 2 and 7 in (b)
2 Incorrectly adjusted carburetor
3 Air leaks at manifold gaskets
4 Leaking vacuum fittings
5 Defective crankcase ventilation system
6 Worn piston rings
7 Worn valve stems or guides
8 Weak exhaust valve springs

(d) Engine misfires

1 Check 1, 2, 3, 4, 8, 10, 11, 12 and 13 in (a); 2, 3, 4 and 7 in (b)
2 Weak or broken valve springs

(e) Engine overheats (see **Chapter 4**)

*(f) Compression test pressure low

1 Check 10, 11 and 12 in (a); 6 and 7 in (c) and 2 in (d)
2 Worn or scored cylinder bores
3 Worn piston rings and grooves
4 Choked air cleaner

(g) Engine lacks power

1 Check 4, 6, 10, 11, 12 and 13 in (a); 2, 3, 4 and 7 in (b); 6 and 7 in (c) and 2 in (d). Also check (e)
2 Leaking cylinder head gasket
3 Distributor automatic advance and retard not operating

(h) Burnt valves or seats

1 Check 10, 11 and 12 in (a); 7 in (b). Also check (e)
2 Excessive carbon around valve or cylinder head

(j) Sticking valves

1 Weak valve springs
2 Bent valve stem
3 Scored valve stem or guide
4 Defective or incorrectly set hydraulic valve lifter
5 Gum deposits on stem or guide

(k) Excessive cylinder wear

1 Check 6 in (a) and also check (e)
2 Lack of oil
3 Dirty oil
4 Piston rings gummed or broken
5 Badly installed piston rings
6 Connecting rod bent or twisted
7 Piston pin worked loose

(l) Excessive oil consumption

1 Check 6 and 7 in (c); 2 and 3 in (f) and 4 and 5 in (k)
2 Oil return holes in pistons blocked
3 Oil level too high in crankcase
4 External oil leaks

(m) Crankshaft bearing failure

1 Check 2, 3 and 6 in (k)
2 Restricted oilways
3 Worn journals or crankpins
4 Loose bearing caps
5 Extremely low oil pressure
6 Dirt particles trapped during installation

(n) Low oil pressure

1 Check 2 and 3 in (k) and also check 2, 3 and 4 in (m)
2 Choked oil filter or inlet strainer
3 Relief valve held open by dirt under seat
4 Weak or broken relief valve spring
5 Very worn oil pump
6 Defective gauge or sender unit

(o) Internal water leakage (see **Chapter 4**)

(p) Poor circulation (see **Chapter 4**)

(q) Corrosion (see **Chapter 4**)

(r) High fuel consumption (see **Chapter 2**)

**A high compression test pressure indicates excess deposits in combustion chambers and on piston crowns.*

Inches	Decimals	Millimetres	Inches to Millimetres		Millimetres to Inches	
			Inches	mm	mm	Inches
1/64	.015625	.3969	.001	.0254	.01	.00039
1/32	.03125	.7937	.002	.0508	.02	.00079
3/64	.046875	1.1906	.003	.0762	.03	.00118
1/16	.0625	1.5875	.004	.1016	.04	.00157
5/64	.078125	1.9844	.005	.1270	.05	.00197
3/32	.09375	2.3812	.006	.1524	.06	.00236
7/64	.109375	2.7781	.007	.1778	.07	.00276
1/8	.125	3.1750	.008	.2032	.08	.00315
9/64	.140625	3.5719	.009	.2286	.09	.00354
5/32	.15625	3.9687	.01	.254	.1	.00394
11/64	.171875	4.3656	.02	.508	.2	.00787
3/16	.1875	4.7625	.03	.762	.3	.01181
13/64	.203125	5·1594	.04	1.016	.4	.01575
7/32	.21875	5.5562	.05	1.270	.5	.01969
15/64	.234375	5.9531	.06	1.524	.6	.02362
1/4	.25	6.3500	.07	1.778	.7	.02756
17/64	.265625	6.7469	.08	2.032	.8	.03150
9/32	.28125	7.1437	.09	2.286	.9	.03543
19/64	.296875	7.5406	.1	2.54	1	.03937
5/16	.3125	7.9375	.2	5.08	2	.07874
21/64	.328125	8.3344	.3	7.62	3	.11811
11/32	.34375	8.7312	.4	10.16	4	.15748
23/64	.359375	9.1281	.5	12.70	5	.19685
3/8	.375	9.5250	.6	15.24	6	.23622
25/64	.390625	9.9219	.7	17.78	7	.27559
13/32	.40625	10.3187	.8	20.32	8	.31496
27/64	.421875	10.7156	.9	22.86	9	.35433
7/16	.4375	11.1125	1	25.4	10	.39370
29/64	.453125	11.5094	2	50.8	11	.43307
15/32	.46875	11.9062	3	76.2	12	.47244
31/64	.484375	12.3031	4	101.6	13	.51181
1/2	.5	12.7000	5	127.0	14	.55118
33/64	.515625	13.0969	6	152.4	15	.59055
17/32	.53125	13.4937	7	177.8	16	.62992
35/64	.546875	13.8906	8	203.2	17	.66929
9/16	.5625	14.2875	9	228.6	18	.70866
37/64	.578125	14.6844	10	254.0	19	.74803
19/32	.59375	15.0812	11	279.4	20	.78740
39/64	.609375	15.4781	12	304.8	21	.82677
5/8	.625	15.8750	13	330.2	22	.86614
41/64	.640625	16.2719	14	355.6	23	.90551
21/32	.65625	16.6687	15	381.0	24	.94488
43/64	.671875	17.0656	16	406.4	25	.98425
11/16	.6875	17.4625	17	431.8	26	1.02362
45/64	.703125	17.8594	18	457.2	27	1.06299
23/32	.71875	18.2562	19	482.6	28	1.10236
47/64	.734375	18.6531	20	508.0	29	1.14173
3/4	.75	19.0500	21	533.4	30	1.18110
49/64	.765625	19.4469	22	558.8	31	1.22047
25/32	.78125	19.8437	23	584.2	32	1.25984
51/64	.796875	20.2406	24	609.6	33	1.29921
13/16	.8125	20.6375	25	635.0	34	1.33858
53/64	.828125	21.0344	26	660.4	35	1.37795
27/32	.84375	21.4312	27	685.8	36	1.41732
55/64	.859375	21.8281	28	711.2	37	1.4567
7/8	.875	22.2250	29	736.6	38	1.4961
57/64	.890625	22.6219	30	762.0	39	1.5354
29/32	.90625	23.0187	31	787.4	40	1.5748
59/64	.921875	23.4156	32	812.8	41	1.6142
15/16	.9375	23.8125	33	838.2	42	1.6535
61/64	.953125	24.2094	34	863.6	43	1.6929
31/32	.96875	24.6062	35	889.0	44	1.7323
63/64	.984375	25.0031	36	914.4	45	1.7717

UNITS	Pints to Litres	Gallons to Litres	Litres to Pints	Litres to Gallons	Miles to Kilometres	Kilometres to Miles	Lbs. per sq. In. to Kg. per sq. Cm.	Kg. per sq. Cm. to Lbs. per sq. In.
1	.57	4.55	1.76	.22	1.61	.62	.07	14.22
2	1.14	9.09	3.52	.44	3.22	1.24	.14	28.50
3	1.70	13.64	5.28	.66	4.83	1.86	.21	42.67
4	2.27	18.18	7.04	.88	6.44	2.49	.28	56.89
5	2.84	22.73	8.80	1.10	8.05	3.11	.35	71.12
6	3.41	27.28	10.56	1.32	9.66	3.73	.42	85.34
7	3.98	31.82	12.32	1.54	11.27	4.35	.49	99.56
8	4.55	36.37	14.08	1.76	12.88	4.97	.56	113.79
9		40.91	15.84	1.98	14.48	5.59	.63	128.00
10		45.46	17.60	2.20	16.09	6.21	.70	142.23
20				4.40	32.19	12.43	1.41	284.47
30				6.60	48.28	18.64	2.11	426.70
40				8.80	64.37	24.85		
50					80.47	31.07		
60					96.56	37.28		
70					112.65	43.50		
80					128.75	49.71		
90					144.84	55.92		
100					160.93	62.14		

UNITS	Lb ft to Kgm	Kgm to lb ft	UNITS	Lb ft to Kgm	Kgm to lb ft
1	.138	7.233	7	.967	50.631
2	.276	14.466	8	1.106	57.864
3	.414	21.699	9	1.244	65.097
4	.553	28.932	10	1.382	72.330
5	.691	36.165	20	2.765	144.660
6	.829	43.398	30	4.147	216.990

CHAPTER 2

THE FUEL SYSTEM

2:1 Description

Twin-barrel carburetors are fitted as standard to all models covered by this manual except the 1968 model fitted with a 302 cu in engine. Gasoline is stored in a tank at the rear of the vehicle. The fuel is drawn from the tank, through a system of pipes, by the mechanical engine-driven fuel pump which then delivers the fuel to the carburetor fuel bowl inlet under slight pressure. A typical fuel system layout is shown in **FIG 2:1**.

A strainer is fitted to the fuel tank sender and pipe assembly to ensure that the fuel to the pump is reasonably clean.

A fuel filter is fitted into the system to ensure that the fuel reaching the carburetor is perfectly clean. On the earlier models the filter takes the form of a cartridge attached to the fuel pump but on later models a line filter is fitted between the pump and carburetor. The fuel pump itself is of the 'sealed' type and very limited servicing can be carried out on it. The pump is driven by the eccentric mounted on the camshaft sprocket.

An air cleaner assembly is fitted to the intake of the carburetor. The cleaner contains an element which filters the air before it enters the carburetor and engine while the housing is designed to act as a silencer to cut out the roar from the carburetor intake, and it also controls the temperature of the air entering the engine.

2:2 Routine maintenance

This is mainly confined to cleaning or renewing the filter elements at the correct intervals and checking that the throttle linkage operates freely and fully.

It is advisable to take the vehicle to a service garage equipped with electronic test equipment so that both the fuel and ignitions systems can be given regular checks for faults or excessive emission from the engine.

Fuel filter:

The sealed Carter fuel pump fitted to 1964 and 1965 models is shown in **FIG 2:2**. Unscrew the filter housing from the fuel pump assembly and discard the old element and gasket. Wash out the housing using cleaning solvent, denatured alcohol or clean gasoline. Ensure that the correct type of element is obtained. Fit the new element to the spout in the fuel pump valve housing cover, making sure that the end labelled 'THIS END MUST BE INSTALLED DOWN' is downwards. Lightly coat the new gasket with engine oil and install it onto the filter housing. Hand-tighten the filter housing until the gasket contacts the pump and tighten the housing by a further $\frac{1}{8}$ turn. Start the engine and check for fuel leaks.

On later models an in-line filter is fitted to the carburetor, shown in **FIG 2:3**. The filter assembly is of one-piece

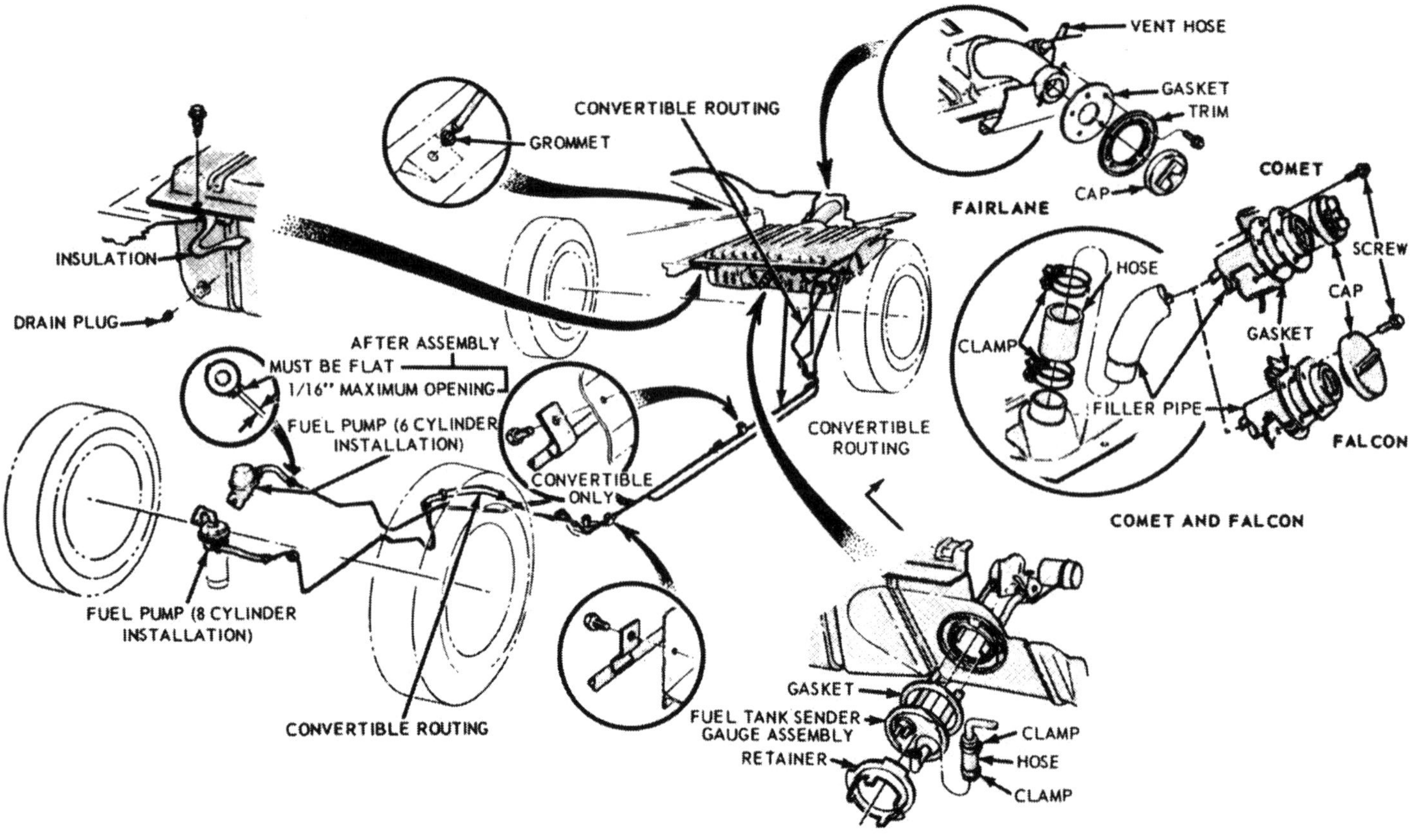

FIG 2:1 Typical fuel line installation (1967 Sedan)

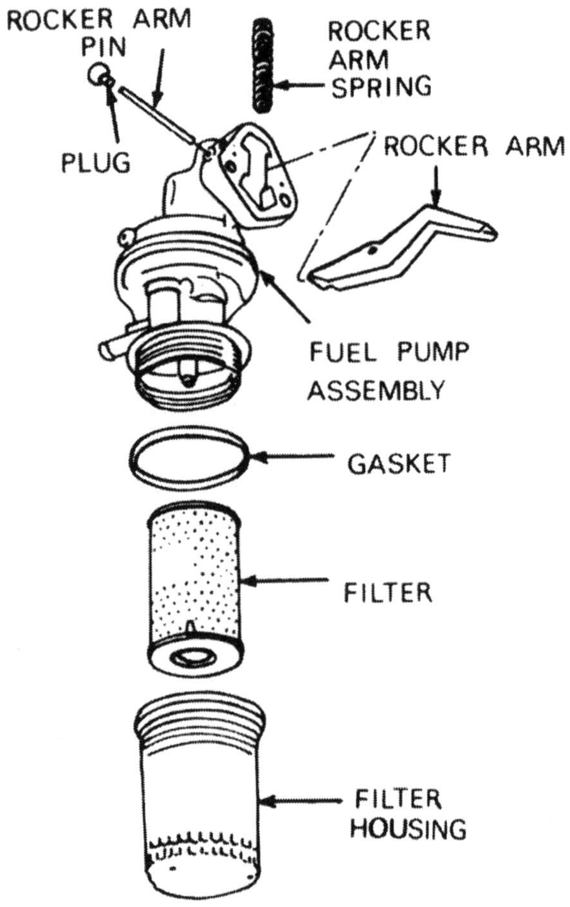

FIG 2:2 The Carter sealed fuel pump fitted to 1965 models

construction and must be renewed complete. Remove the air cleaner to gain access. Loosen the retaining clamp which secures the hose to the filter. Unscrew the filter assembly from the carburetor, disconnect the filter from the hose and discard filter, gasket and retaining clamp. Install a new clamp on the hose.

Place a new gasket on the new filter and screw the assembly back into the inlet port of the carburetor, making sure that the filter is tight. Position the inlet hose onto the filter and secure it by crimping the clamp. Start the engine and check for fuel leaks.

On all models, the fuel filter should be renewed at intervals of 36,000 miles but it should also be renewed if it becomes clogged or choked earlier.

Air cleaner:

Incorrect functioning of the air cleaner or a choked element will make the idle rough, increase fuel consumption and on later models may cause the emissions from the engine to exceed the legal limits.

On the earlier models the filter element should be cleaned at 6000 mile intervals and renewed at 36,000 mile intervals but on later models renew the element at intervals of 24,000 miles. The servicing periods given are the longest that the filter should be left and if the vehicle is used in dirty or dusty conditions the servicing intervals should be reduced.

The air cleaner is fitted with a thermostatically controlled inlet duct, and a typical one is shown in **FIG 2:4**. The thermostat ensures that the air entering the engine is always warmed and, when the temperature is low, the valve moves into the heat-on position so that air is drawn through a muff around the exhaust manifold. Earlier models operate at a lower temperature than the later ones (see **Technical Data**) but the valve should be checked for full and free movement. On later models the duct and valve assembly can be removed, immersed in water and the operating range checked by gradually heating the water. Adjustment is possible by altering the position of the thermostat in its mounting bracket.

On later models an additional vacuum-operated valve is fitted to allow the entry of cold air when the engine is accelerating or under high power. The parts are shown in **FIG 2:5**. With the engine at idle, check that the valve is fully closed. Disconnect the vacuum tube from the vacuum motor assembly and check that the valve opens fully. If the vacuum motor sticks, check that the vacuum tube is not kinked or split and that vacuum is reaching the motor. Check the vacuum motor on a separate vacuum source and if it still fails to operate correctly fit a new valve in its place.

Before cleaning or renewing the filter element, remove the complete air cleaner assembly from the carburetor. Undo the wing-type screws that secure the unit to the carburetor and thermostatically controlled inlet duct. Remove the cover and take out the element. The element may be blown through with an air-line, from the inside to the outside and using low air pressure, or it may be held vertical and tapped onto a firm surface to dislodge loose dirt. Check the element against a bright light and renew it if any cracks or splits can be seen in it. Renew the element if it is choked or has reached the end of its life. Install the element and air cleaner in the reverse order of removal. Ensure that the marking FRONT on the air cleaner faces forwards on the car and that the marking TOP is uppermost on the element, noting that not all models will have these marks.

At the same time as cleaning the air filter element, check the crankcase ventilation system. The hoses can be cleaned with kerosene and a stiff brush but the regulator valve should be renewed.

2:3 The fuel pump

A sealed fuel pump is fitted to all models covered by this manual. The type of fuel pump fitted to the 1965 models is shown in **FIG 2:2**. The type of pump fitted to later models, noting that an in-line filter is used between pump and carburetor, is shown in **FIG 2:6**.

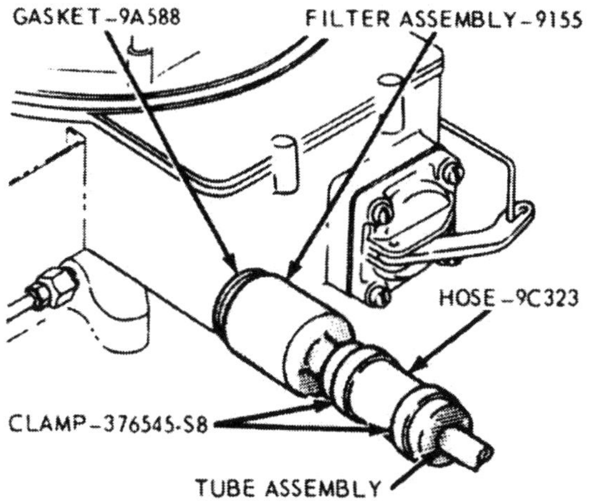

FIG 2:3 Typical in-line fuel filter assembly

Removal:

Disconnect the inlet and outlet lines at the fuel pump. Plug the inlet pipe to prevent fuel from syphoning through. Take out the screws that secure the fuel pump and withdraw the pump from the engine. Discard the gasket fitted between pump and engine.

The pump is installed in the reverse order of removal. Use a new gasket, lightly smeared with sealing compound. Make sure that the fuel pump rocker is riding on the camshaft eccentric. Hold the pump firmly against the crankcase and alternately tighten the attachment screws until the correct torque load is reached. Reconnect the fuel lines, using new clamps and crimping them as required. Start the engine and check for fuel and oil leaks. If the vehicle has stood for some time the engine will have to crank over to prime the carburetor before the engine will fire.

Testing:

The operation of the pump can be briefly checked by cranking the engine with the fuel line disconnected at the carburetor. Point the loose end of the line into a suitable container, or well away from the engine, and crank the motor. At every other turn of the motor there should be a good spurt of fuel from the pipe. Lack of fuel can be caused by a defective pump, blocked lines or air leaks on the suction side of the pump.

For full and accurate tests, a pressure gauge and graduated container must be connected into the fuel

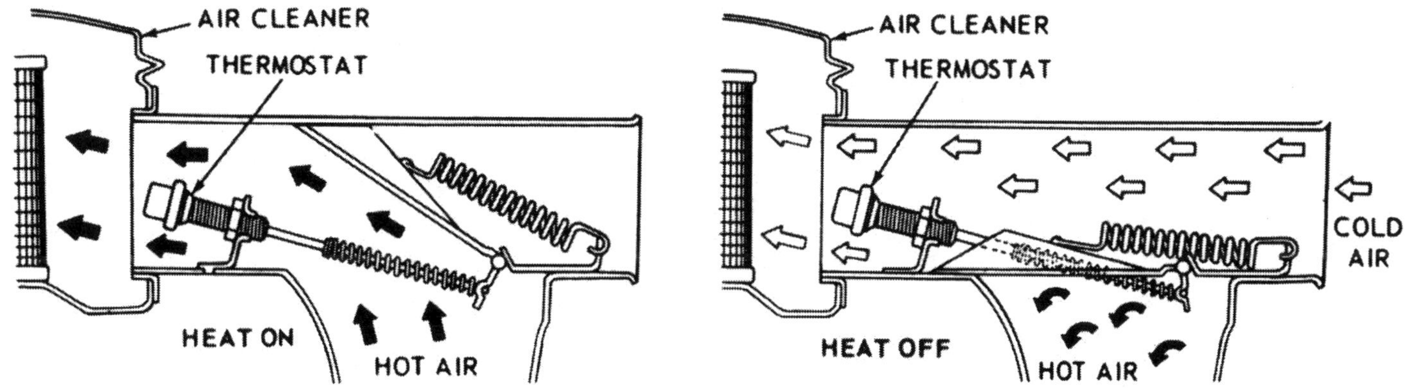

FIG 2:4 The thermostatic duct to the air cleaner. On some models the vacuum motor is fitted to override the thermostat direct

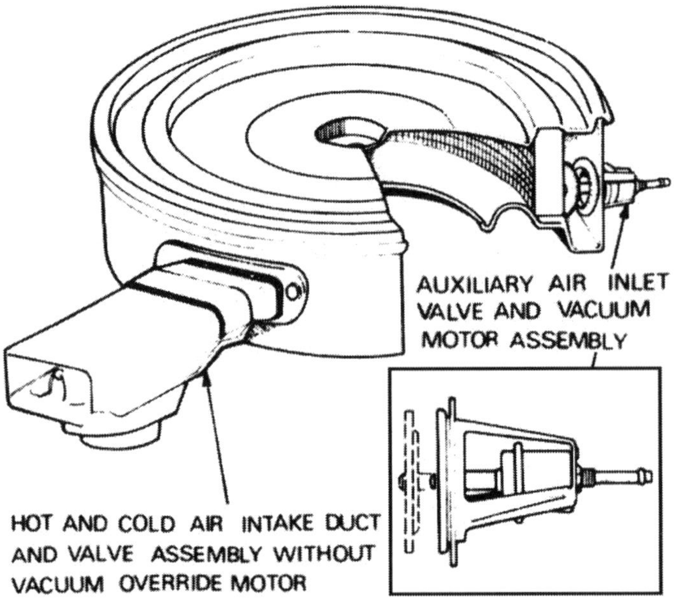

FIG 2:5 Auxiliary cold air inlet valve

system as shown in **FIG 2:7. The fuel lines must be carefully and tightly connected to prevent spillage of fuel which could lead to a fire.** Run the engine until it has reached its normal operating temperature. If the static pressure is low, renew the in-line fuel filter (element on 1965 models), and repeat the test. Both the pressure and volume tests are carried out at engine idle, and the system should be vented of air by opening the hose restrictor momentarily before taking pressure readings.

If the pressure is satisfactory, open the hose restrictor and note the time taken to deliver one pint of fuel. If the time taken to deliver one pint is excessive, stop the engine and connect the fuel pump to a separate fuel supply (a length of hose dipped into a can of fuel). If the time taken to deliver one pint is now within limits, there is a restriction in the fuel lines of the vehicle. If the pressure or volume flow is low in all cases, renew the pump.

Servicing:

The amount of work that can be carried out on the pump is strictly limited. The rocker arm and its spring can be removed and new parts installed but if there are any further defects in the pump, a new pump must be fitted.

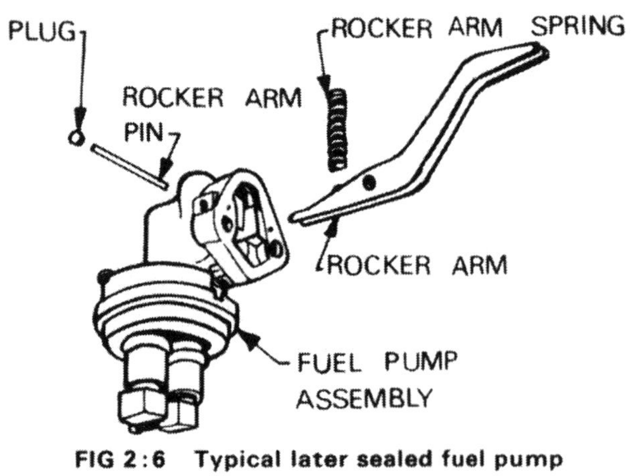

FIG 2:6 Typical later sealed fuel pump

Scrape off the staking that secures the plug for the rocker arm pin. Pry out the plug with a screwdriver, as shown in **FIG 2:8**. Take the pressure of the rocker arm and shake out the rocker arm pin. If the pin sticks, tap the pump against a firm surface until sufficient of the pin is exposed to allow it to be withdrawn using a pair of pliers. Remove the rocker arm and its spring.

Wash out the pump with clean gasoline and check the pin, spring and rocker arm for wear. Renew parts as required. Insert the spring back into its bore. Insert the rocker arm, making sure that it hooks correctly into the diaphragm rod. Position the spring onto the locater of the rocker arm and move the rocker arm so that the pin holes align. Insert the pin to hold the rocker arm in place, and **check that the end of the arm is correctly in the diaphragm and that the spring is correctly located.** Use a new plug to secure the pin and stake the plug in place.

If it is found that the fuel lines are restricted, disconnect them at both ends and blow through in a reverse direction with compressed air. Remove the fuel tank sender unit and clean the strainer on the inlet pipe. In extreme cases it will be necessary to remove the fuel tank, flush it with clean gasoline to remove all sediment and then refit it, or install a new tank if the old one is corroded.

2:4 Carburetor operation

The components of a Ford 2-V carburetor are shown in **FIG 2:9**. A view of the assembled unit is shown in **FIG 2:10** and an underneath view of the unit is shown in **FIG 2:11**. The Ford 2-V carburetor is typical and will therefore be used to explain the operation of a carburetor.

Fuel enters the carburetor at the inlet to the fuel bowl. Before entering the fuel bowl it passes through a filter screen and needle valve assembly. The pressure of the fuel lifts the inlet needle off its seat and allows fuel to pass into the fuel bowl. The bowl has a float fitted in it which rises as the fuel level rises. When the correct level is reached, the float arm presses down the needle and closes it against its seat so that no more fuel can enter the fuel bowl. As fuel is used, the level will drop slightly, and the needle valve will open and allow the pump to replenish the level. The fuel level in the fuel bowl is therefore kept constant within very narrow limits.

Cold start:

Fuel will be deposited on the walls of the intake manifold, as it is cold and a rich mixture is required to initiate ignition. As the engine cranks over on the starter motor, a very rich mixture is required for firing and the mixture strength then requires weakening when the engine starts. As the engine and manifolds warm up the mixture strength needs to be gradually weakened until it is at the correct strength when the engine has reached its normal operating temperature.

An automatic choke is fitted to the carburetor and this ensures a rich mixture at cold start with gradual weakening off as the engine heats. A choke plate is fitted to the inlet of the carburetor barrels and the plate is operated by the automatic choke. For a full rich mixture,

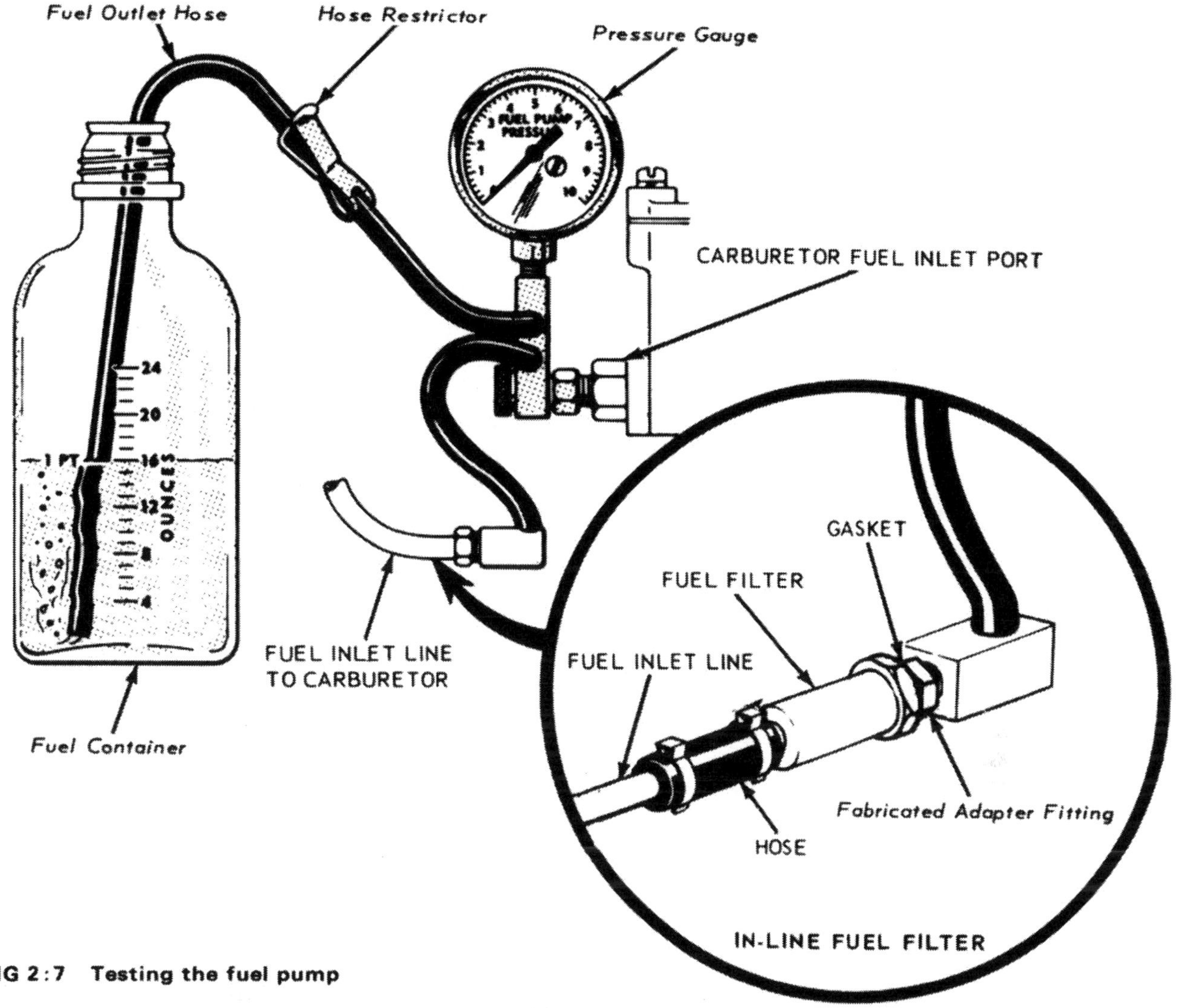

FIG 2:7 Testing the fuel pump

the plate closes the inlets and the suction caused by the the cranking of the engine ensures that fuel is drawn through the jets to give a very rich mixture. The automatic choke, shown in **FIG 2:12**, has a thermostatic bi-metallic spring which is attached to the spindle of the choke plate. The thermostatic spring winds up when cold and slackens when hot. Under cold start conditions, the spring is wound up and holds the choke plate closed. When the engine starts the vacuum piston operates to open the choke plate against the action of the thermostatic spring, and the choke plate opens slightly to allow more air to pass through and partially weaken the very rich mixture required for initial firing. Air drawn past the piston by the manifold suction is also drawn through the heat chamber on the manifold and is therefore heated as the exhaust manifold heats up. The hot air flowing over the spring warms it, and the spring tension is released allowing the choke plate to open further and weaken the mixture still more. When the engine has reached its operating temperature, the spring tension is so reduced that the choke plate is fully open and held there by the vacuum piston. The vacuum piston is at the end of its travel and allows ports to open to ensure that there is a continuous flow of heated air over the thermostatic spring to keep it hot.

The choke rod moves the fast idle cam as the choke plate rotates to ensure that idle speed is increased when the engine is cold. The linkage is so designed that the choke plate will be partially opened at full throttle so that a flooded engine can be dried out by cranking it with full throttle.

The vacuum piston acts as an anti-stall device. If the engine reaches the stall point during warm-up, because the mixture is too lean, the vacuum in the inlet manifold will drop, allowing the spring tension to overcome the vacuum piston pull so that the choke plate closes and the mixture is made richer.

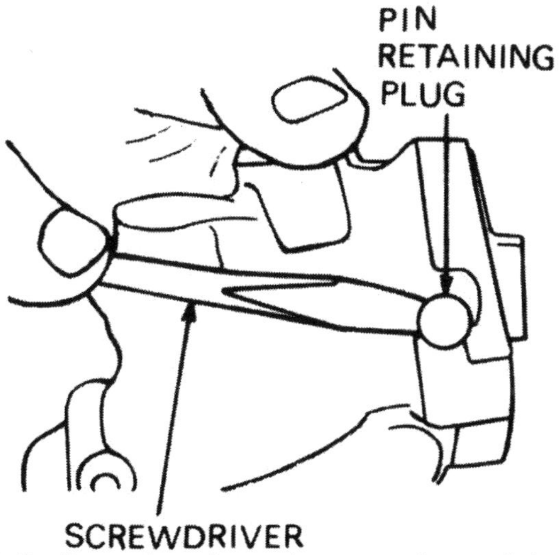

FIG 2:8 Removing the rocker arm pin retaining plug

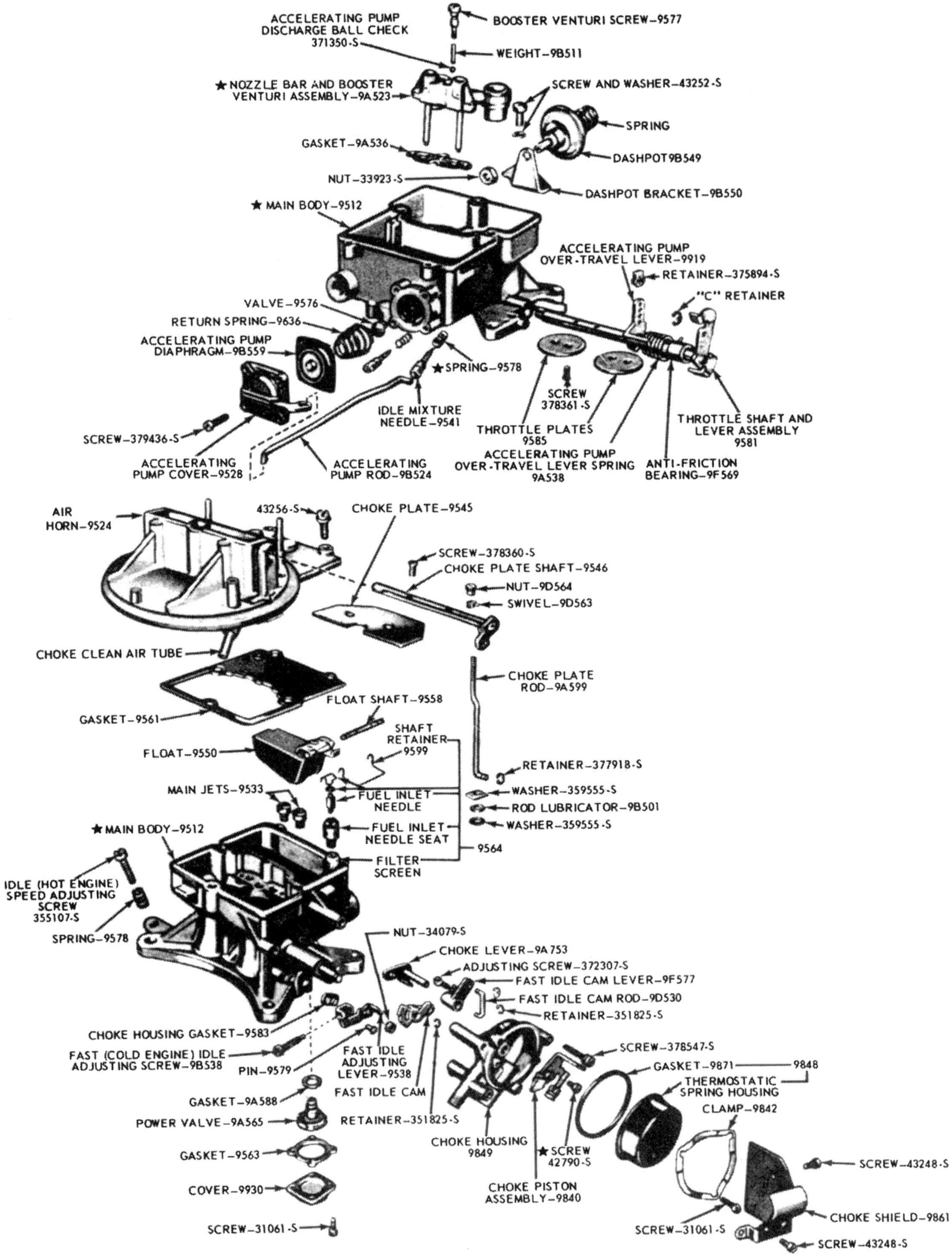

FIG 2:9 The components of the 2-V carburetor (Autolite and Ford are similar)

Idle:

At idle speeds, the air-flow through the carburetor is low and insufficient to cause a large enough pressure drop in the booster venturi in each barrel to draw fuel through the main jets. The throttle butterfly valve in each barrel is closed to the idle position and there is only a small gap between it and the bore of the barrel. The air passing through this narrow gap sets up maximum suction at the side of the valve and the mixture of fuel and some air for idle is drawn through the idle discharge, shown in **FIG 2:13.** Fuel flows through the main jet, from the float bowl into the bottom of the main well. The fuel is then drawn through the idle tube and through internal passages to the idle discharge point. A calibrated orifice in the tip of the idle tube meters the fuel flow. Air is drawn through with the fuel through the calibrated idle air bleed and through a bore as shown in the figure. The air mixes with the fuel to break it down into small droplets and the idle air bleed also prevents syphoning through of fuel when the engine speed is increased beyond idle, or when the engine is stationary.

Idle transfer ports are fitted to the bore above the discharge to ensure smooth transition from idle to main jets. A volume control screw sets the volume of mixture drawn through the idle discharge.

Acceleration:

The fuel has greater inertia than air so that when the throttle is opened for acceleration, the air-flow speeds up immediately while the fuel flow will momentarily lag. An accelerating pump is fitted to inject extra fuel into the flow until the fuel flow can catch up. The system is shown in **FIG 2:14.** The diaphragm is connected by a linkage to the throttle linkage so that the diaphragm is pressed in when the throttle is opened. On the return stroke, throttle closing, the diaphragm is moved by the return spring and fuel is drawn in from the fuel bowl through the Elastomar valve. On the next throttle opening the fuel in the diaphragm chamber is forced through the internal passages and injected into the air-flow through the pump discharge nozzle. The weight and ball prevent air from being drawn back into the chamber when the throttle is released.

A small vent is fitted between the diaphragm chamber and fuel bowl to allow any vapour to escape and ensure that there is no pressure build up in the diaphragm chamber.

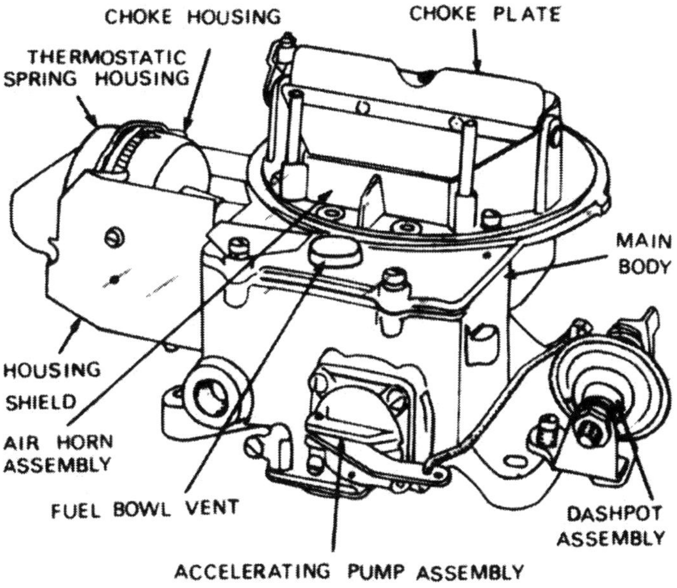

FIG 2:10 Threequarter front view of the carburetor

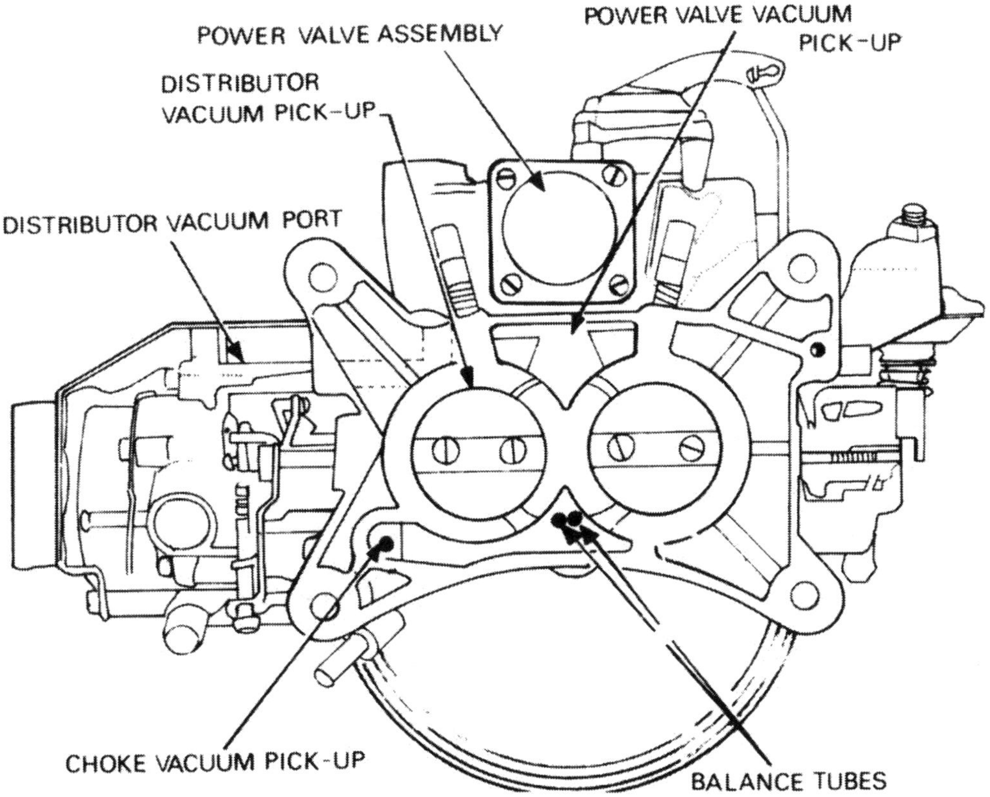

FIG 2:11 Bottom view of the carburetor

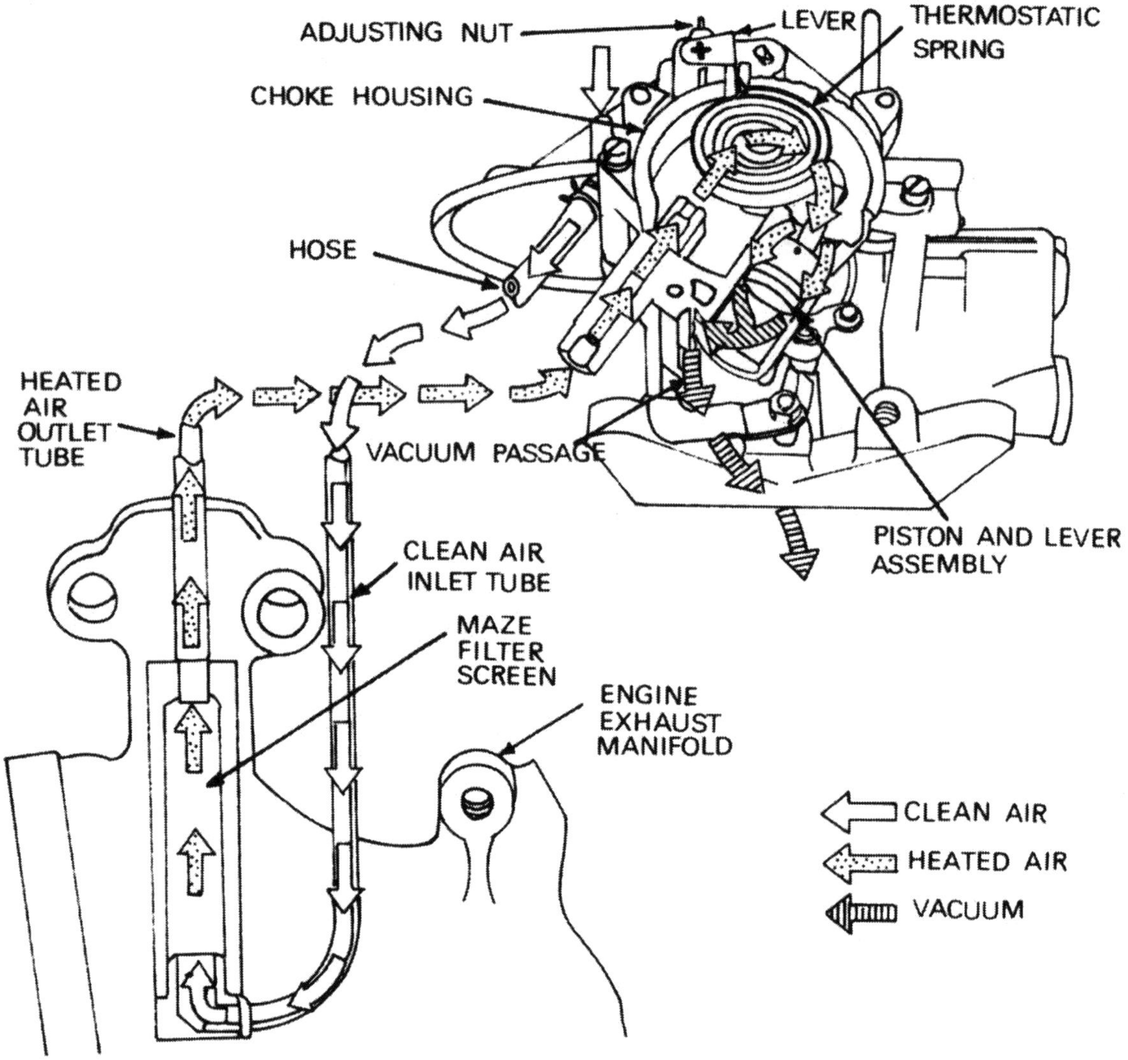

FIG 2:12 The automatic choke system

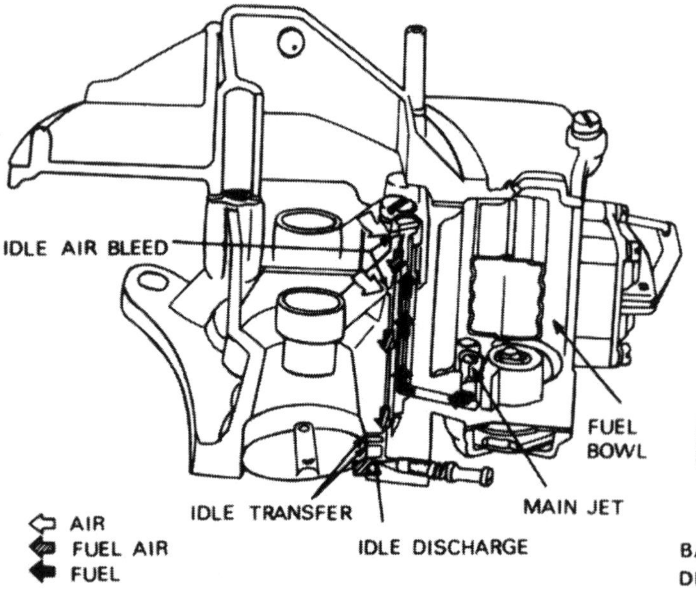

FIG 2:13 The idle fuel system

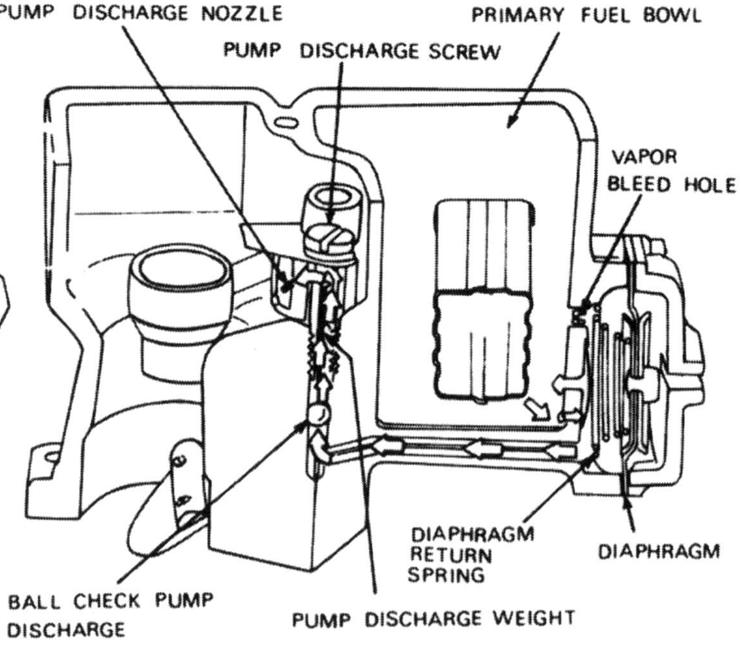

FIG 2:14 The accelerating pump operation

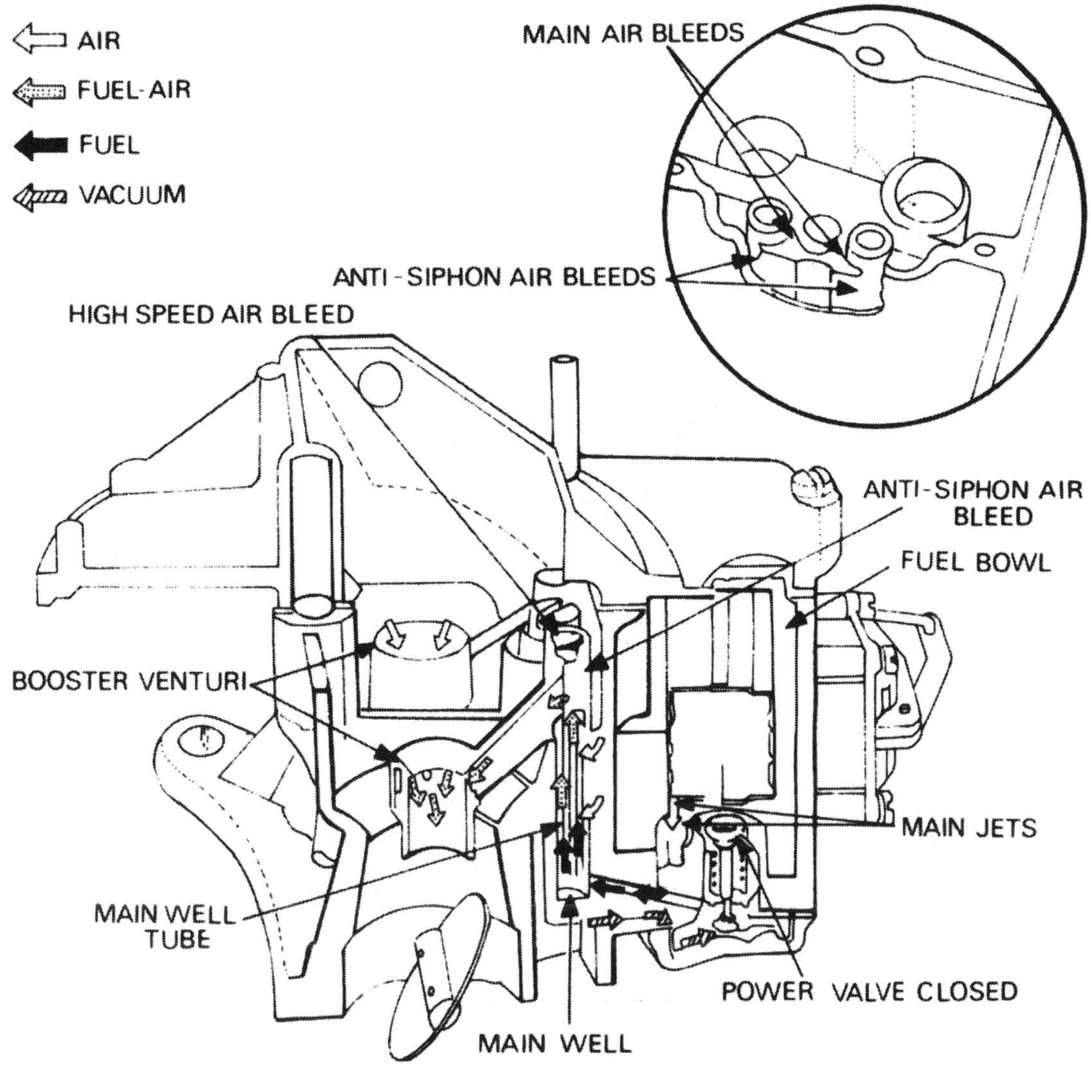

FIG 2:15 The main fuel system

Main fuel flow:

When the air-flow through the carburetor is speeded up the air-flow in the booster venturi causes a drop in static pressure to take place. The faster the air-flow the greater the pressure drop and the air-flow depends on the opening of the throttle valve and speed of the engine. The suction draws the fuel from the fuel bowl, through the main jet and into the booster venturi where it mixes with the main air-flow. Filtered air mixes with the fuel (as shown in **FIG 2:15**) from the high-speed air bleed, and as the engine speed increases the ratio of air drawn in through the high-speed bleed also increases to ensure that the final air/fuel mixture strength does not become too rich at high speeds. The air also helps to emulsify the fuel so that it is already partially broken up into droplets when it joins the main air-stream.

Each barrel also has a balance tube directly below the booster venturi. On deceleration, the balance tube syphons away fuel droplets from the edge of the venturi and leads them to the idle discharge.

Power valve:

For high-speed operation or when the engine is heavily loaded, a power valve is fitted to richen the mixture. The valve operation is shown in **FIG 2:16.** The valve is held open by the spring but the vacuum from the manifold, acting on the diaphragm, holds the valve closed against the spring pressure. When the throttle is wide open, the vacuum in the manifold is insufficient to hold the diaphragm down against the action of spring and the valve opens, allowing extra fuel to bypass the main jet.

Four-barrel carburetor:

A four-barrelled carburetor is fitted to some models made in 1968. The components are shown in **FIG 2:17**. Basically the operation of this carburetor is the same as the twin-barrelled type fitted to other models, though it is more complicated and has additional devices added. Generally servicing will follow the instructions for the standard carburetor, but the owner is advised to take the car to a service garage for full checks, adjustments or servicing.

2:5 Carburetor faults and static settings

The most common cause of faults in the carburetor is dirt or gum blocking jets or passages. Wear does not affect the carburetor excessively as there are few wear points. The most likely wear point is the throttle valve

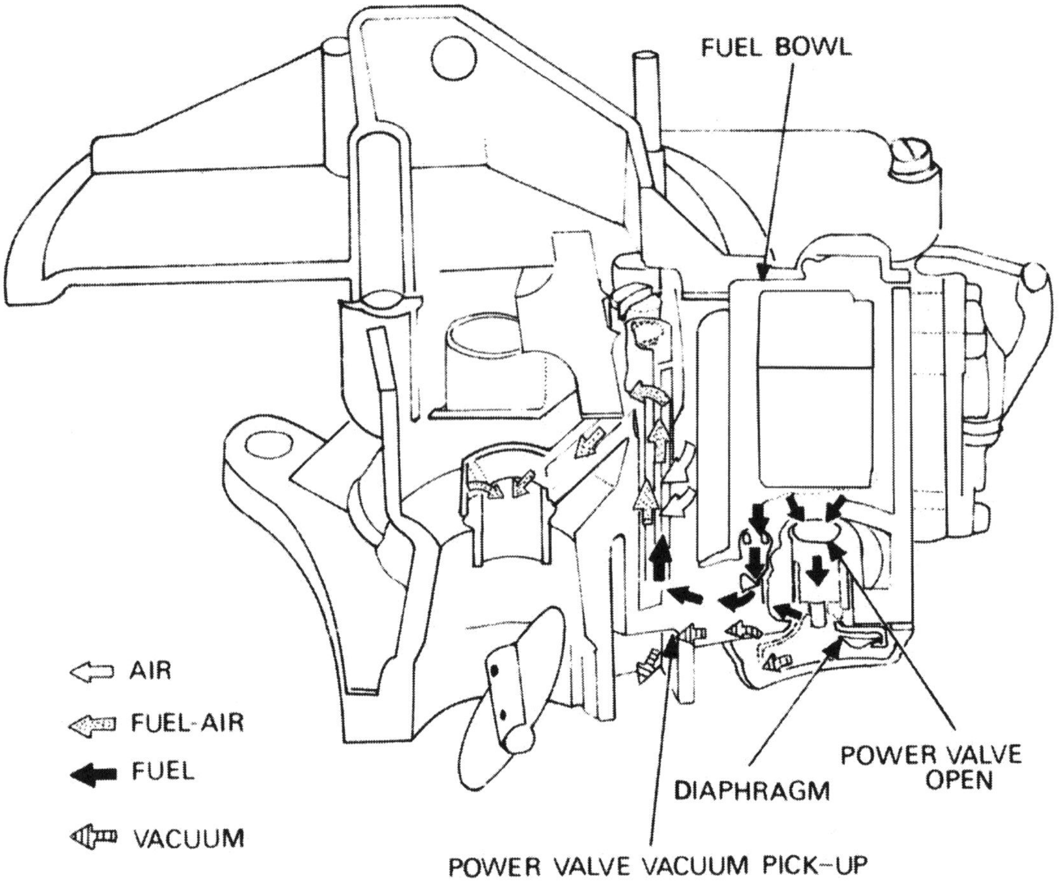

FIG 2:16 The power valve system

spindle or the body where the spindle fits. Anti-friction bearings are fitted but if there is excessive wear, air can leak past the throttle spindle and alter the idling mixture to give poor or lumpy idling.

If wear or dirt are causing faulty operation, the carburetor must be removed, taken apart and reassembled after cleaning and renewing worn parts. When reassembling the carburetor, use all new gaskets and seals—available in service kits.

There is no separate section on disassembling the carburetor but the faults will be laid out in such an order that by following this section through the carburetor will be completely disassembled.

Faults that can be blamed on carburation can also be caused by other components of the power plant. Faulty ignition will cause misfiring and poor idle, so ensure that the ignition system is operating correctly at the right settings and that the points and spark plugs are clean and correctly gapped (see **Chapter 3**). Faults in the crankcase breathing system and Thermactor emission control will also make it impossible to set a satisfactory idle. Check the air cleaner for correct operation. Check all vacuum-operated systems and their hoses. A split hose will allow air to leak directly into the manifold, without passing through the carburetor, and weaken the idling mixture. The intake manifold joints should be checked for leaks for similar reasons.

Only when all the other systems have been checked should the carburetor be examined. Remove the air cleaner assembly to gain access to the carburetor.

Removal:

Disconnect the fuel line from the carburetor inlet. Disconnect the vacuum line to the distributor. Disconnect the automatic choke hoses. Disconnect the accelerator control rod and remove the return spring. Evenly slacken and remove the four nuts that secure the carburetor to the manifold and lift the carburetor up off the studs.

Install the carburetor in the reverse order of removal, using new gaskets. Before installing the carburetor check the spacer, mounting flange on the intake manifold and mounting flange on the carburetor for distortion or damage. High spots or nicks can be smoothed down by careful use of a file. Excessive distortion, usually caused by overtightening the attachment nuts, can be lapped down using a fine grade of emerycloth, or grinding paste, spread over plate glass.

Flooding or fuel starvation:

The appearance of flooding can be given by defective seals and gaskets which allow fuel to leak out, so check for this before taking further steps. The most likely cause of either flooding or starvation in the carburetor is the float needle valve assembly. If starvation is the fault, check that the fuel pump is delivering fuel and that the fuel filters are not clogged (see **Section 2:3**). The carburetor parts are shown in **FIG 2:9**.

Remove the retainer washer that secures the lower end of the choke rod and free the rod. Remove the screws that secure the air horn, lift off the carburetor identification tag and air horn assembly and lay the parts carefully

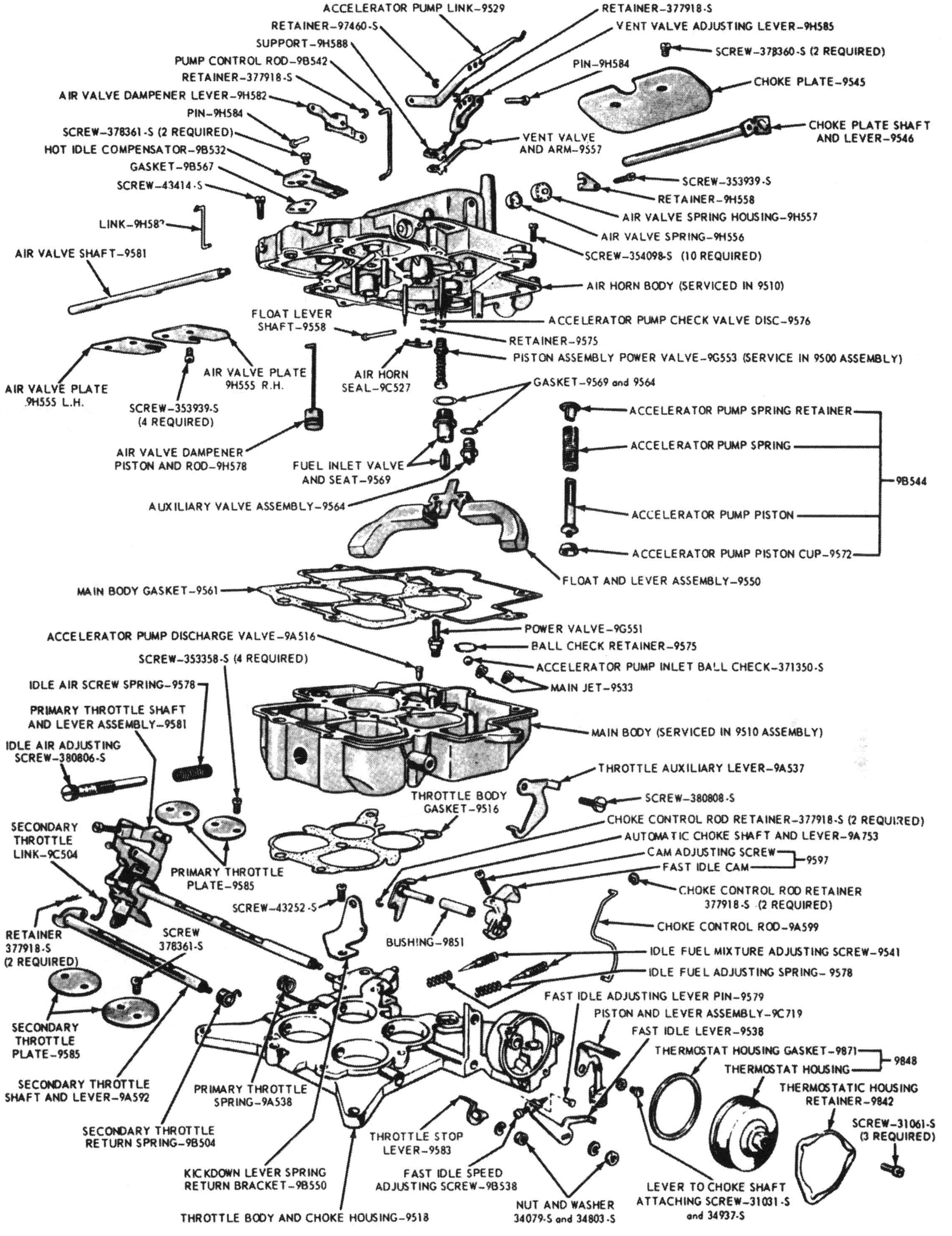

FIG 2:17 The components of the four-barrel carburetor fitted to some 1968 models

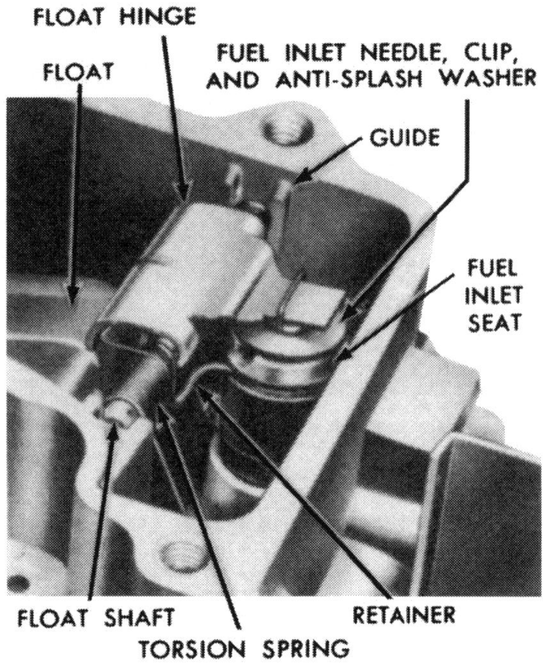

FIG 2:18 The float and needle attachments

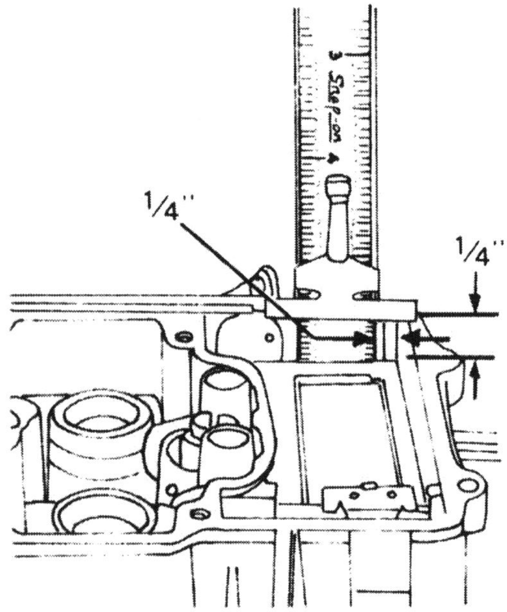

FIG 2:19 Checking the fuel level with the engine idling

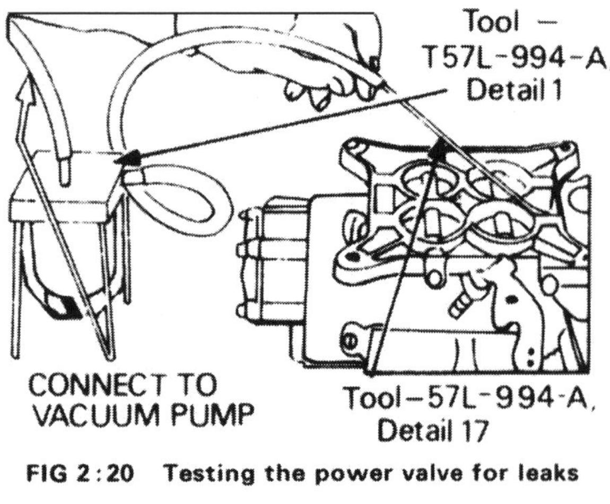

FIG 2:20 Testing the power valve for leaks

to one side. Note that the carburetor need not be removed from the engine.

The float and needle valve attachments are shown in **FIG 2:18.** Pry the retainer free with a small screwdriver so that the float, retainer and needle assembly can be removed from the fuel bowl. Separate the retainer and needle from the float. Unscrew the needle valve seat, preferably using the special jet wrench made for the purpose. Remove the filter screen fitted under the seat.

Check the float itself for damage and renew it if it is defective. Wash the needle valve assembly in gasoline or special carburetor cleaning fluid and check that they do not leak when blown through with the needle held lightly onto its seat. Renew the needle valve assembly complete if it leaks under light air pressure or if the needle has a step worn on its tapered seat. Renew the filter screen if it is clogged.

The parts are installed in the reverse order of removal, ensuring that gasket is fitted under the air horn assembly. Check the dry setting of the float before installing the air horn and check the fuel level after the engine has been run.

Fuel level:

With the air horn removed and the gasket taken off, lightly press on the float arm so that the float is lifted up and the needle valve assembly closed. Check the height of the top surface of the float below the upper surface of the main body, using a steel rule or the special card-board gauge made for the purpose. If the float height is not within limits (see **Technical Data**) gently bend the tab on the float to bring the height within limits. This will give a satisfactory setting for running the engine, but **the actual height of the fuel should still be checked after the engine has been run.**

With the air horn fitted and the car standing on level ground, run the engine until it has reached its normal operating temperature. Stop the engine and remove the air horn. Restart the engine and leave it running at idle. Use a steel rule, with clip as shown in **FIG 2:19** to measure the actual depth from the top of the machined surface on the body to the fuel. Because of surface tension, the fuel will rise up the side of the body slightly and to avoid measuring this meniscus, the steel rule must be at least $\frac{1}{4}$ inch away from any vertical surface of the fuel bowl.

If the level is incorrect, stop the engine to avoid any fire risk. Bend the tab on the float up from its original position to raise the level and down to lower. Restart the engine and let it idle for at least three minutes before taking a further reading. Repeat the sequence until the fuel level is within limits. Too high a fuel level will cause the engine to run rich and can cause flooding, while too low a level will make the mixture lean.

Install the air horn, with its gasket, and air cleaner assembly when the adjustment is completed.

Jet cleaning:

Dirt in the jets will cause erratic running. Remove the air cleaner, air horn and float assembly as described earlier. Unscrew the jets, **preferably using the special jet wrench**, and wash them in gasoline, denatured alcohol or special cleaning fluid. Blow through the jets to remove any dirt. **Do not poke the jets with wire or bristle**

as this will cause wear in the calibrated orifices. Renew any jets that are damaged or appear the slightest bit worn in the calibrated orifice.

If the accelerator pump check ball and weight require removal, the carburetor should be removed from the engine. Take out the screw by the discharge nozzle, invert the carburetor and shake out the weight and ball.

Before reassembling the carburetor, brush and blow out any sediment from the fuel bowl. If there is an excessive amount of gum or dirt in the carburetor, check through the system to find the cause, as the fuel should be well filtered by the time it reaches the fuel bowl. In such cases it is worthwhile draining the fuel tank as there may be water present. It is also worthwhile to disassembly the carburetor and blow through all fuel and air passages with compressed air, after cleaning the parts with special carburetor cleaning fluid. **Note that some carburetor cleaning fluids are caustic and must therefore be handled with care and the instructions carefully followed.**

Hesitation or poor acceleration:

Remove the air cleaner assembly and operate the throttle linkage. There should be a good steady spurt of fuel from the discharge nozzle every time the throttle is opened.

If the fuel spurt is intermittent or poor, remove the air horn assembly. Remove the accelerating pump rod retainer. Press the tab ends of the clip together and at the same time press the rod away from the clip to free it. Take out the screws and remove the accelerating pump cover, diaphragm assembly, and return spring. Check the cover and spring for damage and renew them both if they are defective. Check the Elastomar valve. If the valve appears poor, grasp it firmly and pull it out. **A new valve must be fitted if the old one is removed and the tip of the old valve must be removed from the fuel bowl if it breaks off.** Wash and brush out any sediment from the accelerating pump chamber, noting that the carburetor will have to be removed if the check ball and weight require removal and the fuel way needs cleaning.

Lubricate the tip of the new valve and insert it into place. Pull the valve fully into place by gripping the tip of the valve with needle-nosed pliers from inside the fuel bowl, and pull it firmly into place. Cut off the tip forward of the retaining shoulder and remove the tip from the fuel bowl.

Position the new diaphragm to the cover. Install the return spring and refit the cover to the carburetter. Set the securing screws finger tight, push the accelerating pump plunger through the full length of its travel and fully tighten the screws. Position the operating rod in the inboard hole, closest to the pump plunger.

Power valve:

A defective power valve will make it nearly impossible to set the correct idle and it will be found that the idle mixture screws will have to be nearly on their seats to obtain a reasonable idle.

Remove the carburetor to gain access to the power valve. If fuel is found on the vacuum side of the diaphragm this does not necessarily indicate a defect, as it can have condensed there.

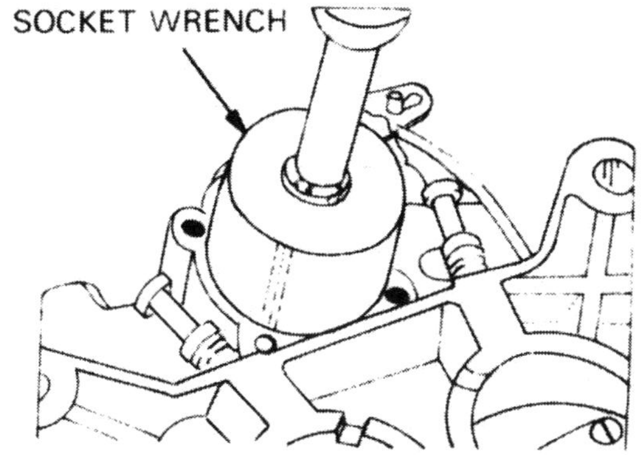

FIG 2:21 Removing or installing the power valve

The method of testing the valve is shown in **FIG 2:20.** Note that the test is shown on a four-barrel carburetor but the principle is the same for the two-barrel carburetors. The glass bowl is half-filled with water and the wand of the tool inserted into the vacuum port for the power valve. Air bubbles through the water will indicate leaks. There will be air bubbles for the first fifteen seconds, as the diaphragm moves and the chamber is exhausted, but the bubbles should then stop. If the bubbles stop correctly, the power valve is not leaking.

To remove the valve, invert the carburetor and take off the cover with its gasket. Use a socket or box wrench to remove the valve, as shown in **FIG 2:21.**
Install the valve in the reverse order of removal.

Idle mixture screws:

These control the mixture at idle by moving a tapered needle in or out of the port. **The screws must never be tightened firmly onto their seats as this wears the tapered needle to a step.** If the idle mixture adjustment is erratic and small movements of the screws make large variations in the mixture strength, stop the engine and remove the idle mixture screws. **Check the tapered needle for wear and renew them if they are worn to a step.** Blow through the ports to remove any dirt.

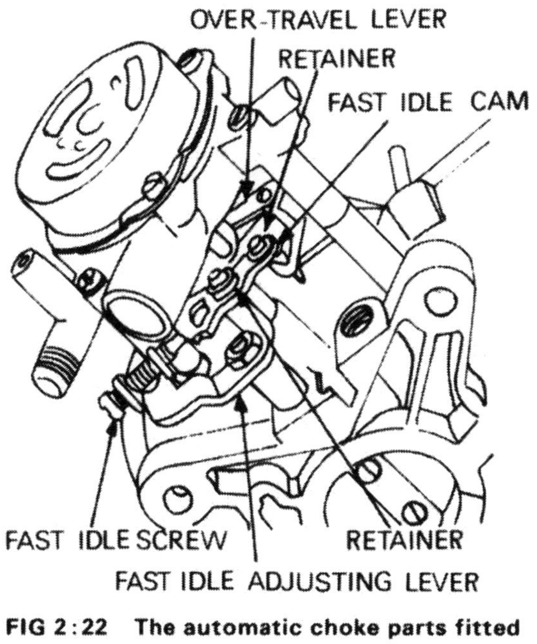

FIG 2:22 The automatic choke parts fitted

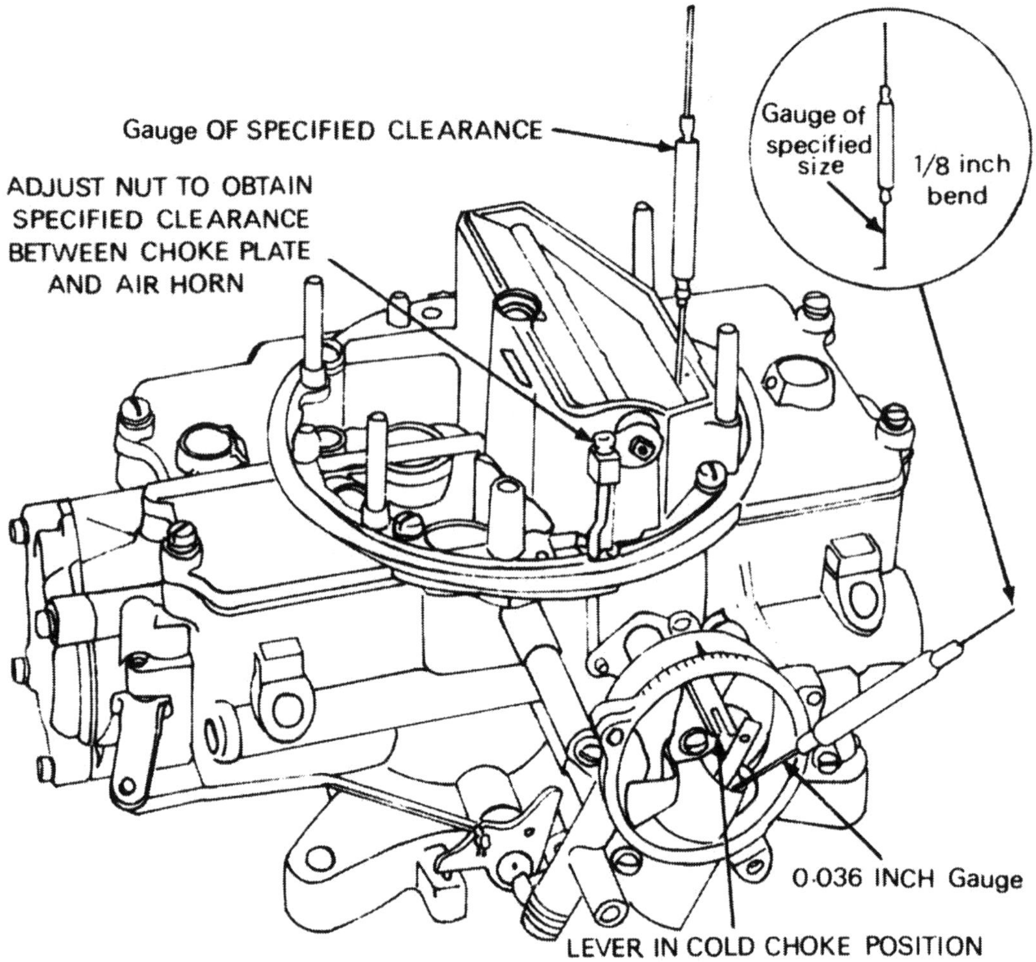

FIG 2 : 23 Adjusting the choke rod

Automatic choke:

The parts fitted are shown in **FIG 2 : 22**. The spring housing is secured to the choke housing by a retainer and three screws. Index marks are made on both the spring and choke housings and the spring housing should be rotated so that the correct index marks align (see **Technical Data**) after the retaining screws have been slackened. Rotating the spring housing alters the tension of the thermostatic spring and arrows show the direction for leaning off the mixture strength when the choke is operating.

Remove the retainer for the fast idle cam, after taking off the air cleaner assembly. Take out the screws and remove the spring housing, its retainer and gasket. **If the thermostatic spring is defective the housing complete must be renewed.**

If the air horn is still fitted, remove the choke rod retainer. Take out the choke housing retaining screws and remove the housing, gasket and fast idle cam. Remove the rod and cam from the fast idle cam lever. The choke lever retaining screw and pin securing the lever link to piston can be taken out to remove the parts from the housing.

If the choke plate has nicked edges or the shaft is worn, new parts should be fitted to ensure smooth operation. Remove the air horn assembly. Disconnect the choke rod from the lever on the shaft, by unscrewing the adjusting nuts. Slide the felt seal and two washers out of the air horn. Remove the staking marks on the screws which secure the choke plate to the shaft. **If the tips of the screws are excessively flared, file them down before attempting to remove the screws.** Take out the screws, slide the plate from its slot and withdraw the shaft from the air horn. Install the parts in the reverse order of removal. The choke plate securing screws should be snug but not tight. Move the choke plate into accurate alignment so that it opens and closes freely without binding. Fully tighten the screws and stake them to prevent them from slackening. When staking the screws, support the shaft on a block of wood to prevent it from bending.

Install the air horn and choke housing parts. Hold the throttle linkage so that it is half open and the fast idle cam does not contact the fast idle adjustment screw. Bend a wire gauge, as shown in **FIG 2 : 23**, and insert the bent end of the gauge between the lower edge of the piston slot and the upper edge of the righthand slot in the choke housing. Turn the countershaft lever clockwise until the gauge is a snug fit in the piston slot and keep it there by applying light pressure on the lever. Insert another gauge between the edge of the choke plate and air horn and adjust the nut until the second gauge is just lightly nipped. Remove the gauges and install the spring housing, making sure that the end of the thermostatic spring is correctly located.

Turn the thermostatic spring housing 90 deg. in the rich mixture direction. Use a gauge to check the gap

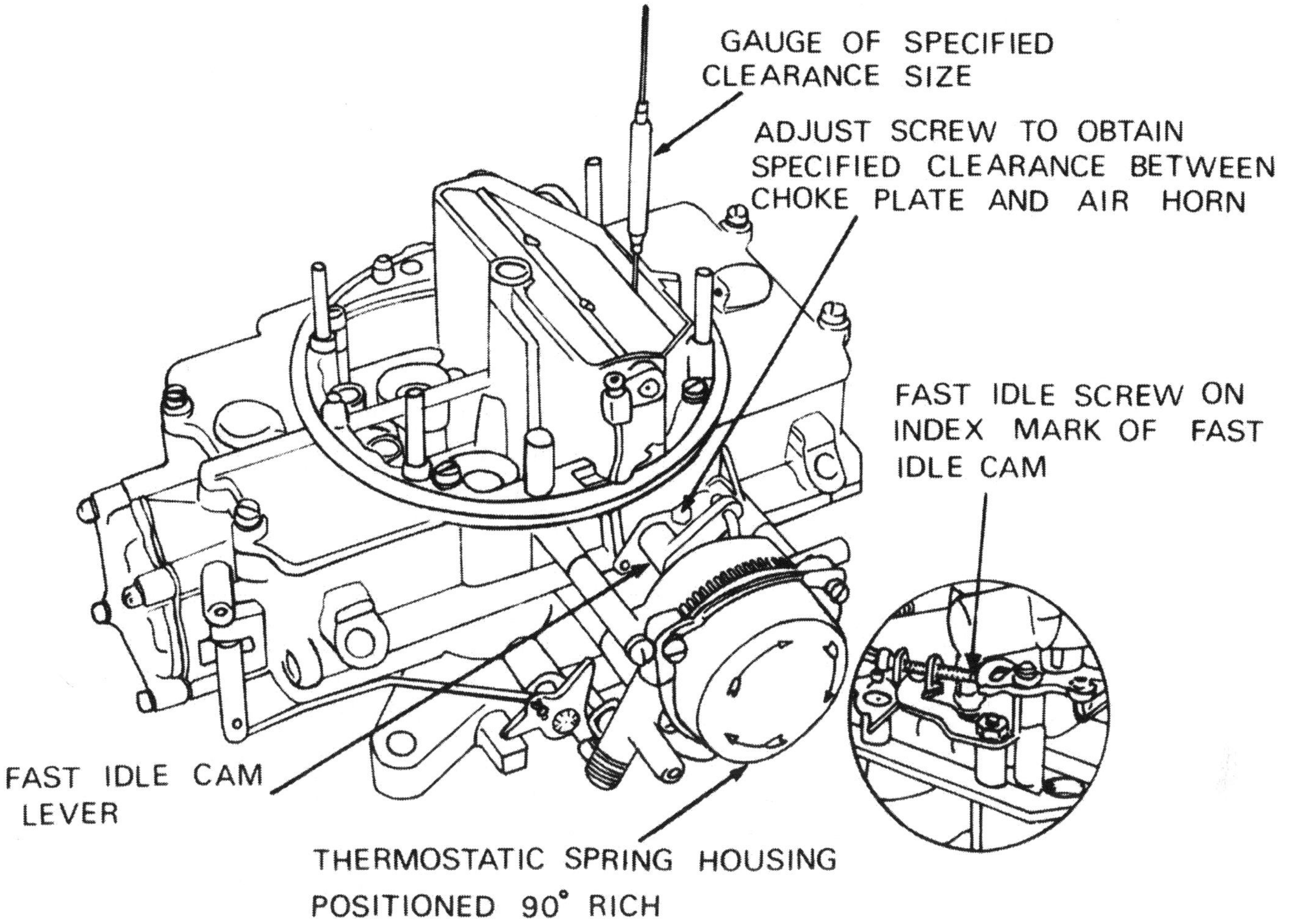

FIG 2:24 Adjusting the fast idle screw and cam

between choke plate and air horn, as shown in **FIG 2:23**. Turn the fast idle cam lever adjusting screw to obtain the correct gap between choke plate and air horn. Set the spring housing to the correct index mark.

The settings and the gauges required are given in **Technical Data**.

Throttle valves and shaft:

The throttle valves and shaft are dealt with in the same manner as the choke plate and shaft. The linkage parts will have to be removed from the shaft. **Before taking out the valve plates, mark both valves and bores with light identification marks (or scribe marks) so that the valve plates will be installed back into their original positions.** When installing the valves, ensure that they are both a free fit in the bores before fully tightening and staking their securing screws.

Anti-stall dashpot:

This will be fitted to models using automatic transmission and the adjustment is shown in **FIG 2:25. The idle settings must be correct before attempting to adjust the dashpot.** Hold the throttle linkage in the closed position and fully depress the plunger of the dashpot with a screwdriver. Measure the gap shown. If the gap is not within limits, slacken the locknut and adjust the position of the dashpot until the gap is within limits.

2:6 Idling settings

Before attempting to set the idling, check that all vacuum operated systems are functioning correctly and that there are no air leaks into the manifold bypassing the carburetor.

On all later model carburetors for the home market limiter stops, shown in **FIG 2:26**, are fitted to the idle mixture control screws to ensure that they can only be adjusted within narrow limits. **These stops must not**

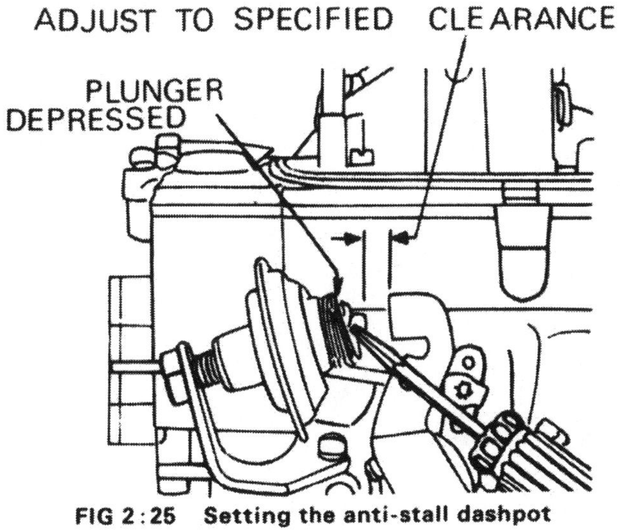

FIG 2:25 Setting the anti-stall dashpot

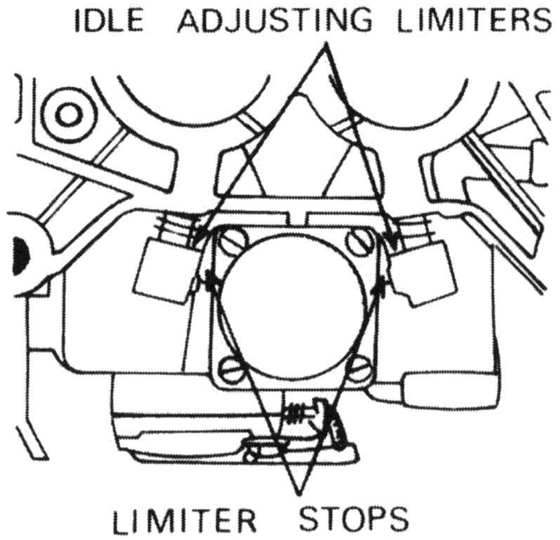

FIG 2:26 The limiters on later idle adjusting screws

be mutilated or damaged. It is most advisable to have the carburetor settings checked using exhaust gas analyser equipment.

Set the static carburetor settings described in the previous section and then set the idle mixture screws and curb idle adjusting screw so that the engine can be started and run.

On models fitted with limiters to the idle mixture screws turn the screws counterclockwise as far as the limiters will allow them to go. On models not fitted with limiters, turn the screws clockwise until they lightly touch their seats **(do not overtighten the screws or the tapered needles will be damaged)** and slacken back each screw by 1½ turns. Back off the curb idle screw, arrowed in **FIG 2:27**, until the throttle valves are fully closed in their bores. Note that it may be necessary to unscrew the dashpot to allow the valves to close fully. Turn the curb idle adjusting screw inwards until it just contacts the stop and then turn it a further 1½ turns inwards to open the throttle valve sufficiently to give a good idle speed.

On models fitted with automatic transmission and a vacuum released parking brake, set the brake while making preliminary adjustments with the engine running, and then disconnect the vacuum line from the manifold and plug the inlet. This ensures that the parking brake will not release when drive is selected.

Start the engine and allow it to idle at 1500 rev/min to ensure that engine and under-hood temperatures have fully stabilized. The fast idle speed can be set using the curb idle speed adjusting screw, positioning the fast idle idle screw on the appropriate step of the fast idle cam or simply clipping the throttle linkage partially open.

Check that the choke plate valve is fully open, but note that the adjustment should be carried out with the air cleaner in position. Turn on the headlights to high beam to ensure that the alternator is under load. On models fitted with automatic transmission, the idle speed adjustment is carried out with 'Drive' selected. On models fitted with air conditioners, the final adjustment is made with the air conditioner ON, with the exception of 302 2-V engines fitted with automatic transmission when the air conditioner must be left OFF.

Adjust the idle speed to the correct limits (see **Technical Data**) using the curb idle adjusting screw. On vehicles with less than 50 miles, set the idle 25 rev/min below specification to allow for the idle increased speed as the engine 'loosens up' with use. Only if difficulty is found in setting the correct idle speed should the air cleaner be removed and the idle setting made.

Turn both idle mixture adjusting screws in by equal amounts until a smooth idle is achieved. The air cleaner must be installed when adjusting the idle mixture.

FIG 2:27 The curb idle adjustment screw

2:7 Trouble shooting guide

(a) Leakage or insufficient fuel delivery

1 Air vent to fuel tank blocked
2 Fuel lines partially blocked
3 Fuel filters partially blocked
4 Defective fuel pump
5 Defective float in fuel bowl
6 Defective needle valve assembly in fuel bowl
7 Loose hose connections
8 Filter screen under needle valve partially blocked
9 Loose carburetor parts or defective gaskets

(b) High fuel consumption

1 Carburetor incorrectly adjusted
2 Dirty air cleaner
3 Automatic choke incorrectly set or failing to open fully
4 Fuel leakage
5 Fuel level in fuel bowl too high
6 Loose or worn jets
7 Incorrect jets fitted
8 Brakes binding
9 Tires under-inflated
10 Engine running too hot
11 Vehicle regularly overloaded

(c) Idling speed too high or difficult to set

1 Defective air cleaner
2 Defective Thermactor
3 Air leaks into manifold from vacuum operated systems
4 Air leaks between carburetor and intake manifold
5 Air leaks between intake manifold and cylinder head
6 Carburetor incorrectly set
7 Throttle linkage sticking
8 Defective automatic choke
9 Worn carburetor throttle valve plates or shaft

(d) No fuel delivery

1 Fuel bowl needle valve jammed closed
2 Vent in fuel tank blocked
3 Fuel filters, strainer or screen blocked
4 Fuel line blocked
5 Defective fuel pump
6 Bad air leak on suction side of pump
7 No fuel in the tank

CHAPTER 3

THE IGNITION SYSTEM

3:1 Description

A schematic layout of the ignition system is shown in **FIG 3:1**. The distributor is fitted with a make-and-break as well as a rotor arm and the distributor shaft is synchronized with the engine so that the points just open and the rotor arm is in the correct position when a cylinder is at the firing point.

Under normal conditions, current from the battery flows through the ignition coil and breaker points. When the contacts open the current is sharply cut off, assisted by the action of the condenser, and the magnetic field induced in the primary windings of the coil collapses rapidly. The collapse of the magnetic field induces a very high voltage in the many secondary coils and this high voltage is led to the center of the distributor cap by a heavily insulated HT lead. From the center of the cap the high voltage passes to the rotor arm, and this leads the high voltage to the electrode in the cap, which is connected to the spark plug of the cylinder at the firing point. The high voltage jumps across the electrodes of the spark plug and ignites the mixture in the cylinder.

A ballast resistor wire is fitted between the supply and ignition coil. When the engine is running, the ballast resistor ensures that full battery voltage is not applied across the coil. The coil is designed to run efficiently on this lower voltage. When the starter motor is operated, a separate lead connects the ignition coil to a terminal on the starter relay and this terminal is live so that full battery voltage is applied to the ignition coil. The coil will therefore be operating at full efficiency even allowing for the drop in battery voltage caused by the starter motor operation.

The actual ignition point required for the most efficient combustion of the mixture will vary with the engine speed and load on the engine. A centrifugal advance mechanism, shown in **FIG 3:2**, is fitted to the distributor shaft. As the engine speed increases, the centrifugal force moves the weights outwards against the pull of the calibrated weight springs. The movement of the weights rotates the cam forward in relation to the mainshaft to advance the ignition. The manifold depression is proportional to the load on the engine and the throttle setting. A small-bore tube connects the vacuum unit to the manifold and the vacuum unit rotates the movable breaker plate about the cam to alter the ignition setting. The vacuum advance mechanism is shown in **FIG 3:3**. This figure also shows a cross-section of the distributor cap giving the means of transferring the HT voltage to the correct side electrode in the cap.

On later models the vacuum advance system is more complicated. A typical vacuum system for a later model fitted with Imco emission control and air conditioning is

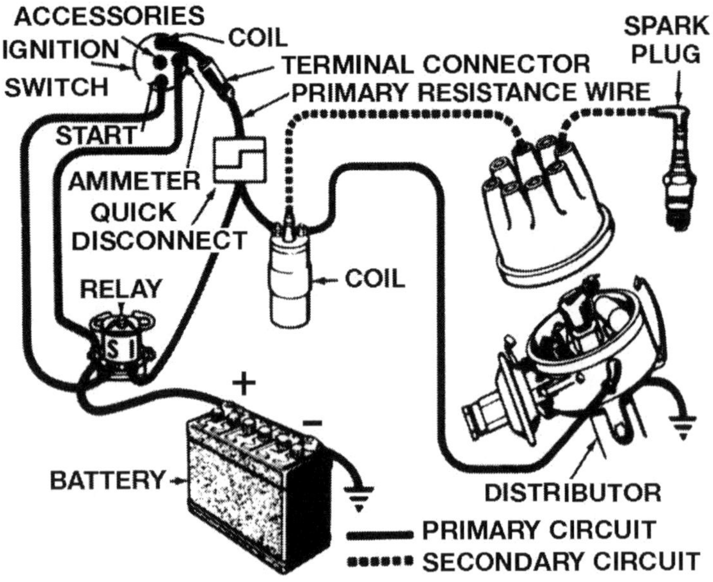

FIG 3:1 Schematic layout of the ignition system

shown in **FIG 3:4**. The exact layout of the system and the color coding will vary depending on accessories fitted—such as manual or automatic transmission, Thermactor instead of Imco emission control, and no air conditioning. **Before disturbing any vacuum hose connections, make a rough sketch of the system to ensure that the hoses will be correctly installed. Incorrect installation will affect ignition timing, carburation and emissions from the exhaust.**

3:2 Routine maintenance

Regularly wipe the HT leads, outside of the distributor cap, and ignition coil clean and dry with a soft piece of cloth. Dirt or moisture build-up will cause arcing and loss of ignition efficiency.

The sparking plugs should be removed, cleaned and tested at intervals of 6000 miles and renewed at intervals of 12,000 miles (see **Section 3:8**).

At 12,000 miles lightly lubricate the distributor, check the breaker points and clean the inside of the distributor

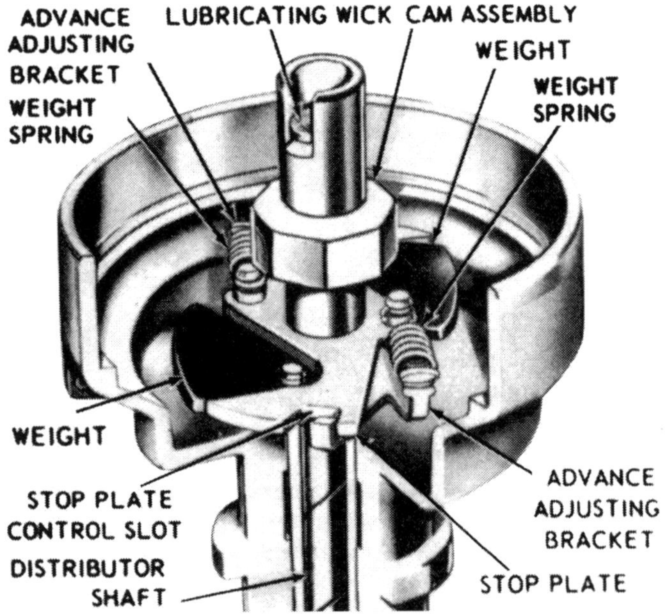

FIG 3:2 The centrifugal advance mechanism

cap. **Excess lubrication must be avoided and under no circumstances allow lubricant onto the breaker points themselves.** Dirty breaker points may be cleaned using a small stiff brush and chloroform or denatured alcohol. Renew the breaker points if they are badly burnt or the build-up of metal has reached the same dimension as the correct breaker gap.

Remove the distributor cap and pull the rotor arm squarely off from its position on the cam. Apply a few drops of oil to the felt pad in the top of the camshaft, exposed after removing the rotor arm. Apply a few drops of oil through the lubricator for the shaft bushing. Inject a few drops through the breaker plate assembly for the centrifugal advance and retard. Lightly grease the surface of the cam, where the breaker foot operates. Wipe away any surplus lubricant.

Use a stiff bristle brush and a mild cleaning solvent, such as denatured alcohol, to clean the rotor arm and distributor cap. Pay particular attention to removing dirt from all the crevices in the distributor cap. Check the cap for cracks in the material, carbon tracking between electrodes or an electrode and the edge of the cap, and also check for badly burnt electrodes. If any defect is found, a new cap must be fitted. Ensure that the carbon button in the cap is in good condition and that the rotor arm makes good contact with it.

Install the parts in the reverse order of removal. The rotor arm lug must be aligned with the slot in the cam and the rotor arm pushed fully home. Slots and lugs on the distributor body and cap align the cap.

3:3 Breaker points

The three main differing versions of breaker point that are fitted are shown in **FIGS 3:5, 3:6** and **3:7**.

Cleaning:

The points should have a frosted slate grey colour for efficient operation. Oil or grease on the contacts will cause them to blacken and there will also be a smudgy dark line on the base plate. Cleaning in such cases is carried out by using a stiff bristle brush and chloroform (or denatured alcohol).

Points which are blue in color or are badly burnt should be renewed. The most likely causes are a defective condenser or radio suppresser. Other causes are radio suppresser incorrectly installed, incorrect alignment of points, voltage regulator in charging circuit incorrectly adjusted, or extended operation at high speeds.

In normal use there will be a transference of metal from one contact to the other, causing a build-up on one and crater in the other. No action need be taken until the build up reaches the same dimension as the correct breaker points gap. When the build up has reached this dimension, renew the contacts. Only in an emergency should the contacts be cleaned with a fine file or ground down.

Renewing breaker contacts:

Remove the distributor cap and rotor arm. Disconnect the primary and condenser wires from the breaker point assembly. Take out the retaining screws and remove the breaker point assembly complete with condenser.

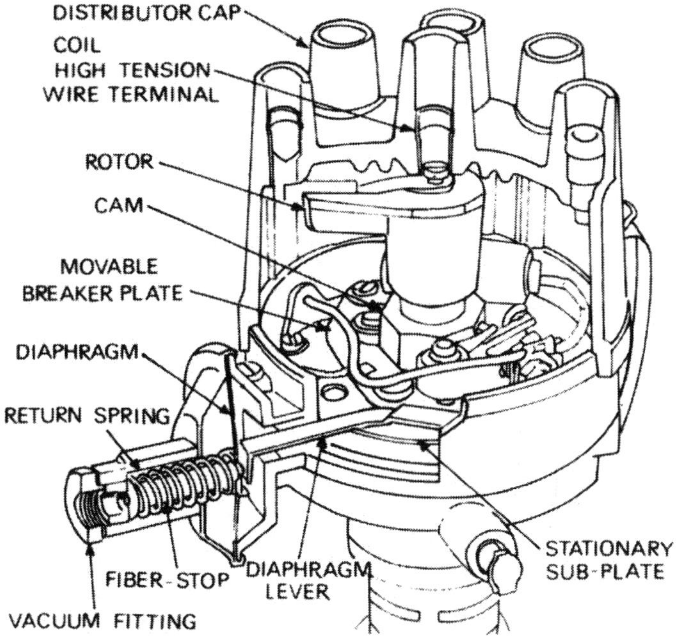

FIG 3:3 The vacuum advance mechanism

New parts are installed in the reverse order of removal, after cleaning off any protective from the new contacts using denatured alcohol. Crank the engine until the contact points are fully closed and check their alignment. When vented type points are used, any misalignment will cause premature wear, pitting and overheating. The correct and incorrect alignments of the points are shown in **FIG 3:8.** The points are aligned by bending the bracket of the stationary breaker point. **Do not bend the moving breaker arm.** Once the parts are installed and aligned, set the breaker point gap correctly.

Adjusting gap:

The method of adjusting new points using feeler gauges is shown in **FIG 3:9. Adjusting used breaker points by this method is not satisfactory as the surface will have worn unevenly giving a false reading on the feeler gauge.** If feeler gauges are used on the old points, make sure that the feeler gauge is inserted between the unworn portion of the points so as not to bridge the gap.

The most accurate method of setting the contacts is to use a dwell motor or oscilloscope, but this will have to be carried out at a service garage.

If feeler gauges are used, crank or turn the engine until the foot of the breaker is on a lobe of the cam and the contacts are at their widest gap. Make sure that the ignition is switched off before carrying out adjustments. The gap is correct when light drag is felt on the correct thickness feeler gauge as it is moved between the points.

Once the correct gap has been adjusted, check the breaker spring tension. Crank the engine until the points are closed. Use a spring balance to measure the force and pull at right angles to the moving breaker until the points just open. On some models the tension can be adjusted slightly by slackening the nut that secures the end of the spring and moving the spring relative to the pivot point. If the tension cannot be adjusted within the correct limits by this method, new points must be fitted.

3:4 Ignition faults

The best method of checking the ignition system is to take the vehicle to a service garage where it can be checked on electronic test equipment. **On models equipped with Imco emission control the vehicle should be taken for regular routine checks to ensure that the ignition system is operating at peak efficiency and preventing excessive emissions.**

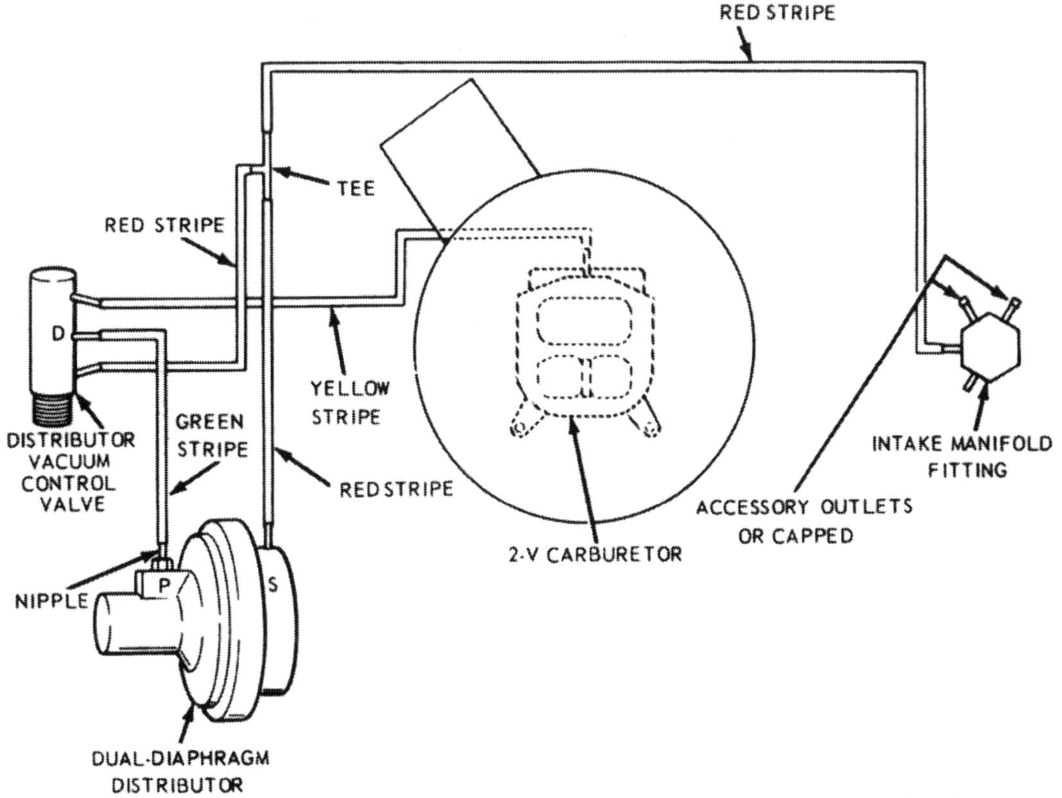

FIG 3:4 A typical vacuum line layout on later models, where manifold suction retards and carburetor suction advances

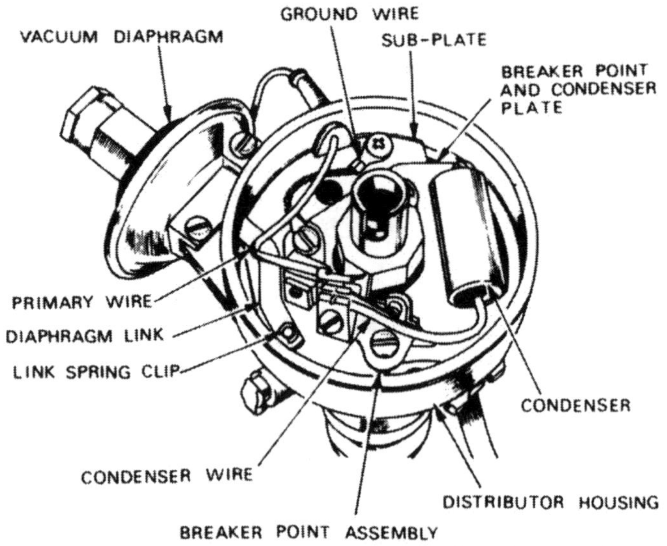

FIG 3:5 The breaker plate assembly fitted with pivot-type points

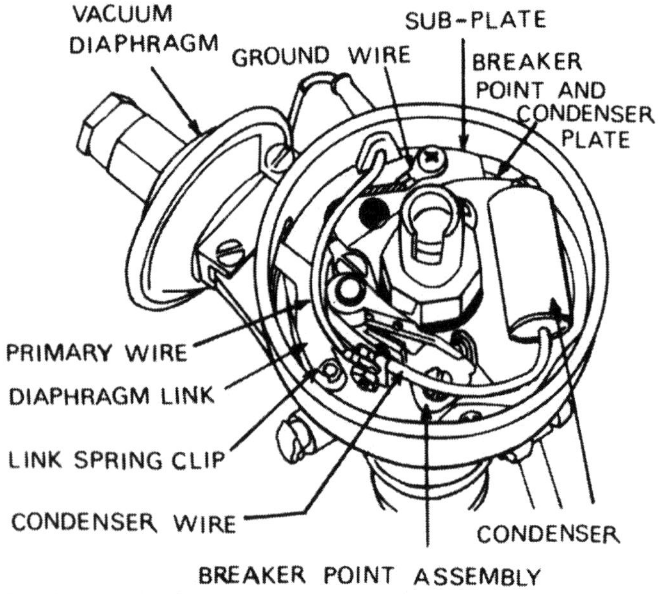

FIG 3:6 The breaker plate assembly fitted with pivotless points

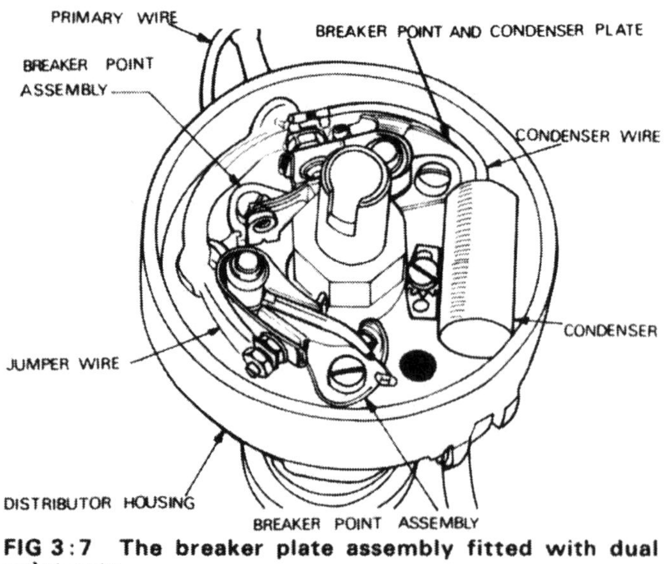

FIG 3:7 The breaker plate assembly fitted with dual point sets

An engine that runs unevenly or has a persistent misfire can have the fault caused by a defect in the ignition system, but first check the carburation and all vacuum operated systems.

Start the engine and set it to idle at a fairly fast tickover. Disconnect each spark plug lead from its spark plug in turn. **Twist and pull on moulded cap only, because by pulling directly on the wire the connection or weather seal may become damaged. Use dry rags or thick gloves to prevent getting shocks from a defective lead.** Disconnecting the spark plug lead from a cylinder that is firing will make the rough running more pronounced while disconnecting the lead from a cylinder that is not firing will make no difference. The exhaust beat will have to be listened to with care as the number of cylinders will tend to mask out failure in any one cylinder. On a four cylinder engine, one cylinder not firing will be immediately obvious but with eight cylinders, even two cylinders out will not be so readily detectable.

When the misfiring cylinder has been found, stop the engine. Fit an old spark plug or piece of rod into the moulded cap and restart the engine. Hold the lead, using gloves or tongs, so that the end of the spark plug or rod is approximately $\frac{3}{16}$ inch away from a convenient grounded metal part on the engine. **Keep the lead well away from the fuel system to avoid any risk of starting a fire.** If the ignition system is satisfactory there will be a regular series of fat blue sparks.

If the ignition system is satisfactory the fault lies either with the spark plug or the cylinder itself. Remove the spark plug for cleaning and testing or install a new spark plug in its place. If this does not cure the fault then the ignition system is not causing the misfire.

If the sparks from the lead are irregular, thin or reddish in color then the fault lies in the ignition system. Check the HT lead for signs of cracking, burn marks and resistance. Renew the lead if it is defective. Repeat the test using the new lead.

If the spark remains weak after renewing the HT lead, or the sparks are weak on all HT leads when they are checked in turn, stop the engine and check the ignition circuit. Remove the distributor cap and rotor arm. Check the breaker points for dirt, burning or excessive wear as well as making sure that the dwell is correct (correct gap when points are open). Clean the cap and rotor arm as described in **Section 3:2**. Examine the distributor cap very carefully for hairline cracks or thin carbon tracking marks between electrodes. Renew the distributor cap if any defect is found. Ensure that the rotor arm is in good condition and makes contact with the carbon button in the distributor cap.

Testing the primary circuit:

On the distributors fitted with dual breaker contact sets, the points are connected in parallel by a jumper lead and one set of breaker points closes the primary circuit while the other set opens it.

A brief field test can be carried out by removing the distributor cap and flicking the points apart with a fingernail, after cranking the engine to the position where the points are closed. If current is flowing through the primary circuit there will be a small low-voltage spark between the contacts as they open. A heavy spark is one indication of a defective condenser.

A better test can be carried out by disconnecting the primary wire from between the distributor low-tension terminal and ignition coil, at the ignition coil, and reconnecting it with a low wattage bulb of the correct voltage in series. The bulb will not glow so brightly as when it is connected directly across the battery terminals because of the resistance in the ballast resistor wire. It is advisable to remove all the spark plugs as this will make the engine easier to turn. Remove the distributor cap so that the action of the points can be watched. Slowly turn the engine. As the points close the test lamp should come on and then go out as the points open.

If the lamp does not come on at all, trace back through the wiring until the fault is found and can be repaired. If the lamp stays on continuously, there is a shortcircuit in the distributor. Disconnect the condenser and repeat the test. If the lamp now operates correctly, the condenser is at fault. Check the wires in the distributor for correct installation and make sure that the insulation is not frayed so that the wires ground.

Resistance wire:

This can be checked using a suitable resistance meter. It can also be checked using a voltmeter. Connect the voltmeter leads as shown in **FIG 3:10.** Connect a jumper lead from the DIST terminal on the coil to a good ground. Make sure that all accessories and lights are turned off. Turn on the ignition switch and note the voltmeter reading. If the reading is outside the limits of 6.6 to 4.5 volts, fit a new resistance wire. A high voltage will cause rapid burning and pitting of the breaker points while a low voltage will cause poor starting and performance.

Condenser:

Condenser failure is not a common fault, though it must not be discounted. Most condensers are made up of metal foil wrapped between paper which acts as an insulator. If the paper fails, the spark erodes away the foil in the area making the capacitor self-healing. A short-circuited condenser can be found using the test lamp procedure outlined earlier.

An open circuit condenser is more difficult to detect without the use of specialized test equipment, though it may be suspected if starting is difficult and the points are blued or badly burnt.

Without specialized equipment, the best check is to install a new condenser and see whether this cures the faults.

3:5 Removing and installing the distributor

Disconnect the primary wire at the distributor. Disconnect the vacuum hoses from the vacuum unit on the distributor. Remove the distributor cap. Make aligning marks on the engine block and distributor case so that the distributor case will be installed back into its original position. Make a mark on the distributor case opposite to the brass end of the rotor arm. Provided that the engine is not turned while the distributor is removed, and the distributor is installed so that the marks again align, the ignition timing will not be disturbed. Remove the distributor hold down cap bolt and clamp. Withdraw the distributor out of the engine block.

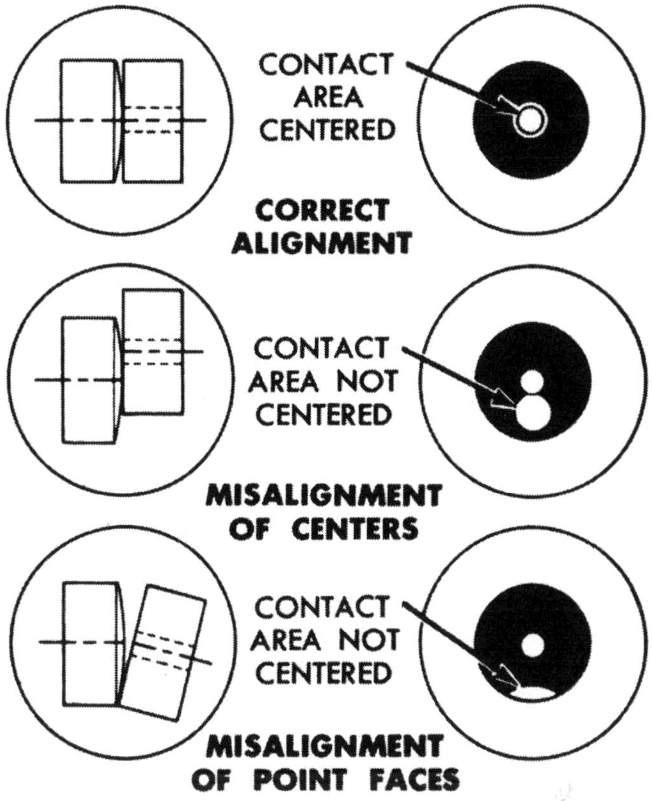

FIG 3:8 Breaker point alignment

Install the distributor in the reverse order of removal. When pressing the distributor back into place, allow for the rotation of the distributor shaft as the gears mesh to ensure that the mark aligns correctly when the distributor is in place. **Ensure that the distributor shaft engages fully and correctly with the oil pump intermediate shaft.** It may be necessary to crank the engine on the starter motor when the distributor is partially engaged in order to engage the oil pump intermediate drive shaft.

If the crankshaft has been rotated with the distributor removed, the ignition timing must be set from scratch.

If the HT leads have to be disconnected from the distributor cap, make sure that they are reconnected in

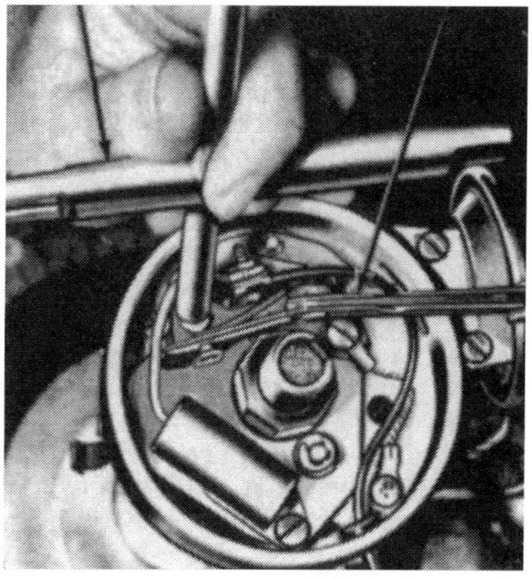

FIG 3:9 Adjusting new breaker points with feeler gauges

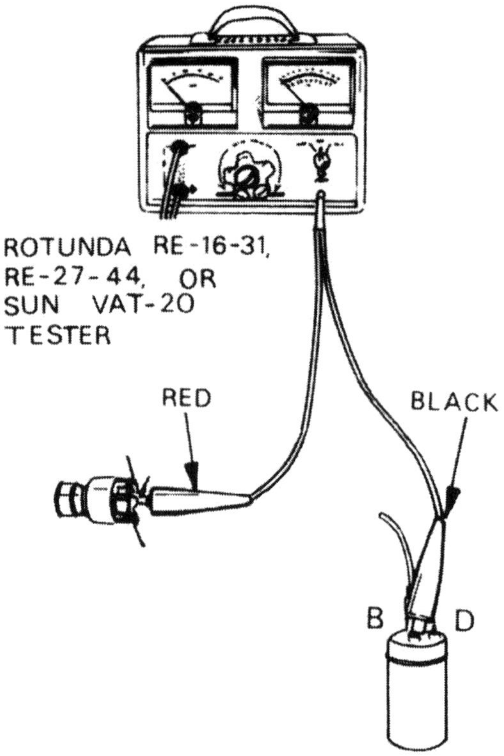

FIG 3:10 Resistance wire test

the firing order. The sequence and runs of the HT leads are shown in **FIG 3:11** for a typical engine. Make sure that the ends of the leads are clean and inserted fully into their sockets. **When disconnecting or installing, push and pull only on the moulded casing at the end otherwise the lead itself may be damaged.**

3:6 Setting the ignition timing

Before removing the distributor, marks should have been made on the engine block and distributor body so that the distributor will be installed into its original position.

Crank the engine until it is at TDC on the end of the compression stroke of No. 1 (front lefthand) cylinder. The compression stroke can be found by turning the engine with the spark plugs removed and a finger or thumb blocking the spark plug hole for No. 1 cylinder. A rise in pressure will be felt as the piston moves upwards on the compression stroke. The position can also be found by watching the rocker arms, if the rocker cover is off, as both valves on No. 1 cylinder will be fully closed. Turn the engine until the timing marks, shown in **FIG 3:12**, are in correct alignment.

With the engine set, install the distributor so that the rotor arm points towards the electrode connected to No. 1 spark plug (marked on the cap). Leave the distributor clamp and cap nut just slack enough to allow the distributor body to be rotated by hand without it moving on its own. Connect a test lamp in series between the ignition coil and distributor in the primary circuit. Apply light pressure against the rotor arm to take up any backlash, and rotate the distributor body slightly in either direction until the test lamp just goes out when the ignition is switched on. The lamp just going out indicates that the breaker points have just opened. Tighten the clamp to prevent the distributor body from turning. Rotate the engine in a forward direction for one complete turn. Slow down the rate of turning just before the end of the turn and stop turning the instant the test lamp goes out. If the timing is correctly set, the timing marks will again be in exact alignment.

If a low grade of fuel is being used, or the individual requirements of the engine dictate it, the ignition setting may be retarded by up to 2 deg. from the standard setting to prevent detonation (spark knock). If retarding is necessary, it should be carried out progressively and while checking the result with a road test before retarding the ignition further. **Do not retard back beyond 2 deg. from the standard setting.**

Performance and economy can sometimes be increased by advancing the timing. **The maximum by which the ignition should be advanced is 5 deg.** Advance the ignition step by step until detonation (spark knock) is heard during hard acceleration from low speeds or when the engine is working hard. Retard the ignition until detonation is just eliminated.

Stroboscopic timing:

A stroboscopic light will ensure that the ignition is accurately set with the engine running. Set the static timing so that the engine can be started and run. Run the engine at a fast idle until the temperatures have stabilized and the choke plate is fully open. Stop the engine and connect the instrument into the ignition circuit as instructed by the instrument manufacturer. Start the engine and set it to run at the correct speed. Disconnect such vacuum hoses as required (see **Technical Data**). Shine the light onto the timing marks. **Avoid wearing loose or flapping clothing and take great care to keep hands, tools or clothes out of the moving drive belts and cooling fan.** Slacken the distributor clamp and rotate the distributor body slightly in either direction until the timing marks are in exact alignment. The timing marks will be easier to see if the correct ones are marked with a thin white line and the edge of the pointer is similarly marked. Tighten the distributor clamp when the timing is correct. Stop the engine and restore the ignition circuit to normal.

3:7 Temperature sensing valve

This is fitted to later models and the valve is shown in **FIG 3:13.** The valve is fitted into the intake manifold of the engine.

To check the valve, attach a tachometer to the engine. Make sure that all vacuum lines are correctly routed and connected. Start the engine and run it until it has reached its normal operating temperature **(do not allow the engine to overheat).** Note the curb idle speed with the transmission in neutral and the throttles closed to the curb idle position. Disconnect the vacuum hose that connects the intake manifold to the valve, at the valve, and plug the end of the hose. There should be no change in idle speed if the valve is satisfactory. If the curb idle speed drops by a 100 rev/min or more, the valve is defective and must be renewed. Reconnect the vacuum hose to the valve, after removing the plug.

Cover the radiator and allow the engine to run until the overheat warning light comes on. **Before making this**

part of the test ensure that the cooling system is satisfactory and that the radiator filler cap is correctly installed. If the valve is satisfactory, the idling speed will have increased by 100 rev/min or more. If the idle speed does not increase then the valve is defective and must be renewed.

Uncover the radiator and allow the engine to cool down to its normal operating temperature before switching it off.

3:8 The spark plugs

The recommended spark plugs and their gap settings are given in **Technical Data.**

Removal:

Slacken the spark plug using a well fitting box wrench or plug wrench. Blow away loose dirt using an air-line or tire pump. Unscrew the spark plug fully by hand, or the wrench if it is stiff. If the spark plug is extremely stiff to unscrew, use the wrench to take it out as far as possible without exerting brute force and then wind the plug in and out to this position several times before attempting to unscrew it further. In extreme cases, wrap the base of the spark plug with rag dipped in kerosene or penetrating oil and leave it to soak overnight. As the plugs are removed they should be stored in order in a rack for later examination of the firing ends.

Examination:

The color and condition of the deposits will give a good guide to the conditions inside the combustion chamber of the cylinder from which the spark plug was removed.

After examination, throw away any spark plugs which have badly burnt electrodes or cracked insulators and fit new ones in their place.

A light powdery deposit ranging in color from brown to greyish tan, coupled with light wear on the electrodes indicates that the conditions are normal. Long periods of constant-speed driving or much city driving will leave the deposits white or yellowish. Cleaning and testing the sparking plugs is all that is required in these cases.

If the deposits are wet and black, they are caused by oil entering the combustion chamber past worn engine components. An engine overhaul is the only full cure but fitting a hotter-running grade of spark plug may help to reduce the problem.

Black fluffy deposits are caused by incomplete combustion. This may be caused by excess idling—particularly if all the spark plugs are similarly affected but it may also be caused by defective ignition or spark plug or running with too rich a mixture.

Overheated spark plugs will have a white blistered look about the central electrode and there may be glints of metal on the insulator if lead-based fuels are used. The electrodes will be excessively burnt away. Some possible causes are engine overheating, very weak mixture, incorrectly set ignition timing, incorrect grade of spark plugs or running at high-speeds with the car overloaded.

Cleaning, adjusting and testing:

Wash oily spark plugs in gasoline before sand-blasting them. Have the plugs cleaned on an abrasive-blasting machine and pressure tested after attention to the

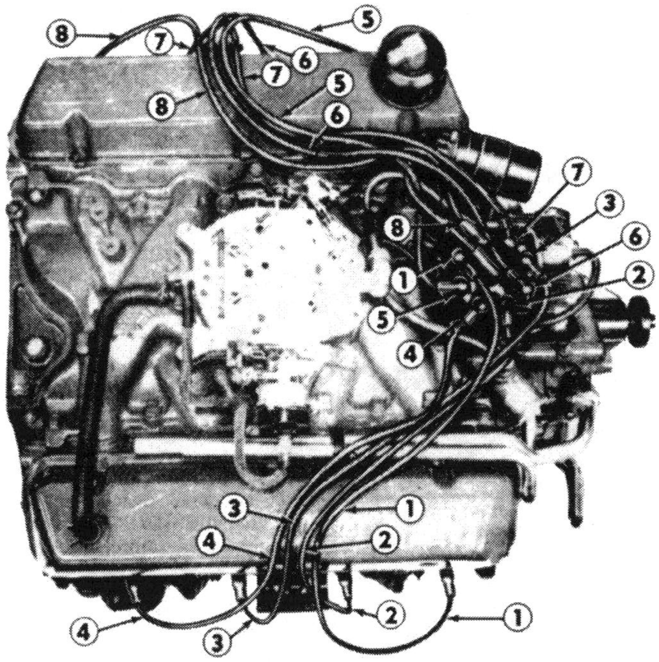

FIG 3:11 Typical ignition high-tension lead layout and routing

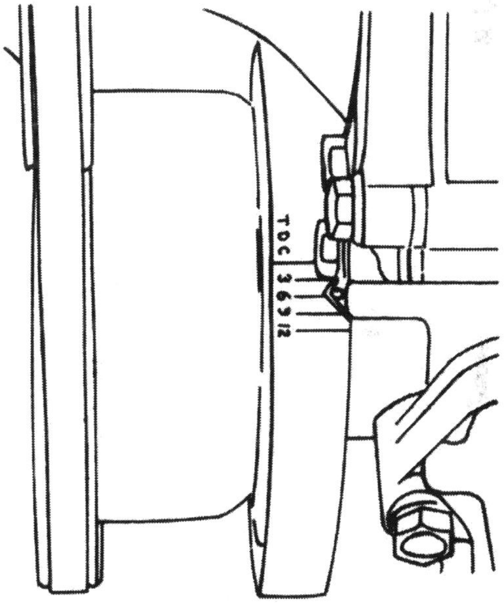

FIG 3:12 Ignition timing marks

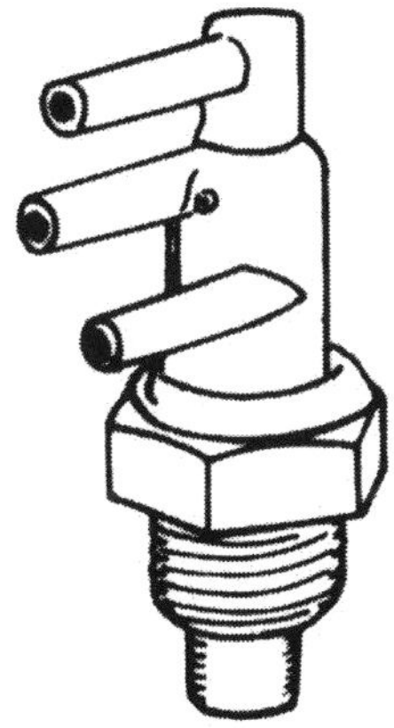

FIG 3:13 Distributor vacuum control valve (temperature sensing valve)

electrodes. In an emergency, a stiff steel-wire brush can be used to clean the plugs though this will not be so effective as abrasive-blasting. Use a wire brush to clean the plug threads.

Trim the electrodes square using a fine file. Adjust to the correct gap by bending the side electrode only. **Never bend the central electrode as this will crack the insulator.**

Apply a little oil around the joint of the insulator and metal body. Pressurize the plug in the testing machine and if air bubbles are seen leaking out through the oil, reject the spark plug.

Clean the external portion of the insulator with denatured alcohol or suitable solvent.

Installing:

If the threads in the cylinder head are dirty, clean them using a well-greased tap. Failing a tap, use an old sparking plug with cross-cuts down the threads.

If grease is applied to the threads use only graphite grease, as any other form of grease will bake hard and lock the plug in the head.

Screw the plugs in by hand and use the wrench for the final tightening only. Half a turn beyond the hand-tight position is quite sufficient and overtightening can cause damage to the spark plugs.

3:9 Trouble shooting guide

(a) Engine will not fire

1 Battery discharged or terminals dirty
2 Breaker points dirty or incorrectly adjusted
3 Distributor cap dirty, cracked or tracking
4 Rotor arm not making contact with carbon button in cap
5 Distributor rotor arm cracked, dirty or omitted on reassembly
6 Defective ingition coil
7 Defective HT lead between coil and distributor
8 Broken breaker spring
9 Breaker contacts stuck open
10 Internal shortcircuit in distributor
11 Damp or dirt on HT leads, distributor cap and ignition coil

(b) Engine misfires

1 Check 2, 3 and 6 in (a)
2 Weak breaker spring
3 Defective HT lead
4 Spark plug(s) loose
5 Spark plug insulation cracked or dirty
6 Spark plug gap incorrectly set
7 Ignition timing too far advanced
8 HT leads installed in wrong order

CHAPTER 4

THE COOLING SYSTEM

4:1 Description

A sectional view of the engine showing the coolant flow is shown in **FIG 4:1**. The coolant circulates around the outside of the cylinder bores and through the passages in the cylinder heads, becoming heated in the process. Hot water normally tends to rise, because of expansion, and the heated coolant passes through the intake manifold (warming it to ensure efficient vaporization of the fuel) before passing out through the outlet elbow on the intake manifold. From the manifold it passes through a hose to the radiator. Two types of radiator can be fitted. The downflow type is shown in the figure and the heated coolant passes into upper header tank. On cross-flow radiators the coolant flows from one side to the other, coming into the righthand side chamber or header tank and being taken out at the bottom of the lefthand side chamber. The passage of air over the fins of the radiator cools the coolant and it is then returned to the engine. To ensure that there is a positive flow of coolant, and a faster flow than would be produced by natural thermo-siphon action, a belt driven centrifugal water pump is fitted into the system. The pump draws the coolant from the radiator and passes it into the engine.

A cooling fan is fitted to the water pump hub to ensure a good air flow through the radiator core when the vehicle is stationary or travelling slowly. Most of the fans fitted are designed not to operate at full efficiency when the vehicle is at speed, as they are not required at speed and will absorb power from the engine.

On vehicles fitted with automatic transmission, a heat exchanger is fitted into the radiator to cool the transmission fluid. On down-flow radiators the heat exchanger is fitted to the lower tank while on cross-flow radiators the unit is fitted to the lefthand chamber.

A thermostat valve is fitted to the outlet elbow from the intake manifold to ensure that the system runs at a constant temperature and also to shorten the warm-up period. When the coolant is cold, the valve stays shut and the coolant is bypassed directly to the water pump inlet. The amount of coolant to be heated is reduced and there are no heat losses through the radiator so the engine rapidly warms up. When the coolant approaches its normal operating temperature, the valve starts to open and coolant can then pass through the radiator to be cooled normally.

The filler cap to the radiator contains two concentric valves. The outer valve seals until the operating pressure of the system is reached and it then opens to vent surplus coolant or air through the overflow tube. By pressurizing the system, the boiling point of the coolant is raised and

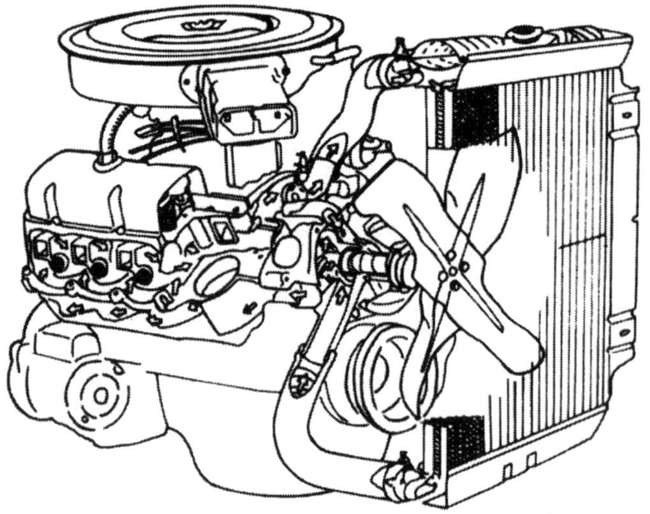

FIG 4:1 The cooling system, showing the coolant flow

this allows a smaller radiator to be used (as the system can be run hotter) and also avoids boiling at localized hot spots in the engine. **The filler cap must never be removed when the engine is overheated, as the pressure will force out scalding water and the subsequent reduction in pressure can cause the coolant to boil, so that the steam forces out even more coolant.** The filler cap is fitted with a double bayonet securing device, whereby the first part of the turn allows the valve to lift and pressure vent overboard and the final part of the turn releases the cap so that it can be removed. When the system cools, the inner valve of the filler cap opens to allow air to return to the header tank and so prevent a vacuum from forming in the system.

A mixture of antifreeze and soft water giving protection to the system down to a minimum temperature of 0°F should always be used in the system, even in climates where there is no danger of the system freezing. The antifreeze, apart from protecting the system against freezing, contains additives which raise the boiling point of the water and also prevent corrosion in the system. A 50-50 mixture is standard and protects to —35°F.

4:2 Maintenance

Routine maintenance is confined to checking that the drive belts are at the correct tension and not deteriorating, that the hoses are secure and not deteriorating, and also checking the coolant level in the radiator.

The life of the antifreeze is two years, so after two years' use the system should be drained, flushed and refilled with fresh coolant.

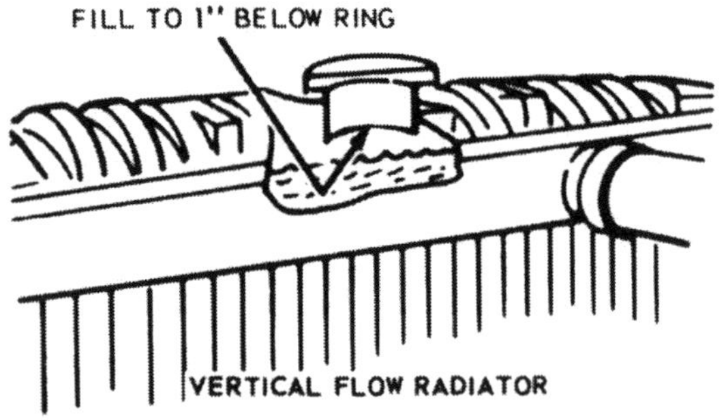

FIG 4:2 Topping up the down-flow radiator

Coolant level:

This should be checked with the engine cold. If losses due to leakage are being replaced, use a mixture of anti-antifreeze and soft water as the use of water alone will dilute the coolant already in the system. On down-flow radiators, top up until the level is one inch below the bottom of the ring, as shown in **FIG 4:2.** On cross-flow radiators the system is topped up to the level marked on the lefthand tank, as shown in **FIG 4:3.**

Draining:

The radiator has a drain cock fitted underneath it. If the system is being drained for any other reason than changing the coolant, attach a length of hose to the drain cock and drain the system into a clean container so that the coolant can be used again.

When completely draining the system, remove the two drain plugs, one fitted on either side of the cylinder block as shown in **FIG 4:4.** Allow all the coolant to drain out.

Flushing:

The use of Rotunda Cooling System Cleanser will remove scale and dirt from the system if the instructions on the tin are followed.

The system can be partially cleaned by flushing through with clean water from a hosepipe inserted in the radiator filler and leaving it to run until the water comes out clear from the drain holes.

If the system is so clogged that Rotunda Cleanser will not remove all the dirt or scale, the car should be taken to a service garage for pressure flushing. When pressure flushing, the thermostat valve should always be removed from its housing. In some cases it will help to remove the radiator, invert it and flush through with a hose in a reverse direction so that water and gravity combined can dislodge scale from the tubes.

Hoses:

Check all the hoses for softening, hardening, collapsing or perishing, and renew any that are defective. Slacken off the hose clamps and pull the old hoses off with a twisting motion. Position the clamp on the hose so that it is at least $\frac{1}{8}$ inch from the end of the hose. Lubricate the inside of the hose with water, soft soap or with an approved water-resistant sealer and slide the hose back into position. Make sure that the hose clamps are beyond the bead and place them centrally onto their sealing surfaces. Tighten the hose clamps sufficiently to prevent leaks without the clamps cutting into the hoses.

Filling:

Make sure that the drain plugs are installed in the cylinder block and that the radiator drain cock is closed. The best mixture to fill the system with is 50 per cent Rotunda ethylene glycol antifreeze and 50 per cent soft clean water. This mixture will protect the system from frost down to —35°F and raise the boiling point to 250°F. A tin of Rotunda Radiator Rust Inhibitor should also be added to the system whenever it is filled with fresh coolant. Fill the system slowly through the radiator

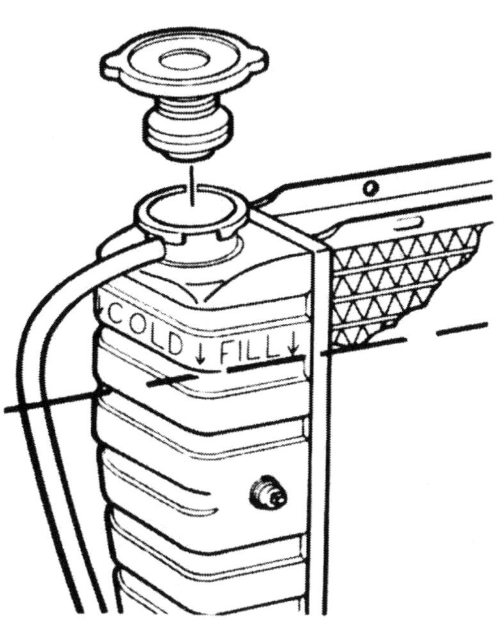

FIG 4:3 Topping up the cross-flow radiator

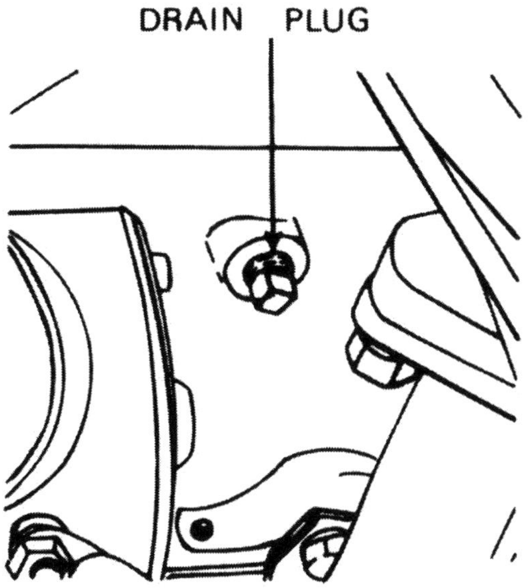

FIG 4:4 Typical engine drain plug

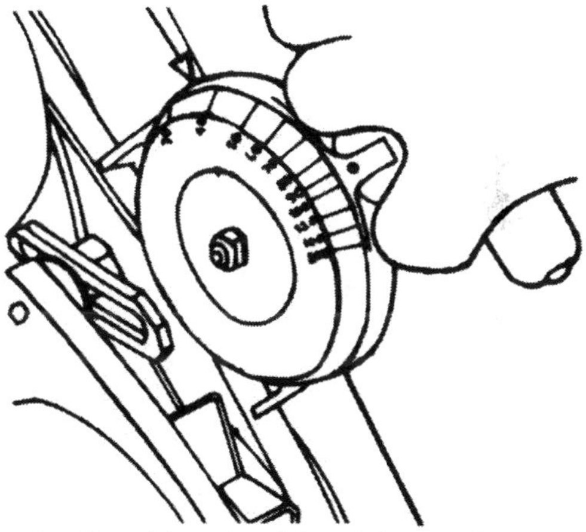

FIG 4:5 Checking the drive belt tension

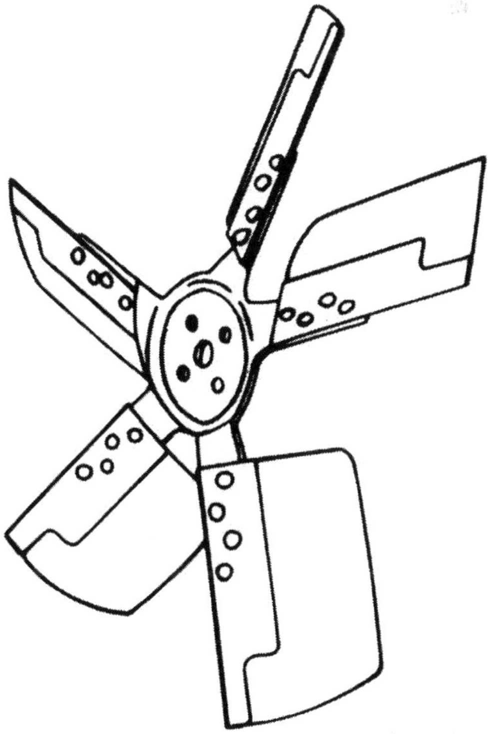

FIG 4:6 The Ford Flex-Blade cooling fan

filler. Disconnect the heater outlet hose at the water pump and wait until coolant flows out through it, making sure that the heater is set to hot, before reconnecting the hose. Fill the engine to the correct level. Start the engine and check for coolant leaks. Leave the engine running for 20 minutes at 2000 rev/min and then allow it to cool. Top up to the correct level. **Up to a quart of coolant may be required to top up the system as entrapped air will have bled out.**

4:3 The radiator

Removal:

Drain the cooling system. Disconnect the upper and lower hoses at the radiator. On models fitted with automatic transmission, disconnect the heat exchanger lines from the radiator and plug both the lines and connectors to prevent the entry of dirt. On vehicles equipped with air conditioning, free the air conditioning radiator and lay it out of the way. Remove the radiator support bolts and supports (if fitted) while supporting the radiator and then lift out the radiator.

Install the radiator in the reverse order of removal.

Cleaning:

Periodically use an air-line or hosepipe to flush accumulations of mud and insects from the radiator cooling fins. Blow from the rear forwards. In extreme cases, remove the radiator so that it can be soaked and scrubbed in a very hot weak detergent solution.

When the radiator is removed, flush it through with a hosepipe. Invert the downflow radiator and stand the cross-flow radiator on its righthand side so that the water will run in the reverse direction to normal. Keep flushing until the water comes out clean and rust-free.

Repairs:

Repairs either to the radiator or the automatic transmission heat exchanger should be left to an authorized dealer or agent specializing in radiator repairs.

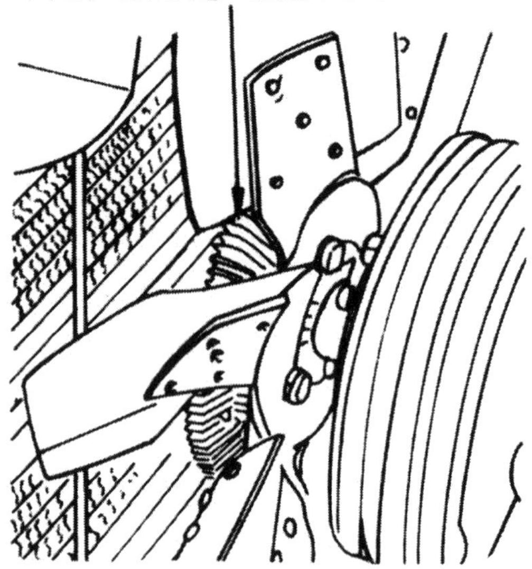

FIG 4:7 Clutch driven cooling fan

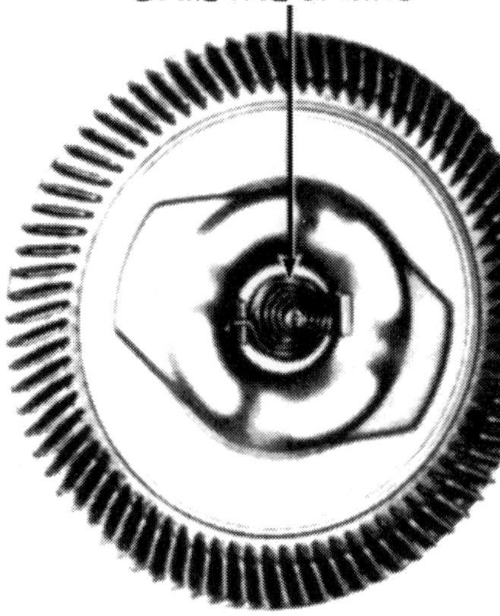

FIG 4:9 The fan drive clutch using a spiral bi-metallic spring

4:4 Drive belts

Belts are used to transmit the drive from the engine to the alternator and water pump, as well as to Thermactor air pump, air conditioning and power steering pumps if these accessories are fitted.

All the drive belts are independently adjustable and they should be set to the correct tension to avoid slip or straining bearings. The most accurate method of checking drive belt tension is to use the gauge shown in **FIG 4:5**. A rough guide is that the belt will deflect by approximately $\frac{1}{4}$ inch from the straight when pressed with moderate thumb pressure. Brand new belts should be set to a tension of 120 to 140 on the gauge but once they have been in use for longer than ten minutes they should be set to a tension of 90 to 110 on the gauge.

The water pump and alternator drive belt(s) are set by slackening the alternator mountings and using a wooden lever on the front drive bracket of the alternator to position the alternator so that the belt is at the correct tension. Tighten the alternator attachments to hold the alternator in place.

The Thermactor pump, if fitted, is driven by a belt between the water pump pulley and Thermactor pump and is set by slackening the Thermactor pump attachments and pulling it away from the engine, then tightening the attachments when the tension is correct.

The power steering pump drive belt is similarly adjusted, but if leverage is required it must be applied to the lug of the pump and not to the reservoir. The air conditioner drive belt is dealt with in a similar manner.

If a belt requires removal, slacken the attachments and move the unit to the fully slack position and ease the belt of a pulley. If several accessories are fitted, front belts will have to be removed first and if the water pump belt(s) are to be removed then all the accessory drive belts must be taken off first. If two belts are fitted to drive the water pump and alternator, they must be renewed as a pair.

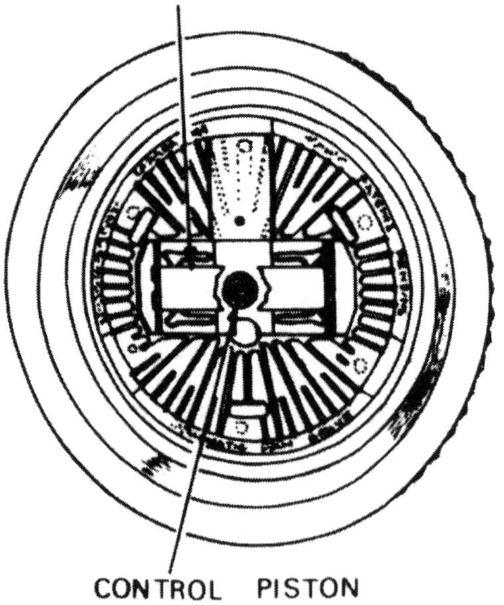

FIG 4:8 The fan drive clutch using a bi-metallic plate spring

4:5 The cooling fan

On some models a Ford Flex-Blade fan, shown in **FIG 4:6**, is fitted. Under heavy load when the engine is running fast, the blades flex and reduce pitch so that the fan absorbs less power.

On other models a clutch is fitted between the pulley and fan blades, and the unit installed is shown in **FIG 4:7**. The unit operates by utilizing the drag of silicone-based fluid between the inner and outer members. Centrifugal force and the pumping action built into the unit forces the silicone-based fluid into the narrow space between the inner and outer members so that drive is transmitted. Immediately the engine is started there will be maximum drive. A bi-metallic spring, either flat as shown in **FIG 4:8**, or spiral as shown in **FIG 4:9**, is fitted to the unit. The bi-metallic spring senses the temperature of the air from the radiator and adjusts the flow of fluid in the unit accordingly. During the warm-up cycle when the air temperature from the radiator is less than 150 to 180°F, the spring moves the plunger to allow the fluid to return to the centre chamber and the clutch is free to slip so that the fan operates at a lower speed. When the air temperature from the radiator rises, the spring senses this rise and allows the fluid to return between the drive members so that the clutch no longer slips and the fan is driven at full speed. The spring will modulate the speed between the high and low settings as the road conditions and engine temperature vary. At high speed the clutch action will limit the maximum speed of the fan irrespective of the engine temperature.

To check the clutch, start the engine and allow it to run for a few minutes. Stop the engine and the fan should turn freely. Restart the engine and allow it to reach its normal operating temperature, blocking off the front of the radiator to speed up the warming up process. Remove the blocking from the front of the radiator and leave the engine running at 1000 rev/min for a short while to ensure that the bi-metallic strip reaches the air temperature. Stop the engine. Protect the hands with rag and try to turn the fan. If the clutch is satisfactory the fan can only be turned with great difficulty. If the fan turns easily, the clutch is defective and must be renewed or repaired.

4:6 The thermostat

The attachment of the thermostat valve on later models is shown in **FIG 4:10.** In all cases, drain the cooling system and remove the outlet elbow from the intake manifold, noting that usually the hose is sufficiently flexible to be left attached to the elbow. Remove the thermostat and gasket. Before installing the thermostat, clean off all traces of old gasket and sealing compound from the manifold and elbow. Smear a new gasket lightly with sealing compound and position it on the manifold. Install the thermostat valve. On earlier models it must be installed with the marking TOP upwards otherwise the engine will overheat. On later models thermostat is installed in the elbow and then turned clockwise to lock it. Attach the elbow to the manifold, fill the cooling system and check for leaks.

Testing:

Suspend the thermostat in a container of water and gradually heat the water while stirring it with a thermometer. If the valve does not start to open within a few degrees of the correct temperature and is not fully open at the correct temperature the thermostat is defective and must be renewed.

Hold the valve up against a strong light and if there are gaps between valve and housing, reject the thermostat and fit a new one in its place.

4:7 The water pump

The pump is attached to the front cylinder cover. A bleed hole is made in the casting for the pump and this hole is for expelling any water that may leak past the seal and is not for lubrication purposes. A defective pump cannot be repaired and either a new or reconditioned pump must be fitted if the old one is damaged, worn or defective.

Removal:

Drain the cooling system. Remove all the drive belts (see **Section 4:4**). Remove the cooling fan and pulley. Disconnect the heater and radiator hoses from the pump. Take out the bolts securing the pump to the front cover and remove the pump and its gasket. Discard the old gasket.

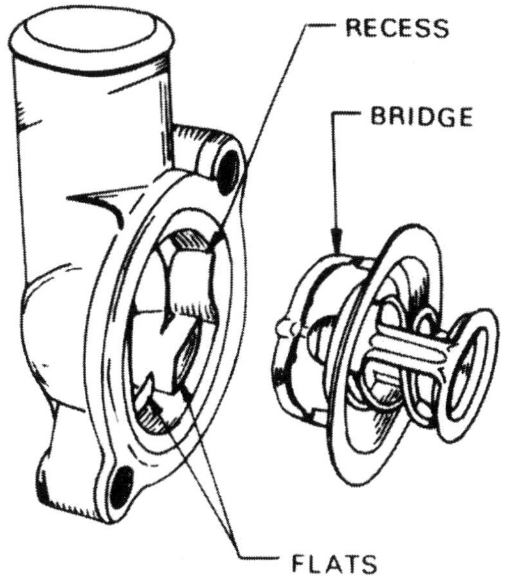

FIG 4:10 The later thermostat valve

The pump is installed in the reverse order of removal. Scrape all old gasket and sealing compound from the mating faces. Smear a new gasket with jointing compound and lay it in position. Attach the pump using its bolts. Install the fan, pulley and belts in the reverse order of removal. Set all the belt tensions correctly.

4:8 Trouble shooting guide

(a) Internal coolant leakage

1 Loose cylinder head bolts
2 Loose intake manifold bolts
3 Cracked cylinder head
4 Cracked cylinder block
5 Defective cylinder head gasket
6 Warped cylinder head

(b) Corrosion

1 Impurities in the water
2 Lack of inhibitor or antifreeze
3 Infrequent draining and flushing

(c) Poor circulation

1 Perished or collapsed coolant hoses
2 Defective thermostat
3 Low coolant level
4 Loose or broken drive belt
5 Engine water passages restricted
6 Radiator tubes restricted

(d) Overheating

1 Check (c)
2 Defective cooling fan clutch
3 Sludge in oil pan and crankcase
4 Low oil level in crankcase
5 Tight engine
6 Choked exhaust
7 Binding brakes
8 Slipping clutch
9 Incorrect clutch adjustments
10 Incorrect valve timing
11 Retarded ignition
12 Mixture too weak

Inches	Decimals	Milli-metres	Inches to Millimetres Inches	mm	Millimetres to Inches mm	Inches
1/64	.015625	.3969	.001	.0254	.01	.00039
1/32	.03125	.7937	.002	.0508	.02	.00079
3/64	.046875	1.1906	.003	.0762	.03	.00118
1/16	.0625	1.5875	.004	.1016	.04	.00157
5/64	.078125	1.9844	.005	.1270	.05	.00197
3/32	.09375	2.3812	.006	.1524	.06	.00236
7/64	.109375	2.7781	.007	.1778	.07	.00276
1/8	.125	3.1750	.008	.2032	.08	.00315
9/64	.140625	3.5719	.009	.2286	.09	.00354
5/32	.15625	3.9687	.01	.254	.1	.00394
11/64	.171875	4.3656	.02	.508	.2	.00787
3/16	.1875	4.7625	.03	.762	.3	.01181
13/64	.203125	5·1594	.04	1.016	.4	.01575
7/32	.21875	5.5562	.05	1.270	.5	.01969
15/64	.234375	5.9531	.06	1.524	.6	.02362
1/4	.25	6.3500	.07	1.778	.7	.02756
17/64	.265625	6.7469	.08	2.032	.8	.03150
9/32	.28125	7.1437	.09	2.286	.9	.03543
19/64	.296875	7.5406	.1	2.54	1	.03937
5/16	.3125	7.9375	.2	5.08	2	.07874
21/64	.328125	8.3344	.3	7.62	3	.11811
11/32	.34375	8.7312	.4	10.16	4	.15748
23/64	.359375	9.1281	.5	12.70	5	.19685
3/8	.375	9.5250	.6	15.24	6	.23622
25/64	.390625	9.9219	.7	17.78	7	.27559
13/32	.40625	10.3187	.8	20.32	8	.31496
27/64	.421875	10.7156	.9	22.86	9	.35433
7/16	.4375	11.1125	1	25.4	10	.39370
29/64	.453125	11.5094	2	50.8	11	.43307
15/32	.46875	11.9062	3	76.2	12	.47244
31/64	.484375	12.3031	4	101.6	13	.51181
1/2	.5	12.7000	5	127.0	14	.55118
33/64	.515625	13.0969	6	152.4	15	.59055
17/32	.53125	13.4937	7	177.8	16	.62992
35/64	.546875	13.8906	8	203.2	17	.66929
9/16	.5625	14.2875	9	228.6	18	.70866
37/64	.578125	14.6844	10	254.0	19	.74803
19/32	.59375	15.0812	11	279.4	20	.78740
39/64	.609375	15.4781	12	304.8	21	.82677
5/8	.625	15.8750	13	330.2	22	.86614
41/64	.640625	16.2719	14	355.6	23	.90551
21/32	.65625	16.6687	15	381.0	24	.94488
43/64	.671875	17.0656	16	406.4	25	.98425
11/16	.6875	17.4625	17	431.8	26	1.02362
45/64	.703125	17.8594	18	457.2	27	1.06299
23/32	.71875	18.2562	19	482.6	28	1.10236
47/64	.734375	18.6531	20	508.0	29	1.14173
3/4	.75	19.0500	21	533.4	30	1.18110
49/64	.765625	19.4469	22	558.8	31	1.22047
25/32	.78125	19.8437	23	584.2	32	1.25984
51/64	.796875	20.2406	24	609.6	33	1.29921
13/16	.8125	20.6375	25	635.0	34	1.33858
53/64	.828125	21.0344	26	660.4	35	1.37795
27/32	.84375	21.4312	27	685.8	36	1.41732
55/64	.859375	21.8281	28	711.2	37	1.4567
7/8	.875	22.2250	29	736.6	38	1.4961
57/64	.890625	22.6219	30	762.0	39	1.5354
29/32	.90625	23.0187	31	787.4	40	1.5748
59/64	.921875	23.4156	32	812.8	41	1.6142
15/16	.9375	23.8125	33	838.2	42	1.6535
61/64	.953125	24.2094	34	863.6	43	1.6929
31/32	.96875	24.6062	35	889.0	44	1.7323
63/64	.984375	25.0031	36	914.4	45	1.7717

UNITS	Pints to Litres	Gallons to Litres	Litres to Pints	Litres to Gallons	Miles to Kilometres	Kilometres to Miles	Lbs. per sq. In. to Kg. per sq. Cm.	Kg. per sq. Cm. to Lbs. per sq. In.
1	.57	4.55	1.76	.22	1.61	.62	.07	14.22
2	1.14	9.09	3.52	.44	3.22	1.24	.14	28.50
3	1.70	13.64	5.28	.66	4.83	1.86	.21	42.67
4	2.27	18.18	7.04	.88	6.44	2.49	.28	56.89
5	2.84	22.73	8.80	1.10	8.05	3.11	.35	71.12
6	3.41	27.28	10.56	1.32	9.66	3.73	.42	85.34
7	3.98	31.82	12.32	1.54	11.27	4.35	.49	99.56
8	4.55	36.37	14.08	1.76	12.88	4.97	.56	113.79
9		40.91	15.84	1.98	14.48	5.59	.63	128.00
10		45.46	17.60	2.20	16.09	6.21	.70	142.23
20				4.40	32.19	12.43	1.41	284.47
30				6.60	48.28	18.64	2.11	426.70
40				8.80	64.37	24.85		
50					80.47	31.07		
60					96.56	37.28		
70					112.65	43.50		
80					128.75	49.71		
90					144.84	55.92		
100					160.93	62.14		

UNITS	Lb ft to kgm	Kgm to lb ft	UNITS	Lb ft to kgm	Kgm to lb ft
1	.138	7.233	7	.967	50.631
2	.276	14.466	8	1.106	57.864
3	.414	21.699	9	1.244	65.097
4	.553	28.932	10	1.382	72.330
5	.691	36.165	20	2.765	144.660
6	.829	43.398	30	4.147	216.990

CHAPTER 5

THE CLUTCH

5:1 Description

A mechanically-operated clutch is fitted to all models covered by this manual. The 1969 model mechanical linkage is shown in **FIG 5:1** and this is fairly typical of all the models in the range.

A sectioned view of the clutch in the flywheel housing is shown in **FIG 5:2**. The mechanical linkage operates the clutch release lever. The clutch cover assembly is bolted to the engine flywheel and therefore rotates with the engine. The clutch disc has a hub with splines in it and it is fitted to the input shaft of the transmission so that they revalve together.

The clutch casing assembly contains springs which act between the case and a pressure plate to force the pressure plate forwards. The clutch disc is gripped, by its friction linings, between the pressure plate and face of the engine flywheel so that the clutch disc is forced to rotate with the engine and transmit drive to the input shaft of the transmission.

When the clutch pedal is depressed, the linkage operates the clutch release lever. The lever moves about its pivot and presses the clutch release bearing against the clutch fingers. The fingers in turn pull the pressure plate back against the pressure of the clutch springs, releasing the clutch disc so that it and the transmission input shaft are free to revolve independently or even come to a stop with the engine running.

As the clutch pedal is released, there is a gradual increase of pressure on the clutch disc and at first there is slip between the clutch disc and pressure faces so that drive is smoothly taken up. Eventually the point is reached where the clutch disc is revolving at engine speed and the drive is again direct. Damper springs are fitted into the clutch disc to help absorb the shock of take-up of drive.

All the components of a 1965 model are shown in **FIG 5:3** and the action of the clutch can be most easily followed through using this figure.

5:2 Clutch adjustments

As the friction linings on the disc wear, the clutch pedal free travel will alter. **It is essential that there is always a clearance between the clutch release bearing and clutch release fingers when the clutch is engaged.** If the release bearing contacts the fingers at all times there will be rapid wear on the release bearing and, in excessive cases, there may be sufficient pressure on the fingers to partially withdraw the pressure plate and allow the clutch to slip.

The method of checking is shown in **FIG 5:4**. Press the clutch pedal right down and measure the total travel. If the total travel is incorrect, adjust the position of the bumper which the pedal hits when it is fully released to set the correct full travel. Start the engine and leave it

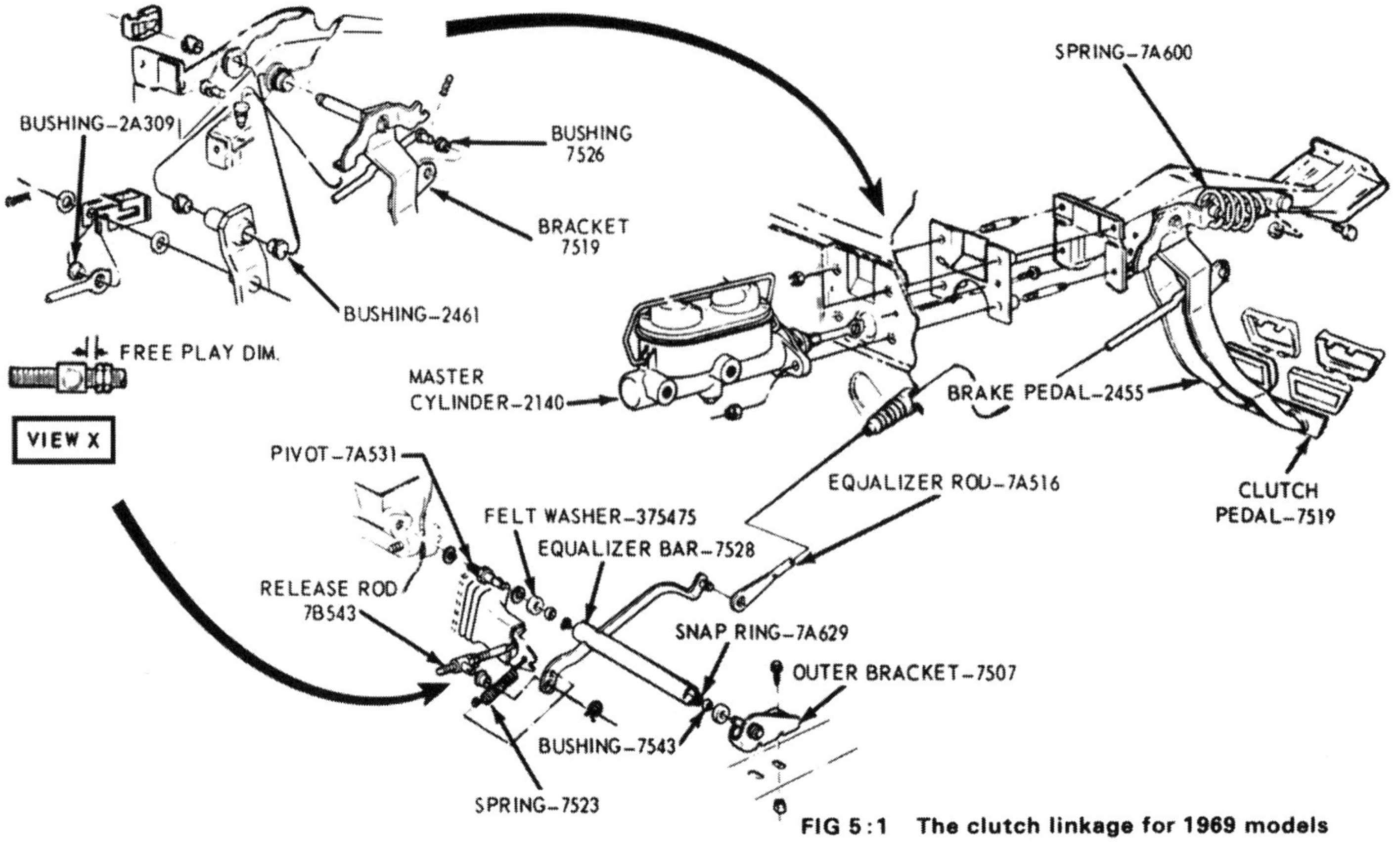

FIG 5:1 The clutch linkage for 1969 models

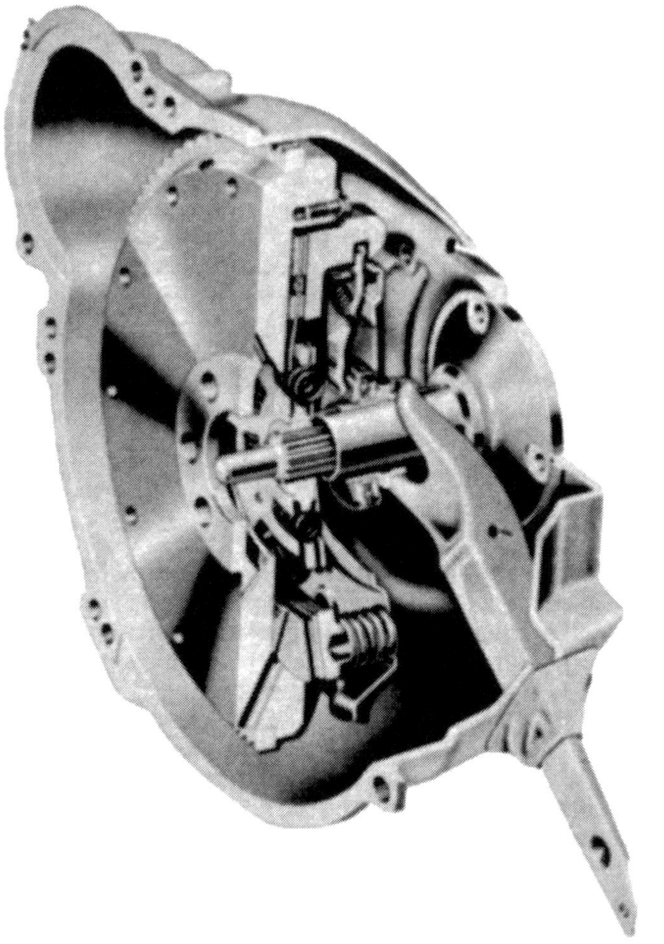

FIG 5:2 Sectioned view through the clutch assembly

running at idling. Press down the clutch pedal by hand until the release bearing can just be felt to contact the fingers in the clutch. This point is readily detectable by hand. On all models the adjustment is correct if the free travel of the clutch pedal is $\frac{7}{8}$ to $1\frac{1}{8}$ inch. The adjustment point on 1964 to 1966 models is shown in **FIG 5:5**. Slacken the nuts and move them in the direction indicated to alter the free play. Tighten both nuts to the trunnion. When the free play at idling is correct, increase the engine speed to 3000 rev/min and check that the minimum pedal free travel is $\frac{1}{2}$ inch. **If the free travel at speed is not sufficient, the free travel at idling must be increased.**

On the later models (see **FIG 5:1**) the check is carried out in a similar manner. To adjust the free play, disconnect the return spring from the clutch release lever. Slacken the nuts on the release lever rod. Slacken off the adjusting nut and pull the release lever rearwards until the release bearing can just be felt to touch the fingers in the clutch. Press the release rod into the release lever so that they are in contact and hold them both in this position. Insert a .136 inch feeler gauge between the adjusting nut and swivel sleeve, and tighten down the adjusting nut until it just nips onto the feeler gauge. Tighten the locknut to hold the adjusting nut in place and remove the feeler gauge. Reconnect the return spring to the release lever. Check that the pedal free travel is correct at idling and 3000 rev/min.

5:3 The flywheel housing

The removal of the flywheel housing, after the transmission has been removed, is dealt with in **Chapter 1, Section 1:11. When removing the housing, carefully collect any shims fitted between it and the engine.**

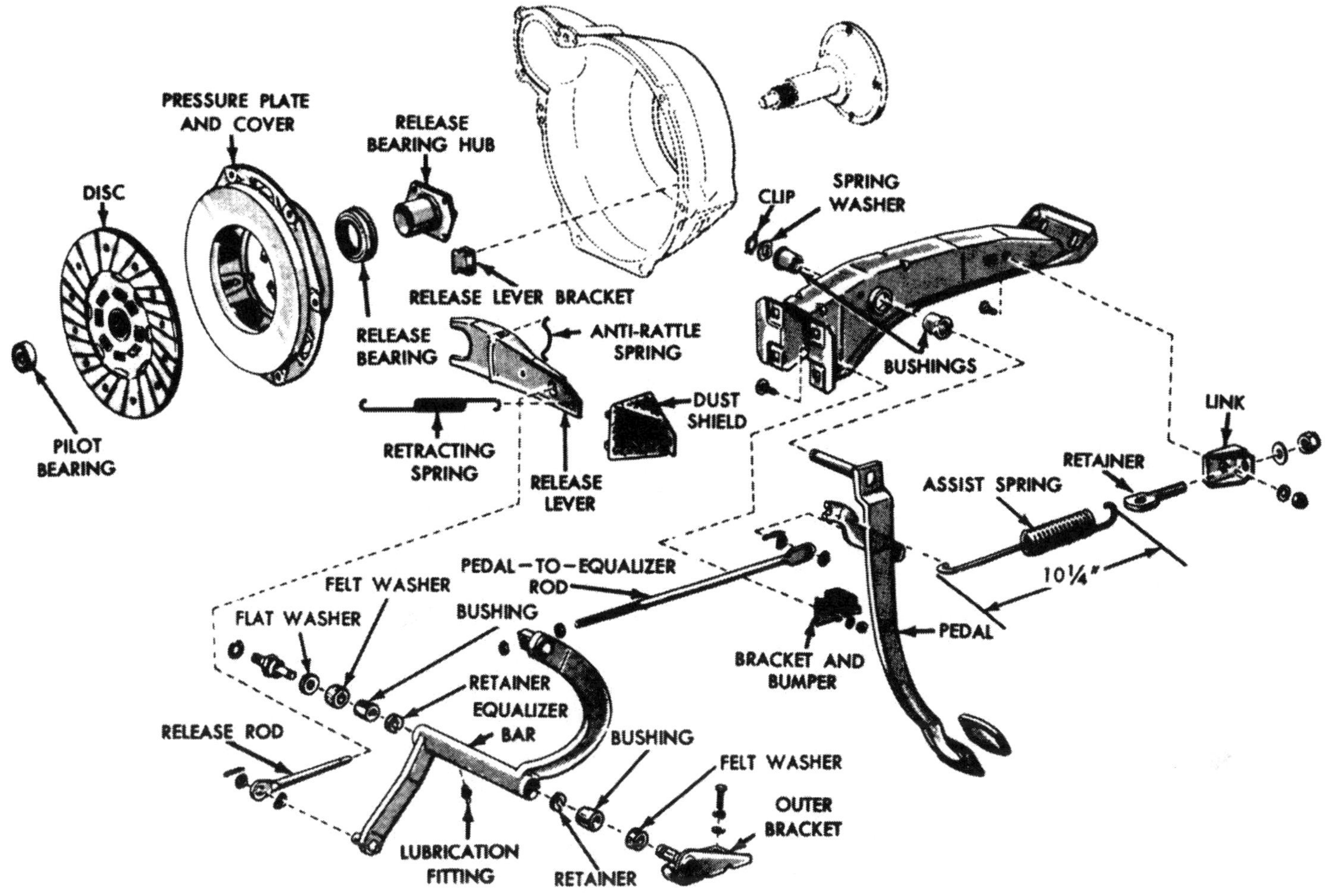

FIG 5:3 The clutch linkage for 1965 models

The shims are fitted to correct the alignment of the housing to the engine, though not more than .010 inch of shims may be used at any one point.

If it is suspected that the alignment of housing and engine is not correct, a dial indicator, pilot tool, and Dia-L-Igner gauge will be required to check the alignment. The clutch release bearing and lever must be removed from the housing and the spark plugs removed from the engine. Full instructions are given on the Dia-L-Igner gauge, and shims made up to the dimensions shown in **FIG 5:6.** If the Dia-L-Igner gauge gives a reading of greater than .014 inch then the housing is not acceptable.

A certain amount of misalignment can be corrected by a service garage, using new locating dowels and drilling faulty mounting holes oversize.

5:4 Release bearing

The release bearing and lever can be removed after the transmission and flywheel housing have been removed from the engine.

The bearing is sealed and no further lubrication can be added to it. **Do not wash the bearing in any solvent otherwise the lubricant will be washed out of it.**

Inspect the bearing for nicks or roughness and clean off any such damage with Crocus cloth. Apply pressure to the bearing and rotate it to check for roughness or noise in operation. If the bearing is rough or noisy, reject it and fit a new bearing in its place. The old bearing can be carefully driven off from the hub but when installing a new bearing use the tools shown in **FIG 5:7.**

Install the parts in the reverse order of removal. The pivot and bearing points on the release lever should be lightly lubricated with Lubriplate and the operating face of the release bearing, where it contacts the release fingers, should be lightly coated with lithium-based grease. **Avoid over-lubrication otherwise grease may find its way onto the clutch disc.**

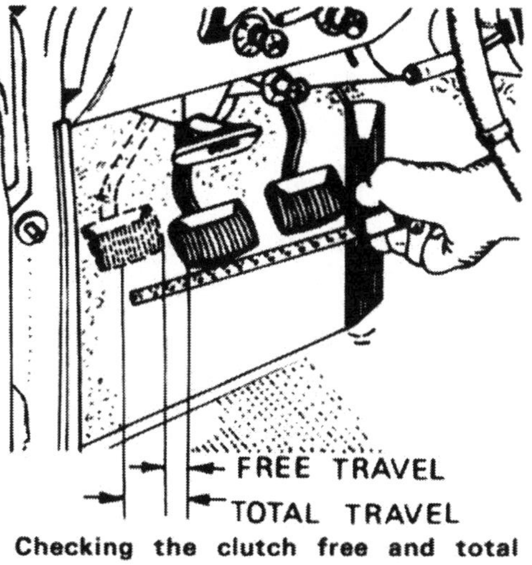

FIG 5:4 Checking the clutch free and total travel

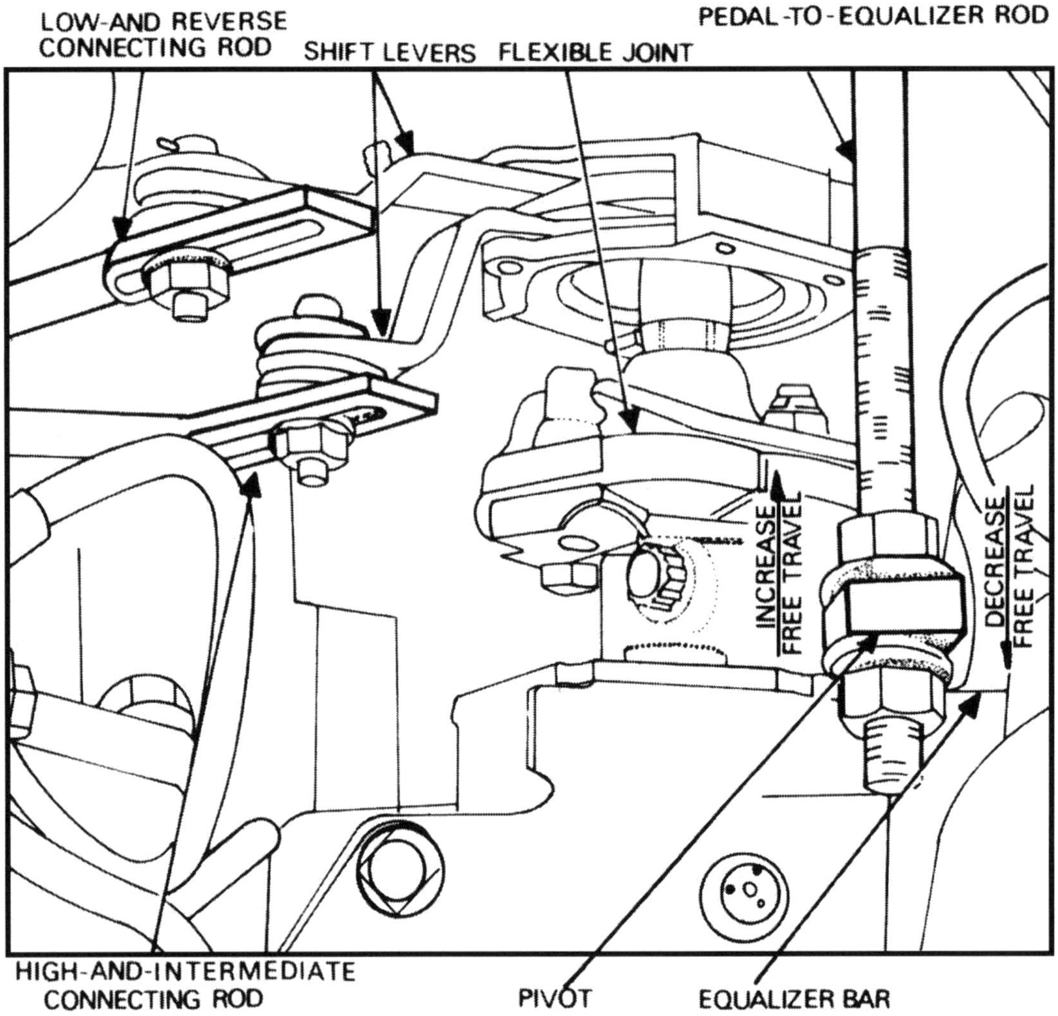

FIG 5:5 Adjusting the clutch pedal free travel on early models

5:5 Servicing the clutch

The removal and installation of the clutch is dealt with in **Chapter 1, Section 1:11. It is essential to use the arbor to centralize the clutch disc while installing the cover assembly,** otherwise the input shaft of the transmission cannot be inserted through the clutch disc into the pilot bushing in the end of the crankshaft.

Damage or wear to either the cover assembly or the clutch disc requires renewal of the component, as neither component can be dis-assembled.

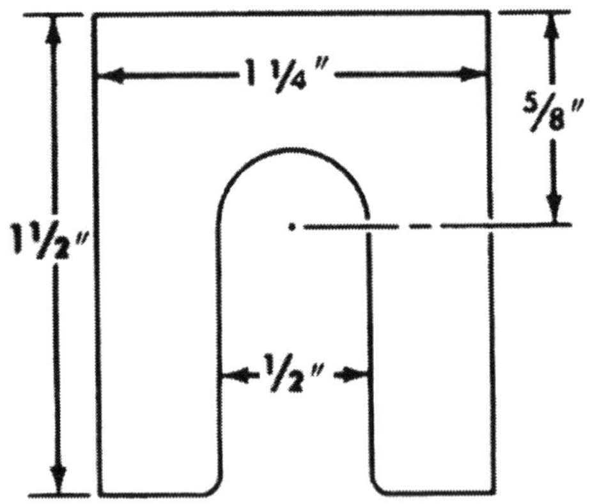

FIG 5:6 Fabricated shims for aligning the flywheel housing with the engine

Cover assembly:

Do not immerse the assembly in solvent. Cleaning should be confined to blowing and brushing out loose dirt and wiping oily surfaces with a cloth moistened with suitable solvent (such as denatured alcohol).

Examine the cover itself for cracks or distortion, paying particular attention to the areas around rivets and the attachment holes.

Inspect the surface of the pressure plate for burn marks, scores, or hairline cracks. Light damage is

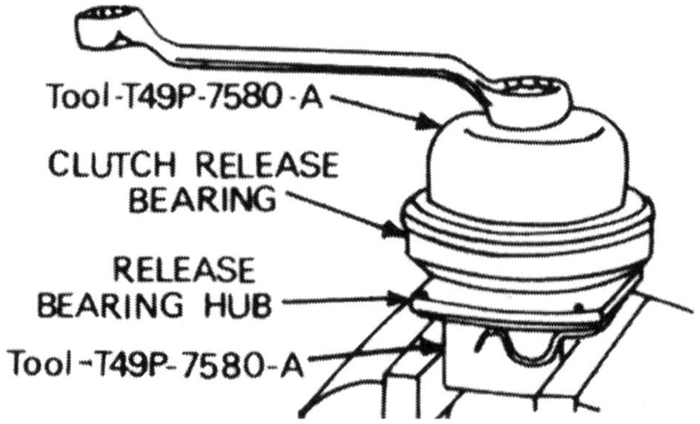

FIG 5:7 Installing a new clutch release bearing onto the bearing hub

removable, though not advisable, and if the damage is deep the cover assembly must be renewed.

Lay the assembly, pressure plate downwards, onto a clean floor or plank of wood. Check that the release fingers are all at approximately the same height and that no finger shows more wear than the others. Press each finger down in turn and release it smartly. If the finger does not return quickly there is a binding condition. If the fingers are at uneven heights or the binding condition cannot be cured, a new cover assembly must be fitted.

Before installing the clutch, lubricate the pivots, as shown in **FIG 5:8**, with a little lithium-based grease and pump the fingers to work the grease well in. Wipe away surplus grease.

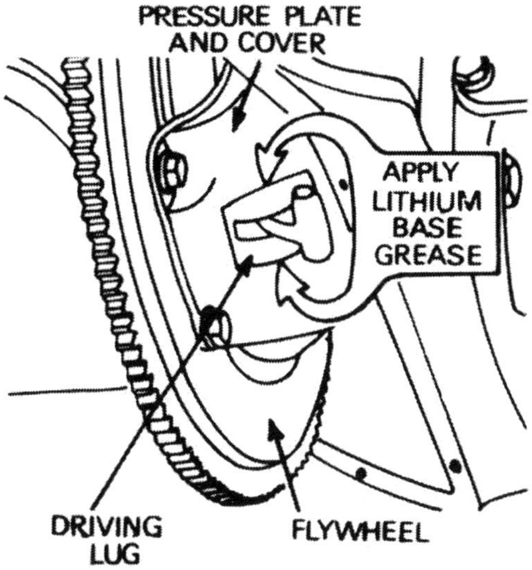

FIG 5:8 The lubrication points on the pressure plate

Clutch disc:

Take great care not to drop the assembly or allow lubricant onto the friction linings.

Check the disc for wear, mechanical damage or contamination and renew it if defective.

The lining surface must be well clear of the rivet heads. Install a new disc if the linings are worn down nearly to the rivet heads as it will save carrying out the work again in a short time.

Check that all rivets are tight and secure. Check that the hub is securely in place and that none of the damper springs are loose or broken. Slide the disc back onto the transmission input shaft and check that it slides freely without excess rotational play or movement in the damper springs.

Any oil leaks into the flywheel housing must be traced and rectified before the clutch is installed. If gaskets and seals are satisfactory the cause can be overfilling of the transmission or a blocked transmission vent.

For maximum efficiency the friction linings should have a smooth polished finish through which the grain of the material is clearly visible. Small amounts of oil or grease will leave dark coloured smears. Larger amounts of oil will leave a dark coloured glaze over the friction surface, hiding the grain of the material, while excessive amounts will be obvious from the free oil in the housing and the oil soaked appearance of the linings. The disc can be considered fit for further use, provided that the linings are not excessively worn or it is not mechanically defective, as long as the grain of the friction material can be seen. **Do not attempt to wash off grease or oil with solvents.**

5:6 Trouble shooting guide

(a) Drag or spin

1 Oil or grease on disc linings
2 Incorrectly adjusted clutch pedal free travel
3 Misalignment between engine and flywheel housing
4 Binding pilot bushing in crankshaft
5 Distorted clutch disc
6 Warped or damaged pressure plate
7 Broken disc linings

(b) Fierceness or snatch

1 Check 1, 3, 4, 5, 6 and 7 in (a)
2 Excessively worn disc linings
3 Worn or loose disc hub

(c) Slip

1 Check 1, 2 and 3 in (a) and 2 in (b)
2 Weak cover assembly springs

(d) Judder

1 Check 1, 2 and 7 in (a)
2 Contact area of friction linings not evenly distributed
3 Buckled disc
4 Faulty or loose engine mountings
5 Defective rear suspensions or drive shaft

(e) Rattles

1 Worn release linkage
2 Worn release lever and bearing
3 Broken spring in disc or cover assembly

Inches	Decimals	Millimetres	Inches to Millimetres		Millimetres to Inches	
			Inches	mm	mm	Inches
1/64	.015625	.3969	.001	.0254	.01	.00039
1/32	.03125	.7937	.002	.0508	.02	.00079
3/64	.046875	1.1906	.003	.0762	.03	.00118
1/16	.0625	1.5875	.004	.1016	.04	.00157
5/64	.078125	1.9844	.005	.1270	.05	.00197
3/32	.09375	2.3812	.006	.1524	.06	.00236
7/64	.109375	2.7781	.007	.1778	.07	.00276
1/8	.125	3.1750	.008	.2032	.08	.00315
9/64	.140625	3.5719	.009	.2286	.09	.00354
5/32	.15625	3.9687	.01	.254	.1	.00394
11/64	.171875	4.3656	.02	.508	.2	.00787
3/16	.1875	4.7625	.03	.762	.3	.01181
13/64	.203125	5·1594	.04	1.016	.4	.01575
7/32	.21875	5.5562	.05	1.270	.5	.01969
15/64	.234375	5.9531	.06	1.524	.6	.02362
1/4	.25	6.3500	.07	1.778	.7	.02756
17/64	.265625	6.7469	.08	2.032	.8	.03150
9/32	.28125	7.1437	.09	2.286	.9	.03543
19/64	.296875	7.5406	.1	2.54	1	.03937
5/16	.3125	7.9375	.2	5.08	2	.07874
21/64	.328125	8.3344	.3	7.62	3	.11811
11/32	.34375	8.7312	.4	10.16	4	.15748
23/64	.359375	9.1281	.5	12.70	5	.19685
3/8	.375	9.5250	.6	15.24	6	.23622
25/64	.390625	9.9219	.7	17.78	7	.27559
13/32	.40625	10.3187	.8	20.32	8	.31496
27/64	.421875	10.7156	.9	22.86	9	.35433
7/16	.4375	11.1125	1	25.4	10	.39370
29/64	.453125	11.5094	2	50.8	11	.43307
15/32	.46875	11.9062	3	76.2	12	.47244
31/64	.484375	12.3031	4	101.6	13	.51181
1/2	.5	12.7000	5	127.0	14	.55118
33/64	.515625	13.0969	6	152.4	15	.59055
17/32	.53125	13.4937	7	177.8	16	.62992
35/64	.546875	13.8906	8	203.2	17	.66929
9/16	.5625	14.2875	9	228.6	18	.70866
37/64	.578125	14.6844	10	254.0	19	.74803
19/32	.59375	15.0812	11	279.4	20	.78740
39/64	.609375	15.4781	12	304.8	21	.82677
5/8	.625	15.8750	13	330.2	22	.86614
41/64	.640625	16.2719	14	355.6	23	.90551
21/32	.65625	16.6687	15	381.0	24	.94488
43/64	.671875	17.0656	16	406.4	25	.98425
11/16	.6875	17.4625	17	431.8	26	1.02362
45/64	.703125	17.8594	18	457.2	27	1.06299
23/32	.71875	18.2562	19	482.6	28	1.10236
47/64	.734375	18.6531	20	508.0	29	1.14173
3/4	.75	19.0500	21	533.4	30	1.18110
49/64	.765625	19.4469	22	558.8	31	1.22047
25/32	.78125	19.8437	23	584.2	32	1.25984
51/64	.796875	20.2406	24	609.6	33	1.29921
13/16	.8125	20.6375	25	635.0	34	1.33858
53/64	.828125	21.0344	26	660.4	35	1.37795
27/32	.84375	21.4312	27	685.8	36	1.41732
55/64	.859375	21.8281	28	711.2	37	1.4567
7/8	.875	22.2250	29	736.6	38	1.4961
57/64	.890625	22.6219	30	762.0	39	1.5354
29/32	.90625	23.0187	31	787.4	40	1.5748
59/64	.921875	23.4156	32	812.8	41	1.6142
15/16	.9375	23.8125	33	838.2	42	1.6535
61/64	.953125	24.2094	34	863.6	43	1.6929
31/32	.96875	24.6062	35	889.0	44	1.7323
63/64	.984375	25.0031	36	914.4	45	1.7717

UNITS	Pints to Litres	Gallons to Litres	Litres to Pints	Litres to Gallons	Miles to Kilometres	Kilometres to Miles	Lbs. per sq. In. to Kg. per sq. Cm.	Kg. per sq. Cm. to Lbs. per sq. In.
1	.57	4.55	1.76	.22	1.61	.62	.07	14.22
2	1.14	9.09	3.52	.44	3.22	1.24	.14	28.50
3	1.70	13.64	5.28	.66	4.83	1.86	.21	42.67
4	2.27	18.18	7.04	.88	6.44	2.49	.28	56.89
5	2.84	22.73	8.80	1.10	8.05	3.11	.35	71.12
6	3.41	27.28	10.56	1.32	9.66	3.73	.42	85.34
7	3.98	31.82	12.32	1.54	11.27	4.35	.49	99.56
8	4.55	36.37	14.08	1.76	12.88	4.97	.56	113.79
9		40.91	15.84	1.98	14.48	5.59	.63	128.00
10		45.46	17.60	2.20	16.09	6.21	.70	142.23
20				4.40	32.19	12.43	1.41	284.47
30				6.60	48.28	18.64	2.11	426.70
40				8.80	64.37	24.85		
50					80.47	31.07		
60					96.56	37.28		
70					112.65	43.50		
80					128.75	49.71		
90					144.84	55.92		
100					160.93	62.14		

UNITS	Lb ft to kgm	Kgm to lb ft	UNITS	Lb ft to kgm	Kgm to lb ft
1	.138	7.233	7	.967	50.631
2	.276	14.466	8	1.106	57.864
3	.414	21.699	9	1.244	65.097
4	.553	28.932	10	1.382	72.330
5	.691	36.165	20	2.765	144.660
6	.829	43.398	30	4.147	216.990

CHAPTER 6

THE TRANSMISSION

6:1 Description

Three or fourspeed transmissions can be fitted. On both types of transmission all forward gears are of the fully synchronized type. Reverse gear is engaged by sliding a spur-cut idler into mesh between the counter-shaft and output shaft gears.

The power flow and internal parts of a threespeed transmission are shown in **FIG 6:1** while those for a fourspeed transmission are shown in **FIG 6:2**. The input shaft is splined to the clutch disc and revolves in a front bearing. The output shaft and input shaft support each other at their junction and needle rollers are fitted to reduce friction. An integral gear on the input shaft is in constant mesh with the mating gear on the countershaft gear cluster. The countershaft gear cluster rotates about the countershaft on needle roller bearings and its end play is controlled by thrust washers. The countershaft gear cluster is in one piece and the gears are in continuous mesh with the gears on the output shaft, all these gears being helical cut. The gears on the output shaft are free to revolve about the shaft so that when neutral is selected, drive is not transmitted to the main shaft. The synchronizer hub is splined to the output shaft internally and the hub has external splines on which the sleeve slides. The sleeve is moved in the appropriate direction by the selector linkage. When the sleeve is moved, pressure is first applied to the blocking ring and friction brings the parts to the same speed. When they are at the same speed further movement of the sleeve engages it onto the teeth of the blocking ring and clutch teeth on the output shaft gearwheel, making the drive positive from the gearwheel to the output shaft. Drive is then passed from the input shaft along the countershaft gear cluster and to the gearwheel on the mainshaft, and from there through the synchronizer assembly to the output shaft. The output shaft is supported in bearings and bush and transmits the drive to the drive shaft.

6:2 Maintenance

Seasonal changes in lubricant are not required. When removing the transmission from the vehicle it is advisable to drain out the old lubricant and discard it, filling with fresh lubricant after installation.

A drain plug is fitted underneath the casing, and the drain plug has a magnet in it to ensure that metallic particles are trapped and not allowed to circulate with the oil.

At regular intervals check the oil level in the transmission. Wipe the area around the combined level and filler plug clean. Stand the car on level ground and

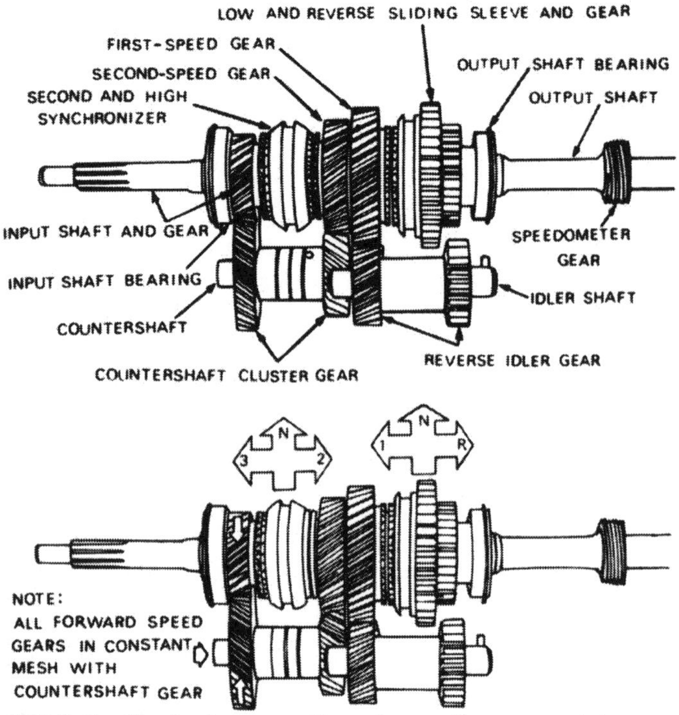

FIG 6:1 Typical power flow for a three-speed transmission

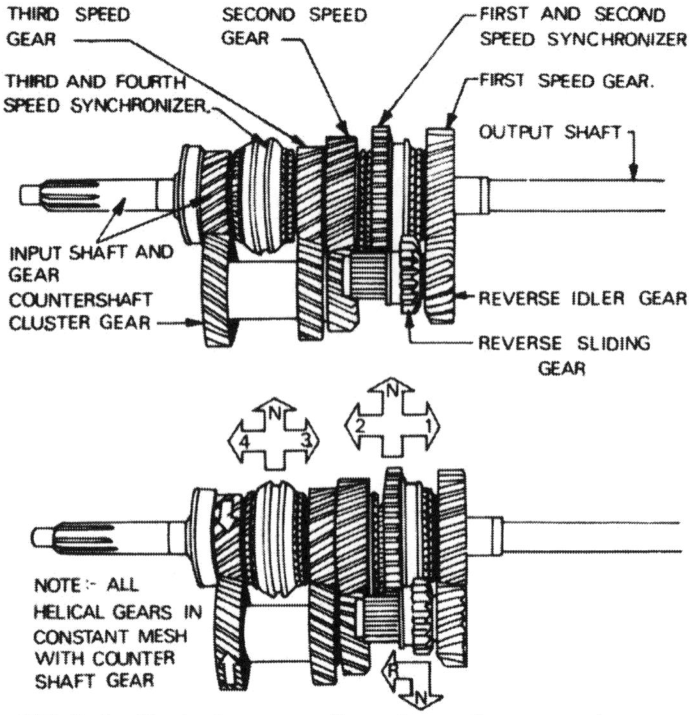

FIG 6:2 Typical power flow for a four-speed transmission

remove the filler plug. If the level is correct the oil will be up to the bottom of the filler hole. If the level is low, fill using a recommended lubricant (Mild EP Gear Oil C3RZ19547-B). Allow the surplus lubricant to drain out from the filler hole before installing the filler plug.

6:3 Transmission removal

Removal on all types of transmission is similar, with only minor variations between types. Full details of the drive shaft are given in **Chapter 8**, and if there is any difficulty in disconnecting the drive shaft this chapter should be consulted.

Removal and installation will be easiest if the car is raised on a hoist or the work is carried out from a pit. If neither of these is available raise the car and support it securely on chassis stands.

1 Drain the oil from the transmission. If a drain plug is not fitted, remove the lower bolt that secures the rear extension housing.

2 On earlier models with a floor mounted shift, remove the dust boot from inside the vehicle and take out the two bolts securing the selector lever so that the lever can be removed, taking care not to lose the two trunnions or springs. The remainder of the shift

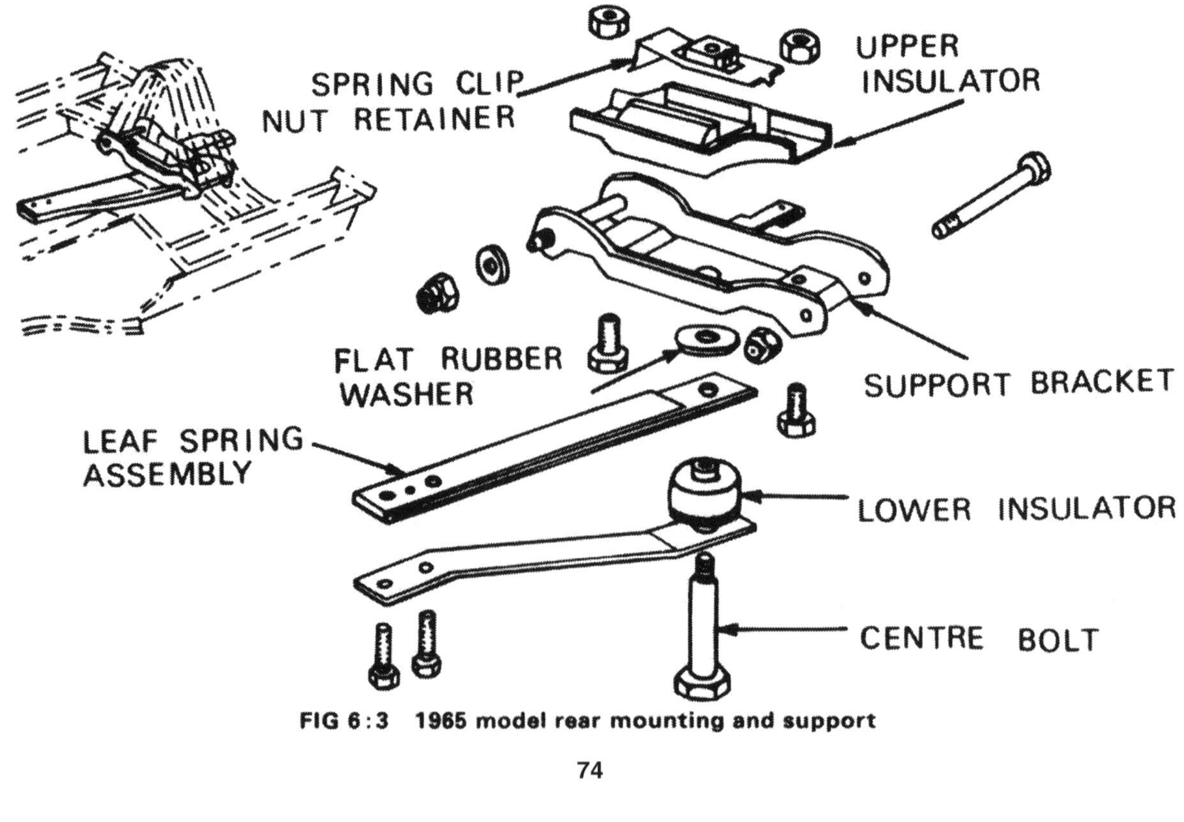

FIG 6:3 1965 model rear mounting and support

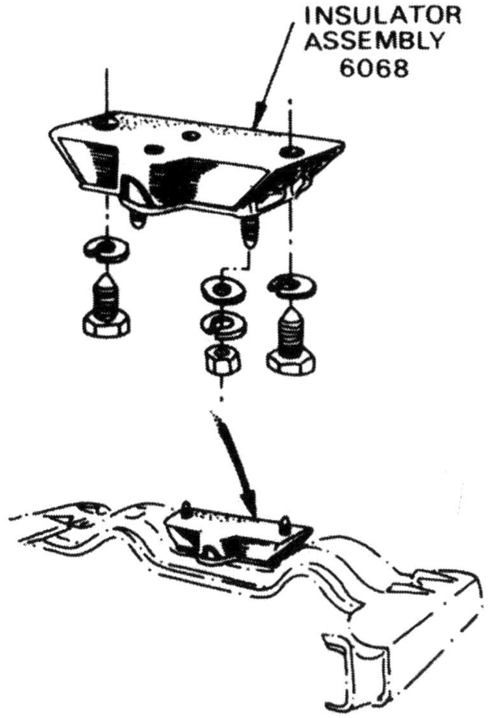

FIG 6:4 Typical later rear mounting and support

linkage can then be left attached to the transmission. This applies only to earlier models.

3 Mark the relative positions of the drive shaft to the flanges on the rear axle and transmission, so that the drive shaft will be installed in its original position. Remove the U-bolts that secure the rear of the drive shaft to the rear axle. Secure the drive shaft bearings in place with tape to prevent them from falling off. Withdraw the front of the drive shaft from the transmission. Plug the extension housing to prevent lubricant from leaking out or dirt entering.

4 Remove the cap screw and lockwasher that secures the speedometer drive cable retainer so that the retainer can be removed and the cable withdrawn from the extension housing. Disconnect the shift rods from the shift levers on the transmission. On later models fitted with a floor mounted shift, remove the three bolts that secure the shift selector assembly to the extension housing and allow the assembly to hang from the shift lever.

5 An early rear engine support is shown in **FIG 6:3** and a typical later rear engine mounting is shown in **FIG 6:4.** Support the power unit with a jack and pad of wood under the engine oil pan. Remove the bolts that secure the rear mounting to the gearbox and raise the engine slightly, about its front mountings, so that the crossmember assembly can be removed from the car. If necessary disconnect the handbrake cable linkage to allow the transmission to be lowered out.

6 Support the transmission on another jack and block of wood. On earlier models, the flywheel housing is removed with the transmission so it will be necessary to disconnect the clutch release mechanism and remove the starter motor. Remove the bolts that secure the transmission to the flywheel housing (flywheel housing to engine on earlier models). Draw the transmission rearwards until the input shaft is clear of the clutch. **Do not operate the clutch and do not allow the weight of the transmission to hang on the input shaft in the clutch.** It may be necessary to lower both jacks, keeping the transmission aligned with the engine, so that the transmission can be withdrawn rearwards. Once the input shaft is clear of the clutch, lower the transmission down and out of the vehicle.

Installation:

The transmission is installed in the reverse order of removal. The task will be made easier if two guide pins are installed in the flywheel housing (engine on earlier models) and the transmissions slid along these to guide the input shaft back into place. If the input shaft is difficult to enter through the hub of the clutch disc,

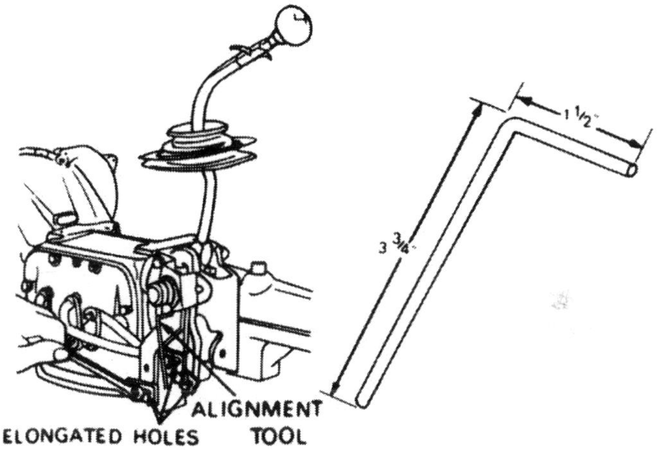

FIG 6:5 Typical method of adjusting the floor shift linkage using a tool fabricated from $\frac{1}{4}$ inch diameter rod

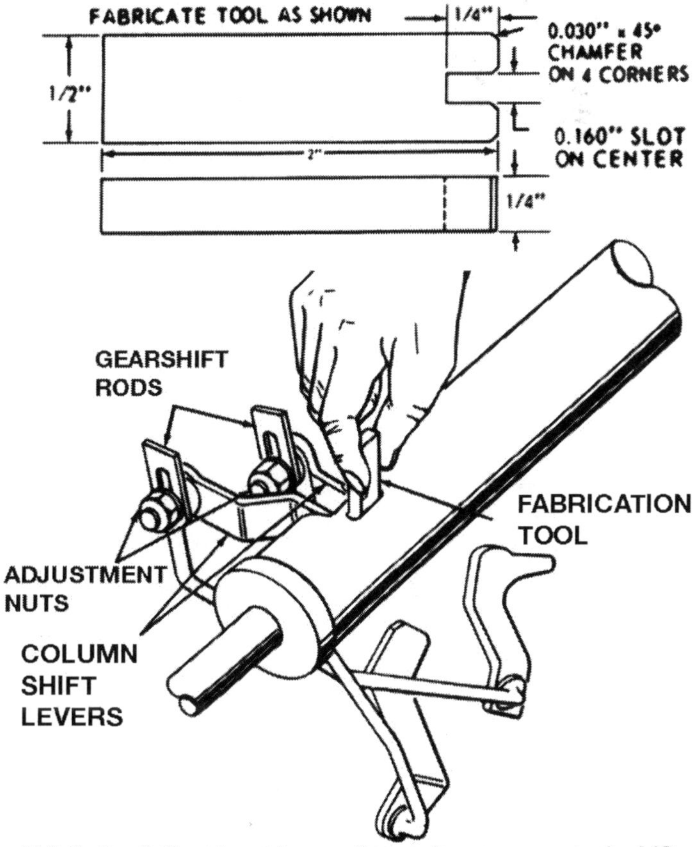

FIG 6:6 Adjusting the earlier column mounted shift linkage using the fabricated tool

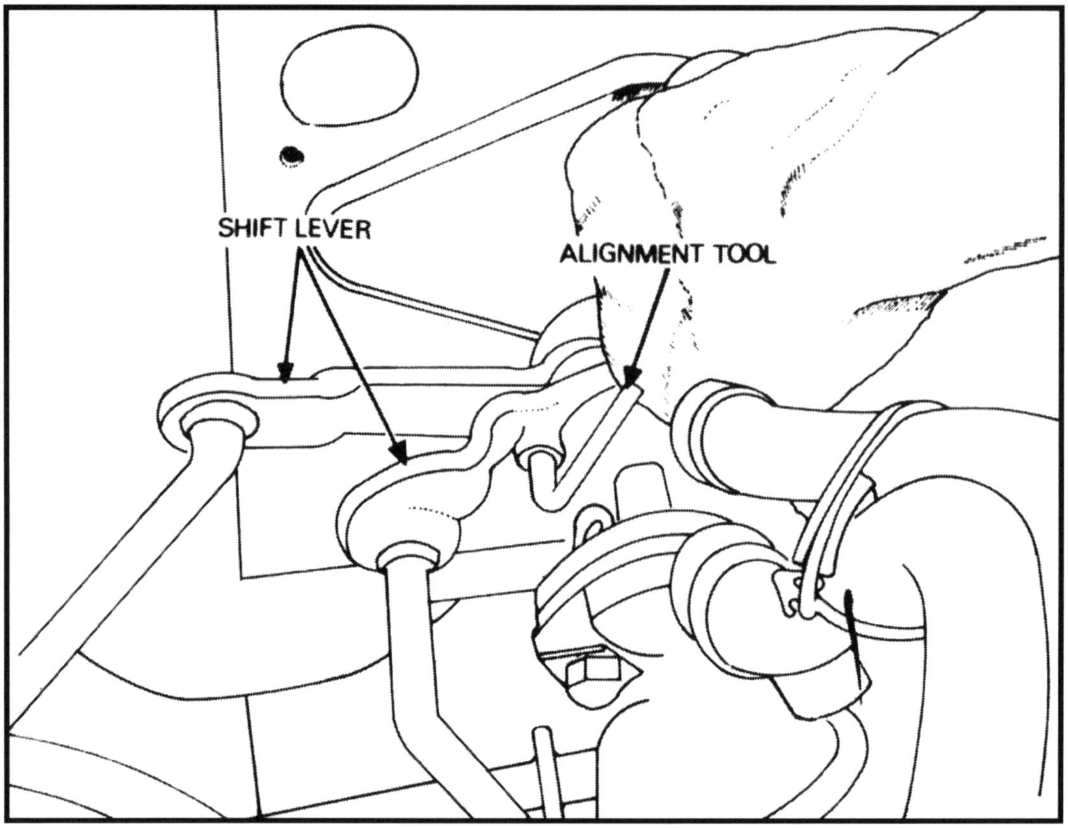

FIG 6:7 Adjusting the column mounted shift linkage on a 1969 model using a tool fabricated from $\frac{3}{16}$ inch diameter rod

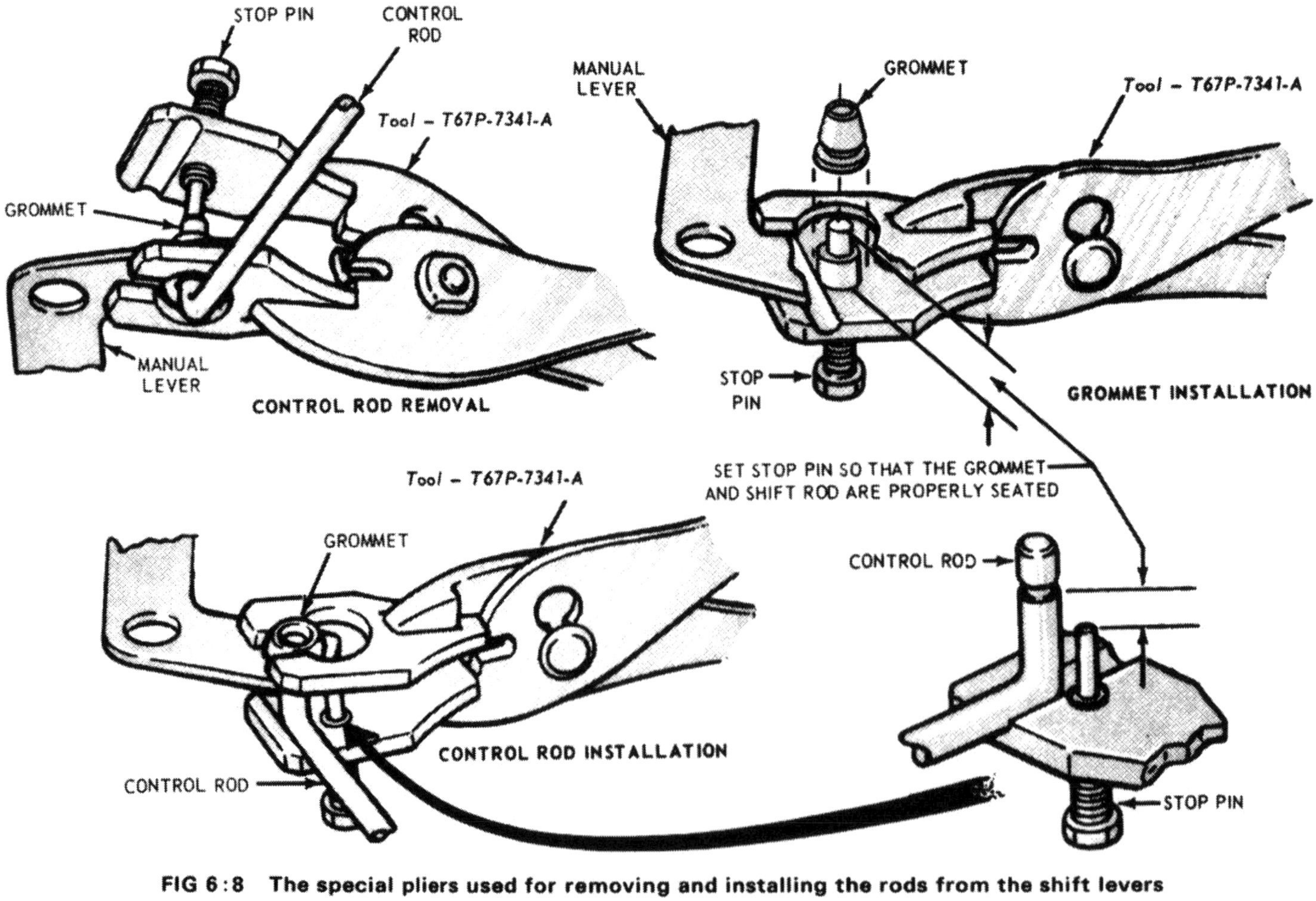

FIG 6:8 The special pliers used for removing and installing the rods from the shift levers

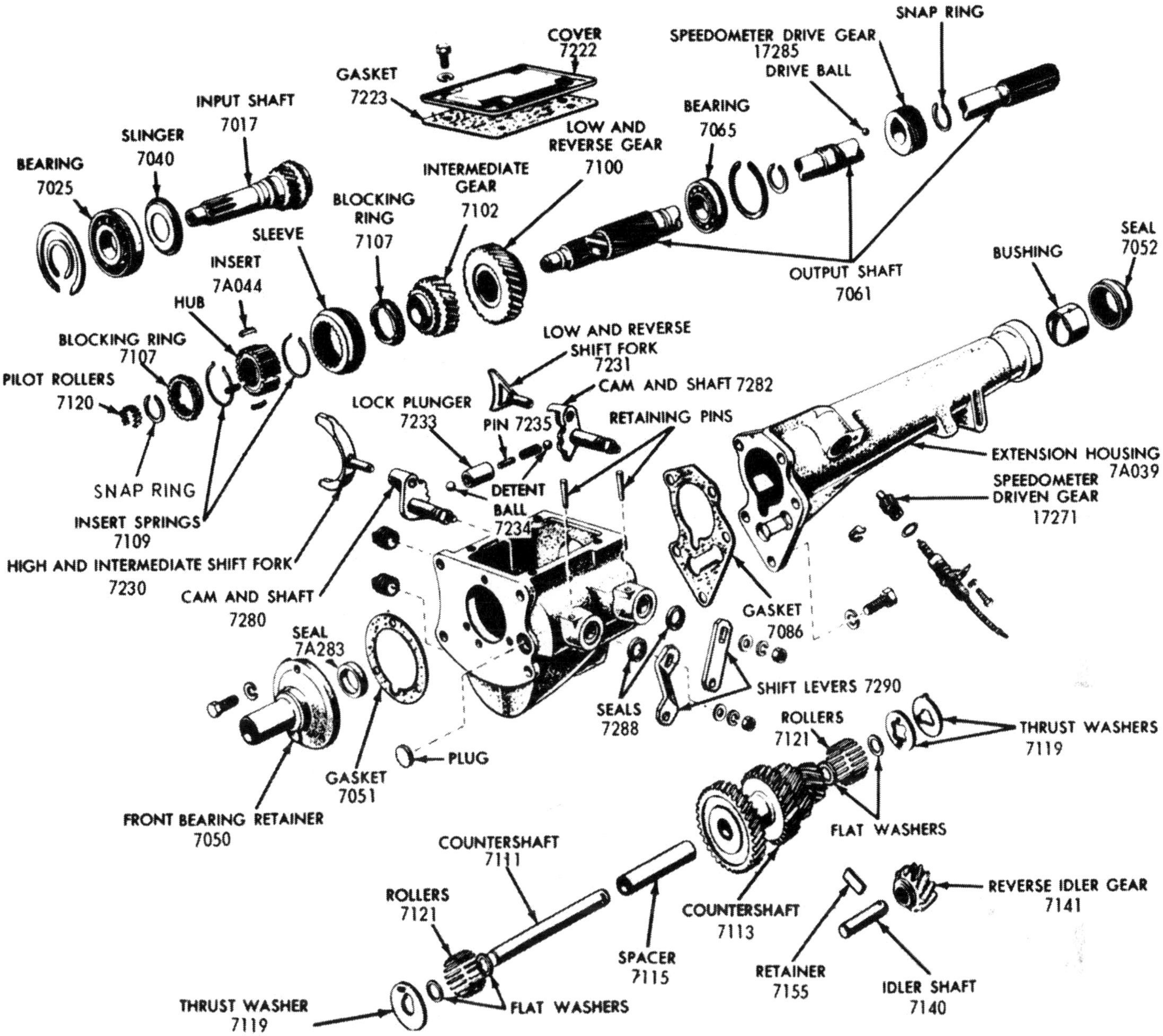

FIG 6:9 The components of a typical three-speed transmission

select a gear and rotate the output shaft to align the splines. **Support the transmission until it is bolted in place and do not allow it to hang on the input shaft in the clutch.**

When the transmission is installed, reconnect the drive shaft so that the previously made marks are again in alignment.

Make sure that the drain plug is fitted and fill the transmission to the level of the bottom of the filler plug aperture with fresh lubricant. Leave the lubricant plenty of time to find its natural level, as rapid filling can allow lubricant to drain from the filler hole before the unit is full.

Check the shift selector linkage adjustments and road test the vehicle.

6:4 Shift linkage adjustment

The method of adjustment for all models fitted with floor mounted shifts is shown in **FIG 6:5**. Make up a tool out of $\frac{1}{4}$ inch rod to the dimensions shown. Set the shift lever to the neutral position and make sure that all the shift levers on the transmission are also in the neutral position. If there is doubt as to the exact position of neutral on the shift levers, disconnect them by taking out the bolts that secure the rods and rotate the levers through their three positions and set them in the central position. Reconnect the rods and tighten the bolts into the elongated holes.

The method of adjustment for earlier models with steering column mounted shift is shown in **FIG 6:6**. Make up a tool to the dimensions shown and fit it into the slot in the steering column. If necessary slacken the adjustment nuts and move the levers so that the tool will fit into place. Set the selector levers on the transmission into the neutral positions and tighten the adjusting nuts in their slotted holes.

On the later models fitted with floor mounted shift, the adjustment is similar to the earlier models. On the standard steering column mounted shift the adjustment is made using a tool fabricated from $\frac{3}{16}$ inch rod, as shown in **FIG 6:7**.

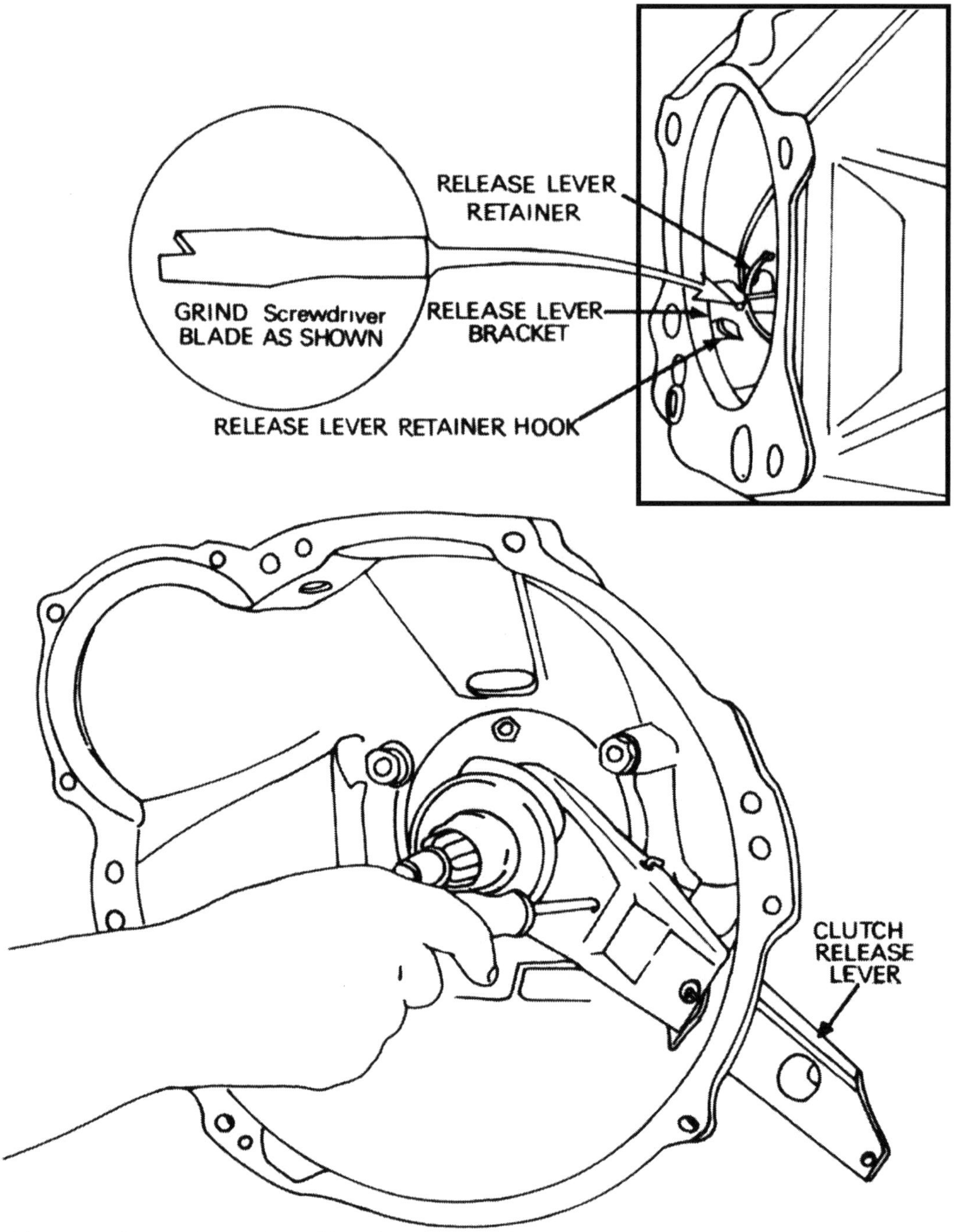

FIG 6:10 Removing or installing the clutch release lever, using a reworked screwdriver

On later models a special grommet is used to secure the rod to the lever on the transmission. Special pliers T67P-7341-A must be used to free or connect the rod, and **once the rod has been disconnected the old grommet must be discarded and a new one fitted in its place.** The method of using the special pliers is shown in **FIG 6:8.** The stop pin must be adjusted to position the grommet, first in the lever and then on the rod.

6:5 Servicing the transmission

Special extractors, tools and an arbor press are required for disassembling and reassembling the transmission. For this reason, and because of the slight variations between different models, the general details of servicing only will be given. The parts must be installed with the correct end plays set, otherwise the transmission will not operate quietly or efficiently. If the owner has any doubts regarding his ability to strip and reassemble the unit, he is advised to take it to a service garage.

Disassembly:

Typical transmission parts are shown in **FIG 6:9**, and these are for a threespeed transmission.

On the models where the clutch housing is removed with the transmission, use a reworked screwdriver to free the clutch release lever, as shown in **FIG 6:10.**

Remove the upper cover with its gasket, and carefully collect the detent spring and plunger on the models where they are held down by the cover. Remove the rear extension and speedometer drive gear.

Remove the selector mechanism, shift rails and shift forks as well as detent springs and plungers. Drive out the roll-pin that secures the countershaft. A dummy countershaft can now be slid in place of the countershaft, and the countershaft gearcluster allowed to slide to the bottom of the case.

Remove the front bearing retainer. Withdraw the mainshaft bearing using a suitable extractor. The input shaft assembly should then be moved forwards. On some models it can be fully withdrawn but on others the gear will prevent it from coming out. Lift out the mainshaft assembly, followed by the input shaft assembly if required.

The remainder of the parts in the casing can then be removed. When disassembling the output shaft assembly a press will be required. Store the parts as they are removed in order, and if there is any doubt mark them with wax crayon or paint for order and position. The synchronizer assemblies must be reassembled so that the parts are in their original positions.

Cleaning:

Wash all the parts except seals and bearings in a suitable solvent. Remove dirt by careful use of a scraper.

Rotate the bearings in a clean bath of solvent until all lubricant and dirt is removed. Hold the bearing and dry it with compressed air. **Do not allow the bearing to rotate with the compressed air as this causes chipping and damage to the races.**

Examination:

Check all faces on the castings for nicks or damage. Remove the damage or distortion using a fine file or stone. Examine the castings for cracks, stripped threads or any other damage which could affect the operation of the transmission.

Examine the teeth on all gears for chipping, wear or broken teeth. Renew any gears that are defective and examine the teeth of the gear that mates with it with extra care.

All shafts should be true without bends or distortion and they should show no chatter or fret marks. Check bushes for wear or fretting. Inspect all splines on the shafts for wear, damage or nicks.

Renew any bearings that run roughly or show any fault on the surfaces of the races and balls. Faults that can be expected on bearings are; pitting, wear, cracking, and corrosion.

Check the synchronizer units for wear in the index slots. Slide the hubs into the sleeves, with the etched index marks aligned, and check that the sleeve slides freely on the hub. Fit the blocker ring on the cone and check that there is at least .010 inch between the face of the blocker ring and the teeth on the gear.

It is advisable to discard all old seals and gasket, installing new ones on reassembly. Check the bush in the extension housing and renew it if it is worn or defective. It should be noted that both the bushing and seal can be renewed without having to remove the extension housing from the transmission or the transmission from the vehicle. The drive shaft must be removed (see **Chapter 8**) and renewal of the bushing and seal is then similar to the method used on the automatic transmission (see **Chapter 7, Section 7:6**).

Reassembly:

The transmission is reassembled in the reverse order of disassembly. Selectively fit thrust washers to set the correct end plays. Lubricate the parts as they are installed.

6:6 Trouble shooting guide

(a) Jumping out of gear

1 Shift mechanism incorrectly adjusted
2 Broken or weak detent spring
3 Worn detent grooves in shift rails
4 Worn blocking rings or synchronizer units
5 Shift fork loose on rail

(b) Noisy transmission

1 Insufficient or incorrect grade of lubricant
2 Excessive end plays on shafts and parts
3 Worn or damaged bearings
4 Worn or damaged gear teeth

(c) Difficulty in engaging gear

1 Check 1, 4 and 5 in (a)
2 Stiff shift mechanism
3 Incorrect clutch adjustment
4 Defective spigot bushing in crankshaft

(d) Lubricant leaks

1 Damaged joint faces
2 Defective gaskets
3 Defective oil seals

Inches	Decimals	Millimetres	Inches to Millimetres: Inches	Inches to Millimetres: mm	Millimetres to Inches: mm	Millimetres to Inches: Inches
1/64	.015625	.3969	.001	.0254	.01	.00039
1/32	.03125	.7937	.002	.0508	.02	.00079
3/64	.046875	1.1906	.003	.0762	.03	.00118
1/16	.0625	1.5875	.004	.1016	.04	.00157
5/64	.078125	1.9844	.005	.1270	.05	.00197
3/32	.09375	2.3812	.006	.1524	.06	.00236
7/64	.109375	2.7781	.007	.1778	.07	.00276
1/8	.125	3.1750	.008	.2032	.08	.00315
9/64	.140625	3.5719	.009	.2286	.09	.00354
5/32	.15625	3.9687	.01	.254	.1	.00394
11/64	.171875	4.3656	.02	.508	.2	.00787
3/16	.1875	4.7625	.03	.762	.3	.01181
13/64	.203125	5·1594	.04	1.016	.4	.01575
7/32	.21875	5.5562	.05	1.270	.5	.01969
15/64	.234375	5.9531	.06	1.524	.6	.02362
1/4	.25	6.3500	.07	1.778	.7	.02756
17/64	.265625	6.7469	.08	2.032	.8	.03150
9/32	.28125	7.1437	.09	2.286	.9	.03543
19/64	.296875	7.5406	.1	2.54	1	.03937
5/16	.3125	7.9375	.2	5.08	2	.07874
21/64	.328125	8.3344	.3	7.62	3	.11811
11/32	.34375	8.7312	.4	10.16	4	.15748
23/64	.359375	9.1281	.5	12.70	5	.19685
3/8	.375	9.5250	.6	15.24	6	.23622
25/64	.390625	9.9219	.7	17.78	7	.27559
13/32	.40625	10.3187	.8	20.32	8	.31496
27/64	.421875	10.7156	.9	22.86	9	.35433
7/16	.4375	11.1125	1	25.4	10	.39370
29/64	.453125	11.5094	2	50.8	11	.43307
15/32	.46875	11.9062	3	76.2	12	.47244
31/64	.484375	12.3031	4	101.6	13	.51181
1/2	.5	12.7000	5	127.0	14	.55118
33/64	.515625	13.0969	6	152.4	15	.59055
17/32	.53125	13.4937	7	177.8	16	.62992
35/64	.546875	13.8906	8	203.2	17	.66929
9/16	.5625	14.2875	9	228.6	18	.70866
37/64	.578125	14.6844	10	254.0	19	.74803
19/32	.59375	15.0812	11	279.4	20	.78740
39/64	.609375	15.4781	12	304.8	21	.82677
5/8	.625	15.8750	13	330.2	22	.86614
41/64	.640625	16.2719	14	355.6	23	.90551
21/32	.65625	16.6687	15	381.0	24	.94488
43/64	.671875	17.0656	16	406.4	25	.98425
11/16	.6875	17.4625	17	431.8	26	1.02362
45/64	.703125	17.8594	18	457.2	27	1.06299
23/32	.71875	18.2562	19	482.6	28	1.10236
47/64	.734375	18.6531	20	508.0	29	1.14173
3/4	.75	19.0500	21	533.4	30	1.18110
49/64	.765625	19.4469	22	558.8	31	1.22047
25/32	.78125	19.8437	23	584.2	32	1.25984
51/64	.796875	20.2406	24	609.6	33	1.29921
13/16	.8125	20.6375	25	635.0	34	1.33858
53/64	.828125	21.0344	26	660.4	35	1.37795
27/32	.84375	21.4312	27	685.8	36	1.41732
55/64	.859375	21.8281	28	711.2	37	1.4567
7/8	.875	22.2250	29	736.6	38	1.4961
57/64	.890625	22.6219	30	762.0	39	1.5354
29/32	.90625	23.0187	31	787.4	40	1.5748
59/64	.921875	23.4156	32	812.8	41	1.6142
15/16	.9375	23.8125	33	838.2	42	1.6535
61/64	.953125	24.2094	34	863.6	43	1.6929
31/32	.96875	24.6062	35	889.0	44	1.7323
63/64	.984375	25.0031	36	914.4	45	1.7717

UNITS	Pints to Litres	Gallons to Litres	Litres to Pints	Litres to Gallons	Miles to Kilometres	Kilometres to Miles	Lbs. per sq. In. to Kg. per sq. Cm.	Kg. per sq. Cm. to Lbs. per sq. In.
1	.57	4.55	1.76	.22	1.61	.62	.07	14.22
2	1.14	9.09	3.52	.44	3.22	1.24	.14	28.50
3	1.70	13.64	5.28	.66	4.83	1.86	.21	42.67
4	2.27	18.18	7.04	.88	6.44	2.49	.28	56.89
5	2.84	22.73	8.80	1.10	8.05	3.11	.35	71.12
6	3.41	27.28	10.56	1.32	9.66	3.73	.42	85.34
7	3.98	31.82	12.32	1.54	11.27	4.35	.49	99.56
8	4.55	36.37	14.08	1.76	12.88	4.97	.56	113.79
9		40.91	15.84	1.98	14.48	5.59	.63	128.00
10		45.46	17.60	2.20	16.09	6.21	.70	142.23
20				4.40	32.19	12.43	1.41	284.47
30				6.60	48.28	18.64	2.11	426.70
40				8.80	64.37	24.85		
50					80.47	31.07		
60					96.56	37.28		
70					112.65	43.50		
80					128.75	49.71		
90					144.84	55.92		
100					160.93	62.14		

UNITS	Lb ft to kgm	Kgm to lb ft	UNITS	Lb ft to kgm	Kgm to lb ft
1	.138	7.233	7	.967	50.631
2	.276	14.466	8	1.106	57.864
3	.414	21.699	9	1.244	65.097
4	.553	28.932	10	1.382	72.330
5	.691	36.165	20	2.765	144.660
6	.829	43.398	30	4.147	216.990

CHAPTER 7

THE AUTOMATIC TRANSMISSION

7:1 Description

A C4 automatic transmission can be fitted as an optional extra in place of the standard manually shifted transmission described in the previous chapter.

A sectioned view of the unit is shown in **FIG 7:1** and the main components in **FIG 7:2**. The converter is bolted to the engine drive plate and takes the place of the mechanically-operated clutch. The drive from the converter is passed through an epicyclic gearbox. The drives are selected by engaging clutches and brake bands using hydraulic pressure from the control system.

Converter:

A sectional view of the converter is shown in **FIG 7:3**. The unit is filled with automatic transmission fluid and there is a constant flow of fluid passing through the unit from and to the gearbox. The fluid flow is to ensure that the converter is always filled when the engine is running and also transfers away heat produced by slip in the converter. The impeller is attached to the casing and revolves with it. As the impeller revolves it drags the fluid around with it. Because of centrifugal force there will be a greater fluid pressure at the outlet from the impeller vanes (outside) than at the inlet and the pressure differential sets up a rotational fluid flow in the plane of the section. The combined drag and centrifugal flow sets up a flow around the unit which follows the path of an imaginary spiral spring laid around between impeller and turbine. The stator is fitted to a one-way clutch. When the turbine is stationary or rotating at a low-speed compared with the impeller, the oil flow from the turbine is guided by the stator so that it impinges on the impeller at a favourable angle and assists the impeller to turn. As a result the torque on the turbine is greater than that produced by the impeller. At the stall point, when the turbine is stationary, the torque multiplication factor is approximately 2:1. As the turbine speeds up, the torque multiplication factor drops towards 1:1 until this factor is reached at a turbine speed of $\frac{9}{10}$ of impeller speed. At this speed the fluid flow impinges on the back of the stator blades, but power is not lost as the one-way clutch allows the stator to revolve freely without altering the fluid flow direction.

When the car is stationary and the throttle opened, with a drive selection made, the turbine will have large torque applied to it and the torque multiplication factor will decrease until the turbine is rotating at the same speed as the impeller and the unit acts as a fluid coupling. There will always be a slight speed differential because of frictional losses but under optimum conditions this loss is usually in the order of 2 per cent.

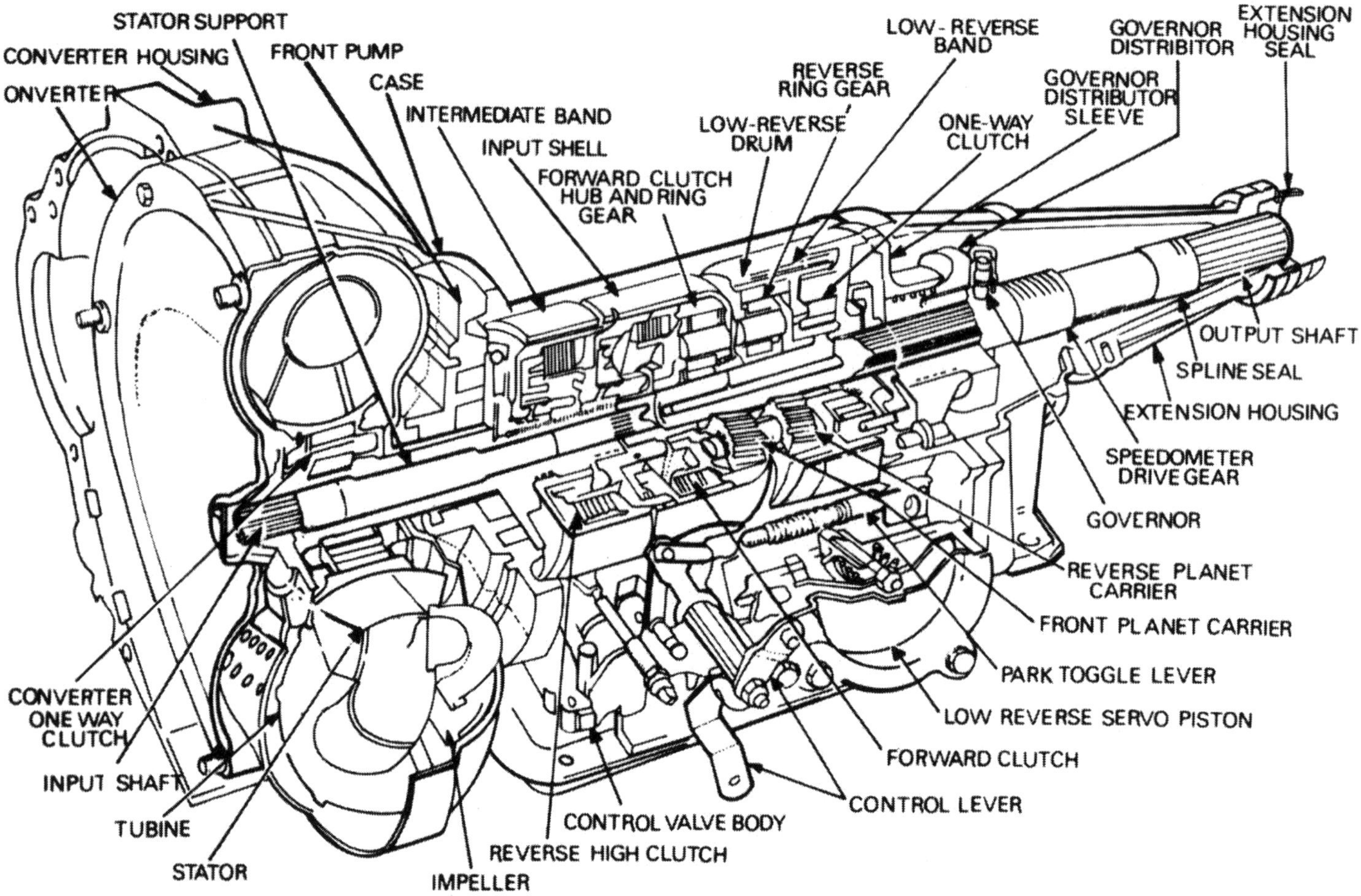

FIG 7:1 Sectional view of the automatic transmission

The frictional losses are dissipated as heat. **If the turbine is held stalled with the engine running at speed, the resultant churning of the fluid will cause rapid overheating in the unit.** The turbine can only be stalled by selecting drive and holding the car stationary.

Epicyclic gearbox:

An epicyclic gear train consists of a sun gear and outer gear, with planet gears revolving on a planet carrier between the outer and sun gears. By holding any one component and driving another, drive can be taken from the third component. The ratio can be varied and the drive reversed using a single epicyclic train.

Two epicyclic gear trains are used in the gearbox to produce the three forward speeds and one reverse. The power flow through the transmission in the various selections is shown in **FIG 7:4**. The clutches are applied by hydraulic pressure from the control system. A sectional view through an intermediate servo and band is shown in **FIG 7:5**. The adjustment screw is external on the unit so that the band can be set without dismantling the transmission. When hydraulic pressure is applied to the piston, the ends of the band are squeezed together and the band grips the drum to prevent it from rotating.

The gearbox should not be disassembled by the owner as special tools are essential. On reassembly all the parts must be set to the correct clearances and a great deal of skill is required.

Hydraulic control system:

A schematic layout of the system is shown in **FIG 7:6**. It must be emphasized that the layout is schematic and that the fluid is led through integral passages in the castings.

The fluid is contained in the oil pan under the unit. A pump is fitted to the input shaft and draws the oil from the oil pan through an oil screen. Under normal conditions the pressure of the oil from the pump is limited to 60 lb/sq inch by the main regulator valve. When the pressure from the pump rises above this level, the regulator valve allows the surplus oil to pass through to the converter and if the pressure is still high, the valve opens further to allow excess fluid to return directly to the oil pan. When the operation of the unit requires pressures higher than 60 lb/sq inch, the pressure booster valve holds back the regulator valve.

The manual valve is connected to the selector linkage and is operated by the selector in the vehicle. This valve selects the direction of drive and also overrides the controls to lock out the higher gears as required.

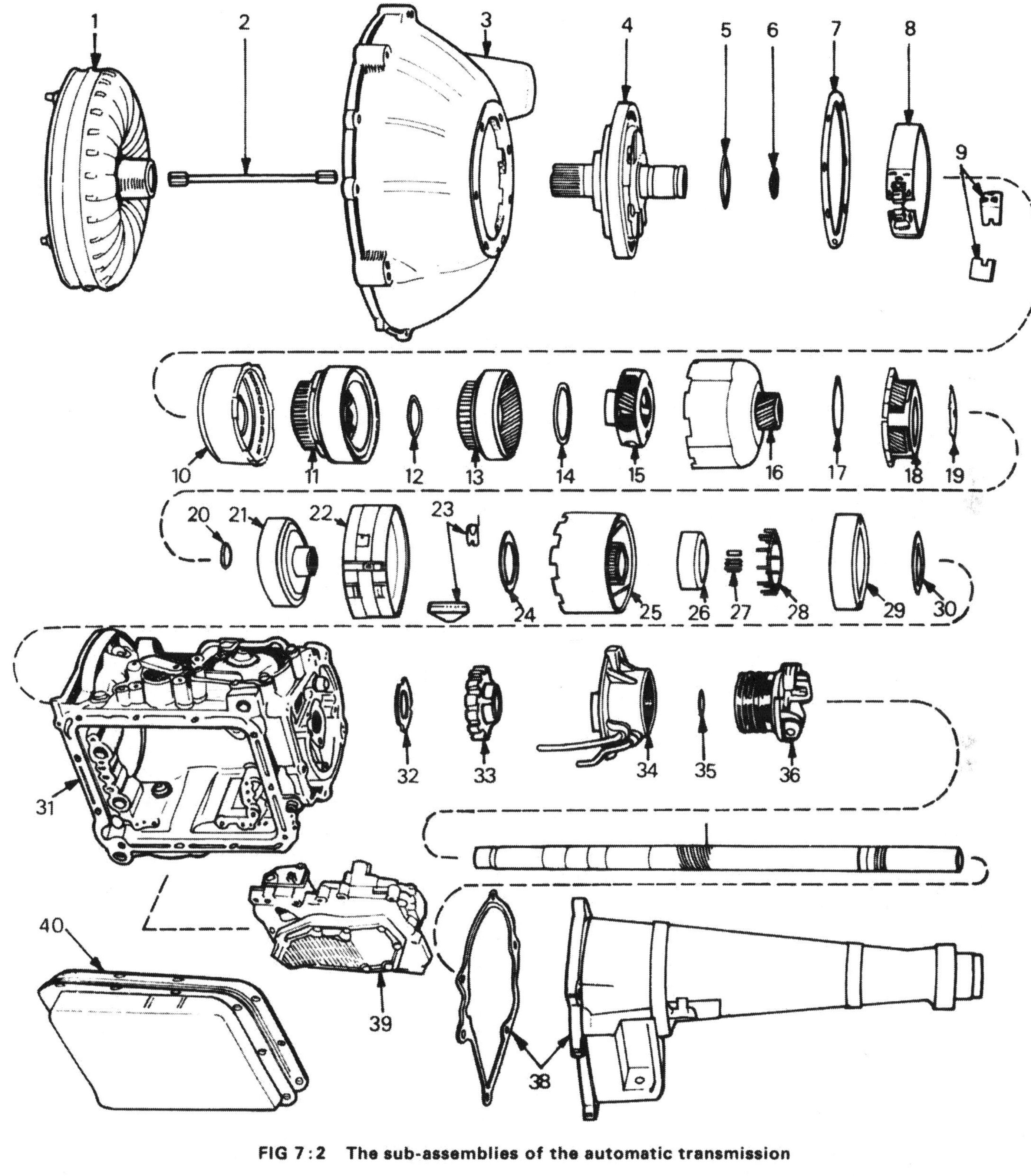

FIG 7:2 The sub-assemblies of the automatic transmission

Key to Fig 7:2 1 Converter 2 Input shaft 3 Converter housing 4 Front pump 5 and 6 Thrust washers 7 Pump gasket 8 Intermediate band 9 Band struts 10 Reverse and high clutch drum 11 Forward clutch and cylinder 12 Thrust washer 13 Forward clutch hub and ring gear 14 Thrust washer 15 Front planet carrier 16 Input shell, sun gear and thrust washer 17 Thrust washer 18 Reverse planet carrier 19 Thrust washer 20 Snap ring 21 Reverse ring gear and hub 22 Low and reverse band 23 Band struts 24 Thrust washer 25 Low and reverse drum 26 One-way clutch inner race 27 Twelve rollers and springs 28 Spring and roller cage 29 One-way clutch outer race 30 Thrust washer 31 Case 32 Thrust washer 33 Parking gear 34 Governor distributor sleeve 35 Snap ring 36 Governor valves and distributor 37 Output shaft 38 Extension housing and gasket 39 Control valve assembly 40 Oil pan and gasket

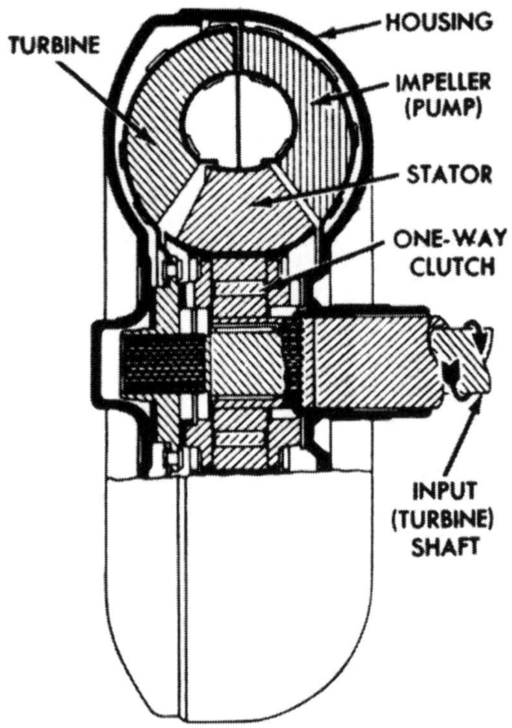

FIG 7:3 Sectional view of converter

The vacuum manifold moves the primary throttle valve dependent on the intake manifold suction. The governor is attached to the output shaft.

Tests on the operation of the hydraulic control system or any repair work, beyond that given in this chapter, must only be carried out by experts.

7:2 Routine maintenance

This is confined to ensuring that the fluid level is correct, cleaning dirt from the casing and at regular intervals adjusting the brake bands.

Cleanliness:

All operations on the automatic transmission must be carried out with scrupulous cleanliness. This includes wiping the filler tube and dipstick clean before removing the dipstick.

The outside of the casing should be periodically cleaned and air vents checked for plugging. The automatic transmission operates at a high temperature and dirt on the casing will act as an insulator, causing the unit to overheat—particularly in areas of high ambient temperatures.

Fluid level:

Run the engine at 1200 rev/min until the automatic transmission has reached its normal operating temperature. Leave the selector at the PARK position. Reduce the engine speed to normal idle and move the selector through each selection in turn before finally returning it to park. Wipe the top of the filler tube clean and withdraw the dipstick. Wipe the dipstick clean on a piece of leather or lint-free cloth. Insert the dipstick back into the filler tube and press it fully home before withdrawing it again. If the level is low, top up with automatic transmission fluid to the upper mark on the dipstick. **The level must always be taken with the transmission hot and the engine running at idle. Do not overfill the transmission.** Overfilling will lead to frothing of the oil and incorrect operation of the unit.

Linkage:

Check that the linkages operate freely and correctly and lightly lubricate pivot points.

Draining:

Routine or seasonal fluid changes are not required. **If the fluid is drained while the transmission is hot, take great care as the fluid will be hot enough to cause serious scalding.** Remove the cover from the bottom of the converter housing and crank the engine until the two converter drain plugs can be removed. One of the drain plugs is shown in **FIG 7:7**. Allow the fluid to drain out into a suitable container before discarding it.

If the remainder of the unit requires draining, the oil pan must be removed from under it.

Oil cooler:

The oil cooler does not require routine maintenance. If there has been internal damage in the automatic transmission, the oil cooler will require flushing through.

If transmission fluid is found in the radiator header tank when checking the coolant level, there is a defect in the oil cooler. Such a defect will also allow water to enter the transmission. Water in the automatic transmission will cause a rise in the level and frothing as the water boils, and it may even cause frothing fluid to overflow from the filler tube.

The oil cooler is built into the header tank and cannot be repaired, except by specialists, so if the oil cooler is leaking a new radiator assembly must be fitted.

Brake band adjustment:

These normally only require adjustment at intervals of 36,000 miles. A special tool, No. T59P-77370-B or 7345, is made for adjusting the bands and the tool is tightened until the handle clicks and overruns. If the special tool is not available a torque spanner can be used instead. The method of adjusting the intermediate band is shown in **FIG 7:8** and that for the low-reverse band is shown in **FIG 7:9**. Slacken the locknut and back it off several turns so that it is well clear. Use the special tool and tighten it until the handle clicks (or a load of 10 lb ft is reached on a torque spanner). For the intermediate brake band slacken off the adjusting screw by exactly 1½ turns and for the low-reverse band slacken back the adjusting screw by exactly 3 turns. Hold the adjusting screw and tighten the locknut to secure it in place.

7:3 Removing the transmission unit

The automatic transmission weighs considerably more than the manually operated transmission and a jack is essential for lowering and moving the unit out from under the car. A special

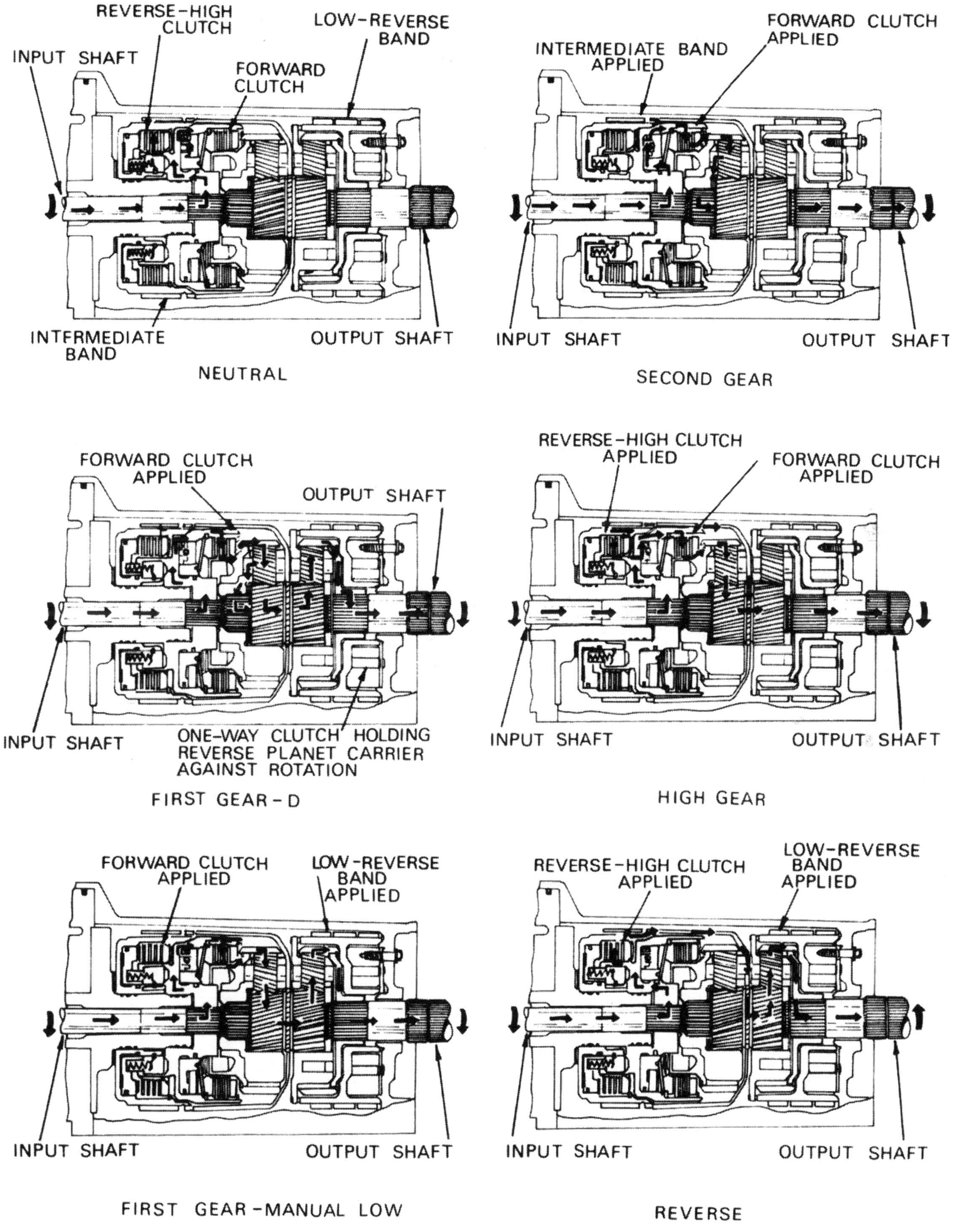

FIG 7:4 Power flow through the transmission

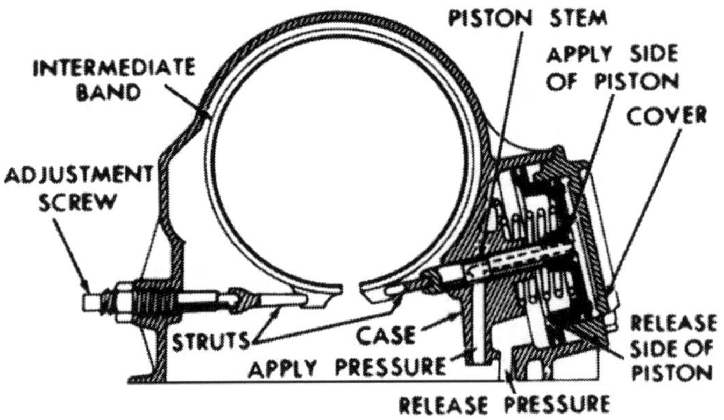

FIG 7:5 Intermediate servo and brake band

support for the transmission is shown in **FIG 7:10** and it should be noted that the transmission is held down to the support using a chain.

1 Drain the transmission by removing the two drain plugs from the converter (see **FIG 7:7**). Slacken off the oil pan bolts and allow the fluid to pour out by lowering one corner of the oil pan. Install the drain plugs and tighten the oil pan attachment bolts after draining out the fluid. Disconnect the oil cooler lines from the transmission, plugging both lines and ports to prevent dirt from entering them. Remove the dipstick and oil filler tube from the case.

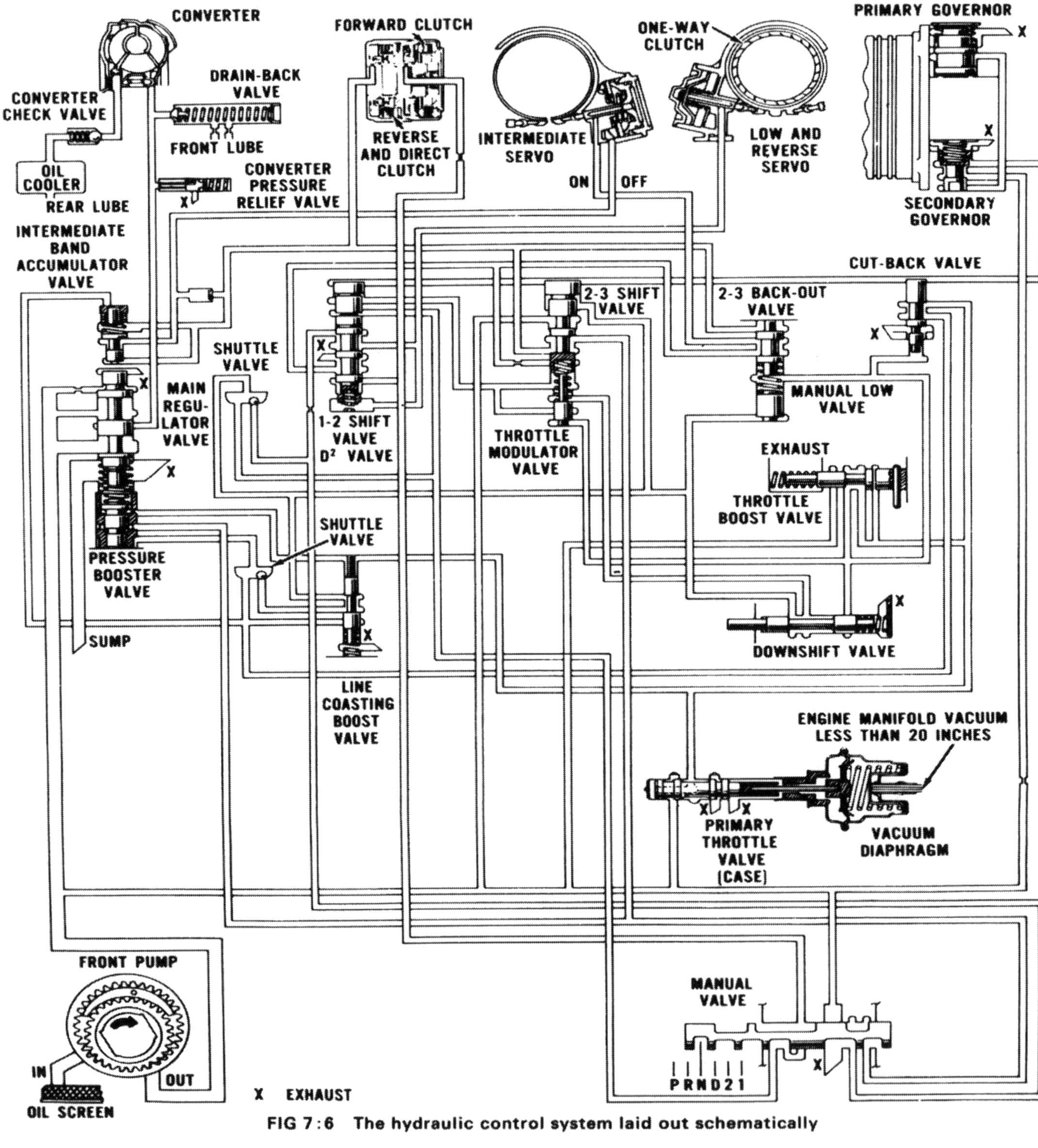

FIG 7:6 The hydraulic control system laid out schematically

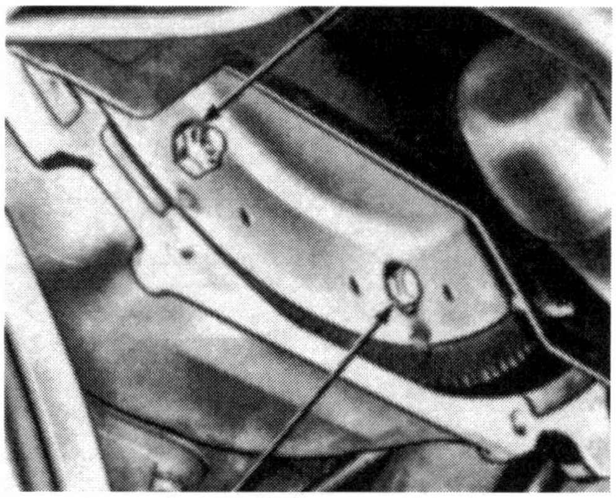

FIG 7:7 A converter drain plug and attachment bolt

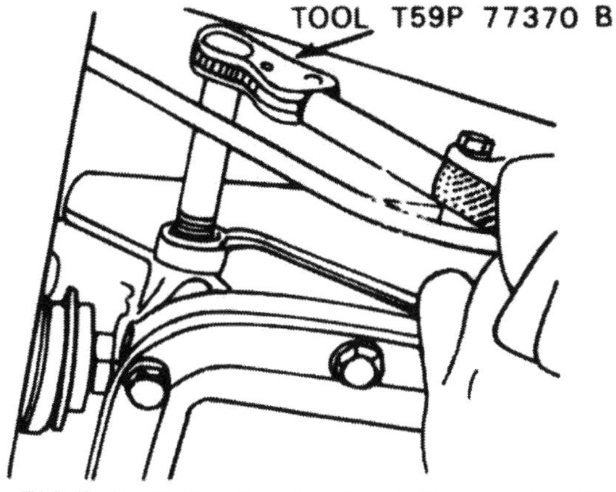

FIG 7:9 Adjusting the Low-Reverse band

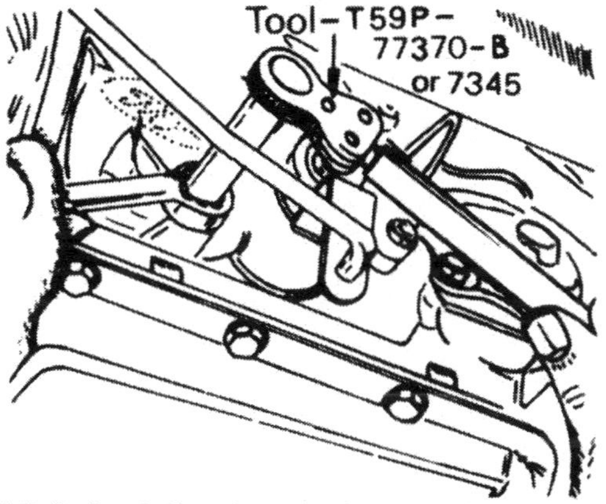

FIG 7:8 Adjusting the intermediate band

FIG 7:10 The transmission support jack

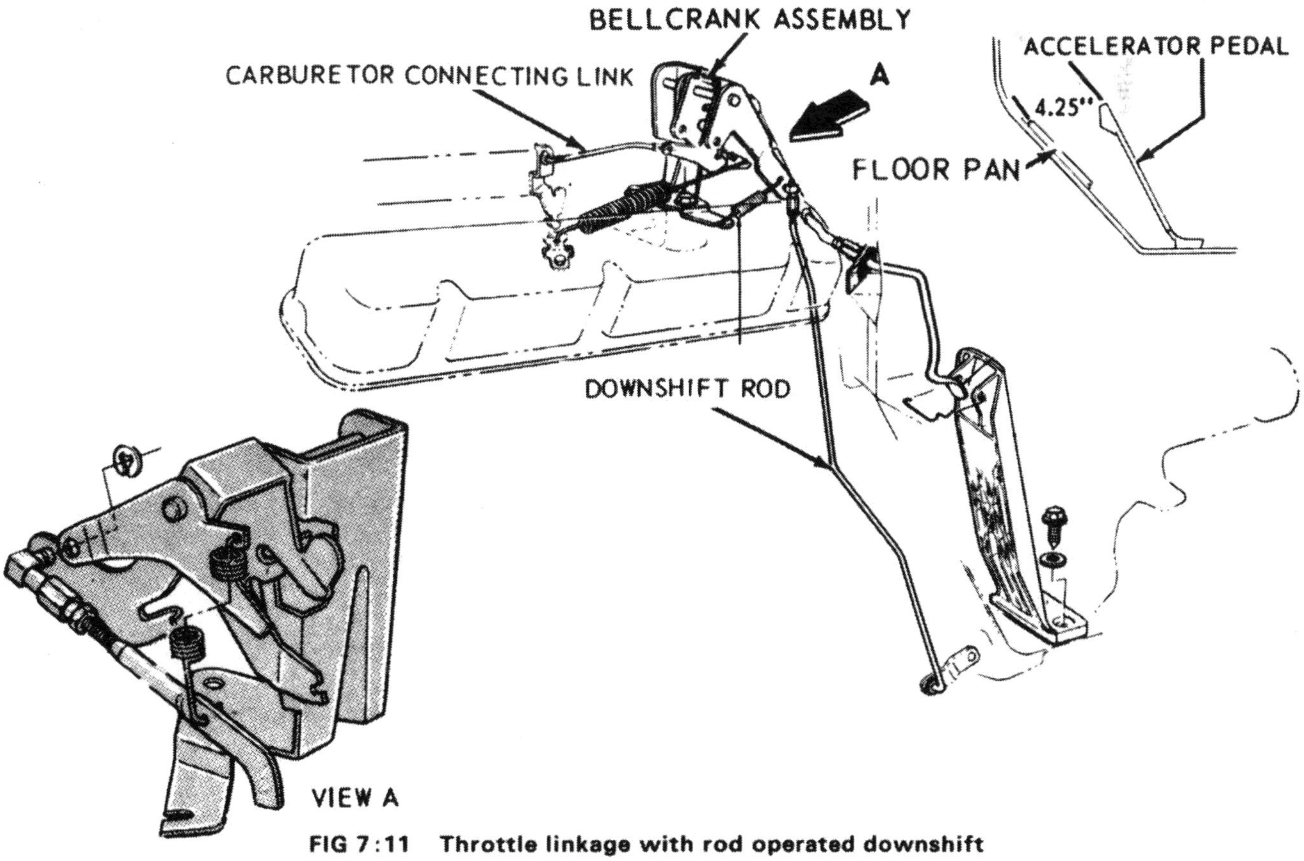

FIG 7:11 Throttle linkage with rod operated downshift

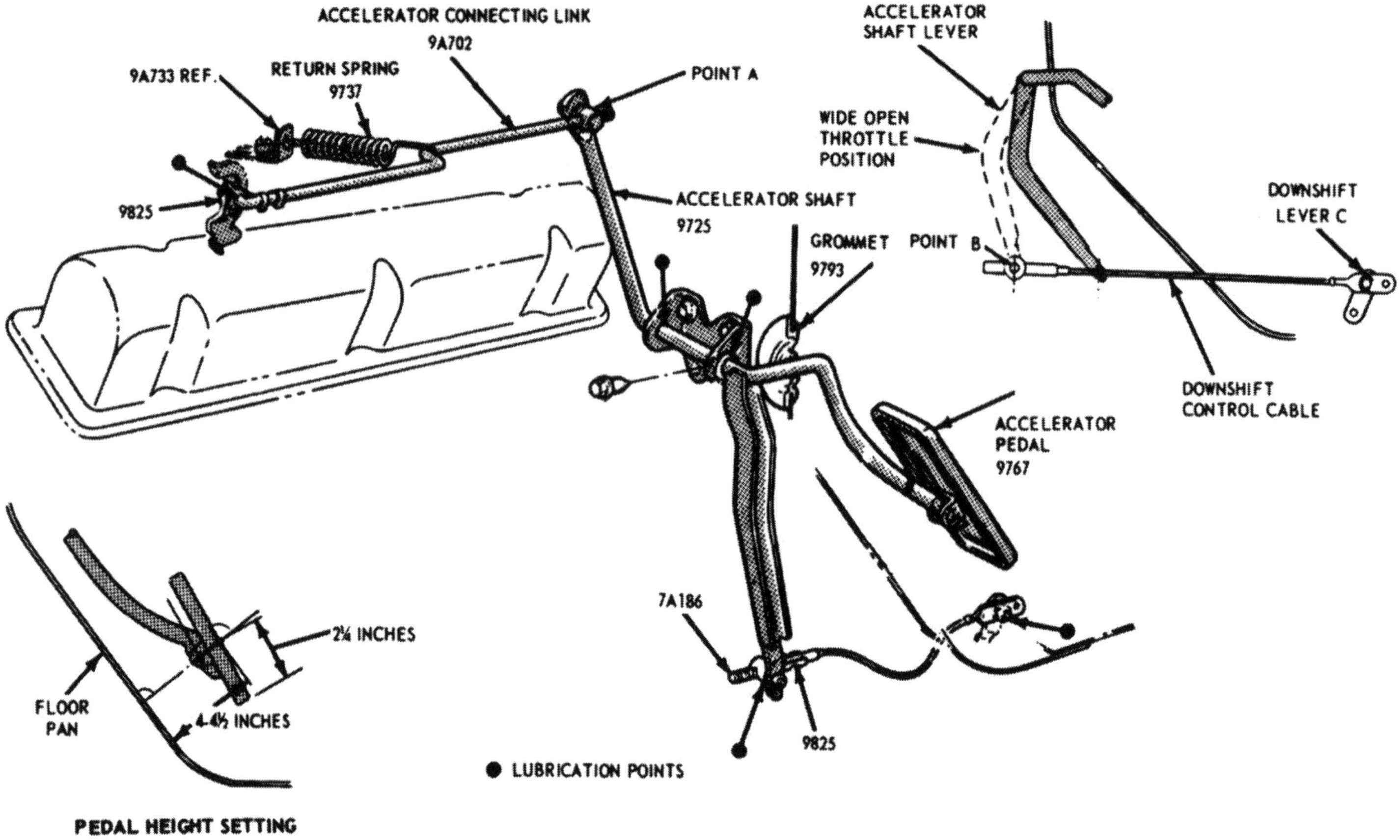

FIG 7:12 Throttle linkage with cable operated downshift

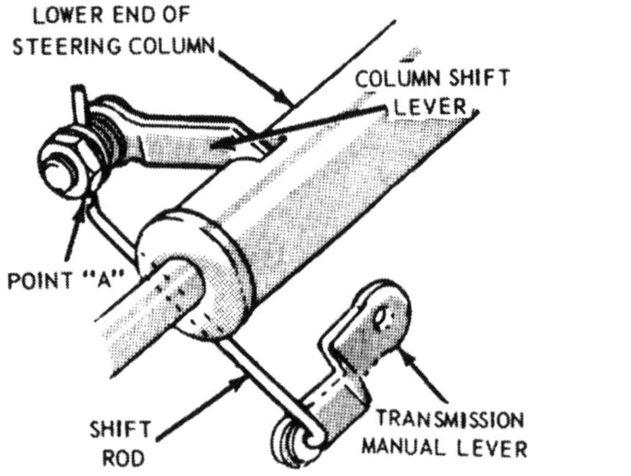

FIG 7:13 Adjusting the column mounted selector linkage

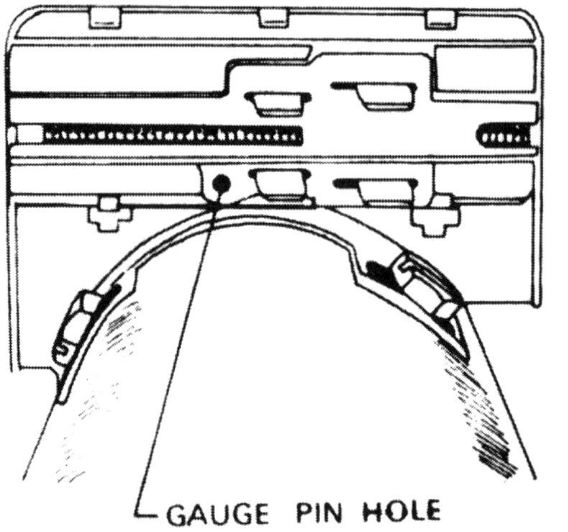

FIG 7:14 Adjusting the neutral start switch

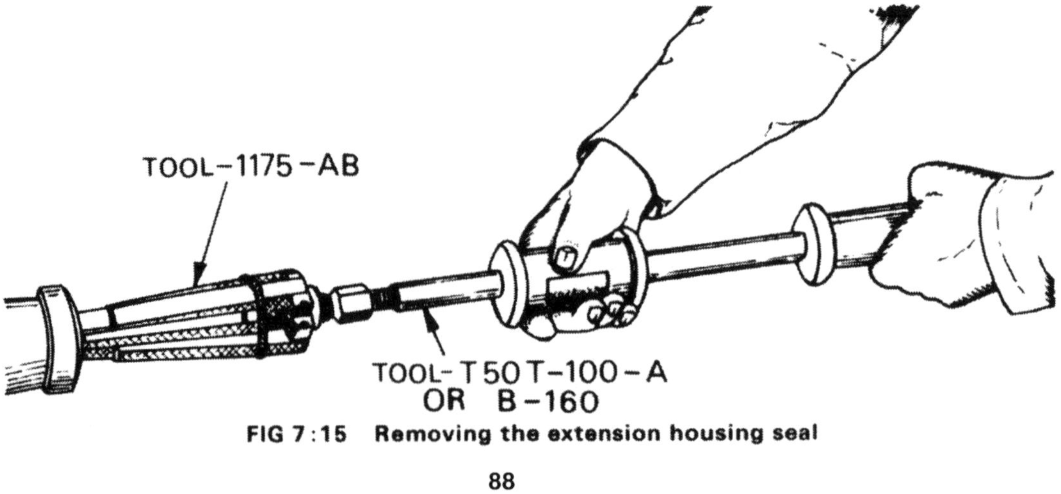

FIG 7:15 Removing the extension housing seal

2 Remove the drive shaft (see **Chapter 8**) after making aligning marks so that it will be installed into its original position. Plug the extension housing on the rear of the transmission to prevent dirt from entering the unit. Disconnect the shift and kickdown rods.

3 Disconnect the speedometer drive cable, by taking out its retaining clip. Disconnect the vacuum line at the transmission vacuum unit. Disconnect the exhaust pipe at the flange on the manifolds. Free the parking brake cable from the equalizer lever. Remove the starter motor, after disconnecting the battery and the starter motor cables. Disconnect the neutral start switch wires, labelling them if necessary.

4 Remove the two bolts that secure the extension housing to the rear crossmember mounting. Remove the four bolts that secure the converter to the engine drive plate (see **FIG 7:7**). Support the engine on a jack and pad of wood under the oil pan. Support the transmission on the transmission jack and secure it with a length of chain. Remove the bolts that secure the rear mounting crossmember to the frame. Remove the five bolts that secure the converter housing to the engine. Draw the transmission slightly rearwards and then lower it down and out of the car.

Installation:

The transmission is installed in the reverse order of removal. All bolts should be torqued to the correct specifications. Check and adjust the shift and kickdown linkages. Fill the transmission to the correct level with fluid.

Oil cooler flushing:

If any dirt has entered the oil cooler or the transmission has been removed because of internal faults, the oil cooler should be flushed. With the transmission filled to the correct level with fluid, disconnect the return line from the oil cooler at the rear of the transmission. Start the engine and drain out at least two quarts of fluid from the return line. If the fluid does not flow freely from the cooler, shut off the engine and disconnect the other line to the oil cooler at the transmission. Blow through the lines and cooler using compressed air at a maximum pressure of 100 lb/sq inch in a reverse direction to the normal flow. Disconnect the lines from the oil cooler if necessary so that they can be blown through separately. If the blockage cannot be removed using compressed air, new lines or radiator assembly must be fitted. **Do not attempt to cure leaks by bypassing the cooler or plugging the lines.**

Torque converter:

Once the transmission has been taken out of the vehicle, the converter can be removed from the transmission. The converter is sealed on manufacture so if it is found to be defective it must be renewed as a unit.

Special tools are made for checking the converter. The tools are used for pressure testing the unit, as well as for to be checked. If it is suspected that faults in the transmission are caused by a defective converter, take the unit to a service garage holding the special tools so that the unit can be checked.

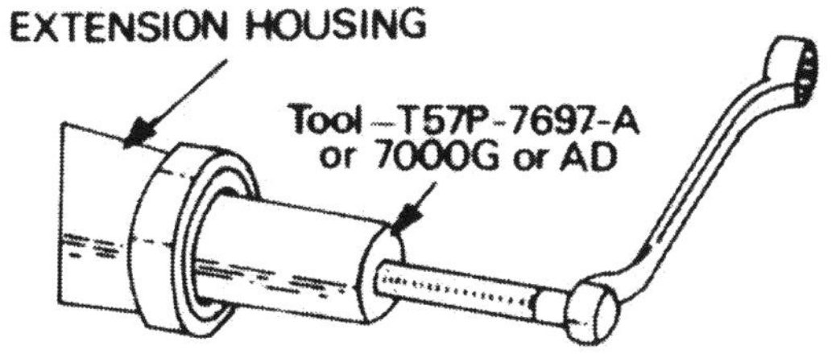

FIG 7:16 Removing the extension housing bushing

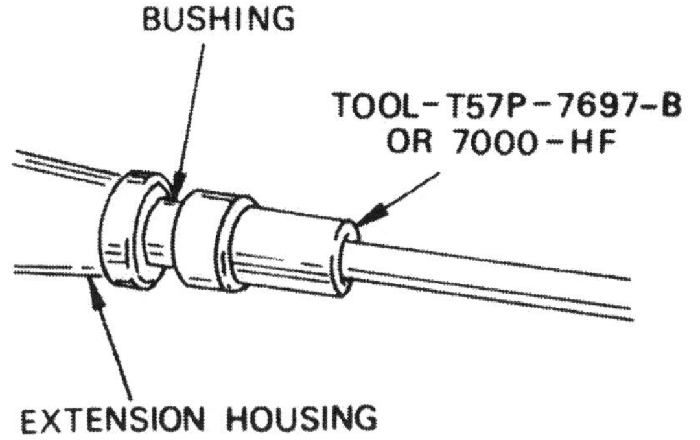

FIG 7:17 Installing the extension housing bushing

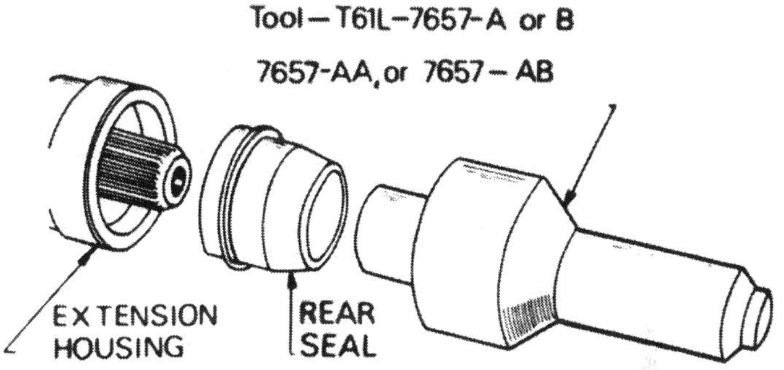

FIG 7:18 Installing housing seal

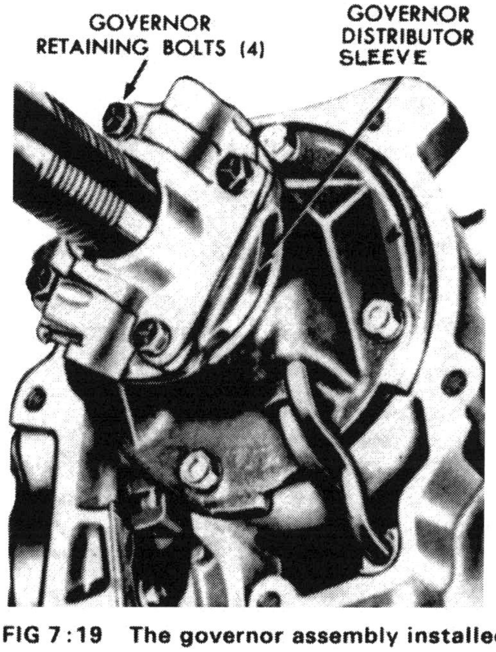

FIG 7:19 The governor assembly installed

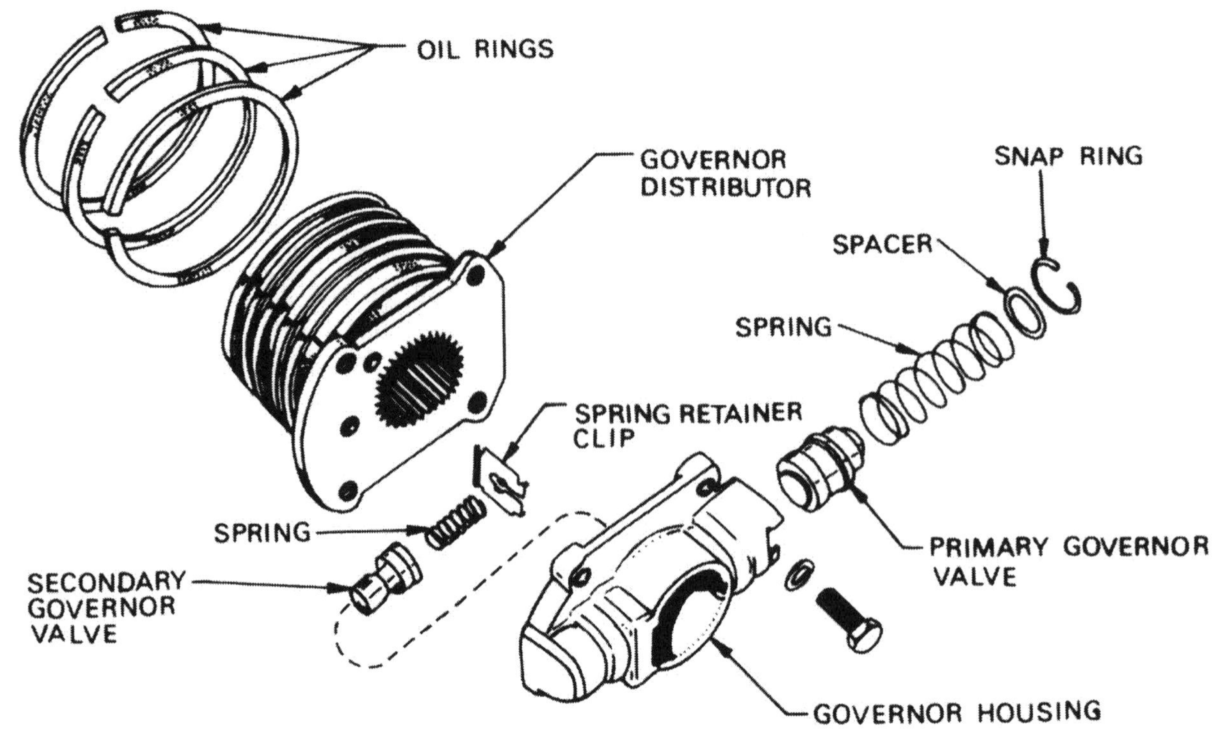

FIG 7:20 The governor components

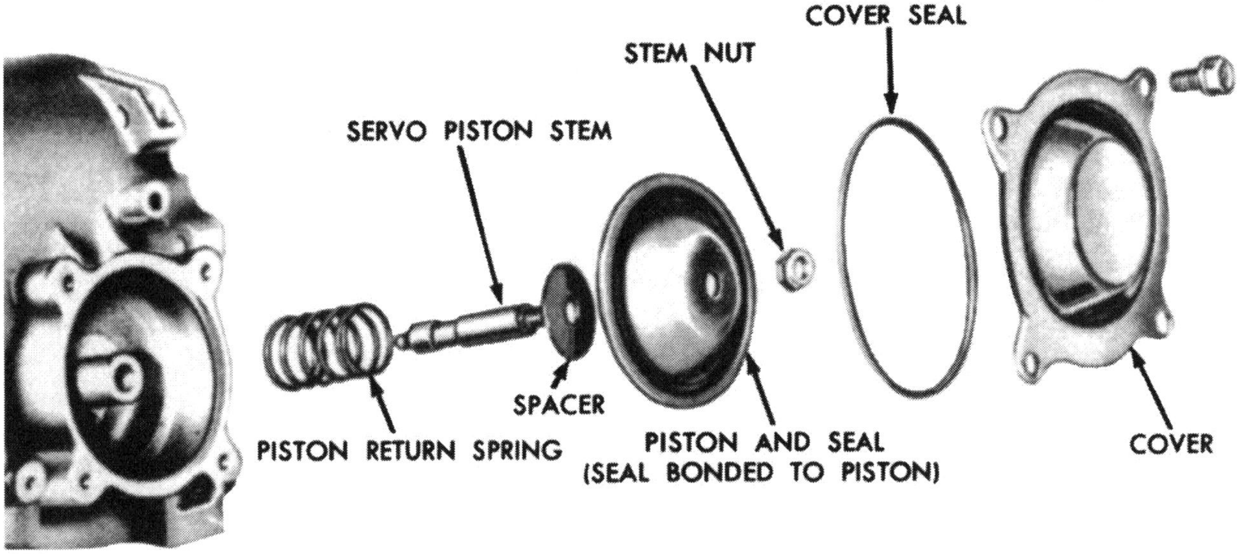

FIG 7:21 The Low-Reverse servo components

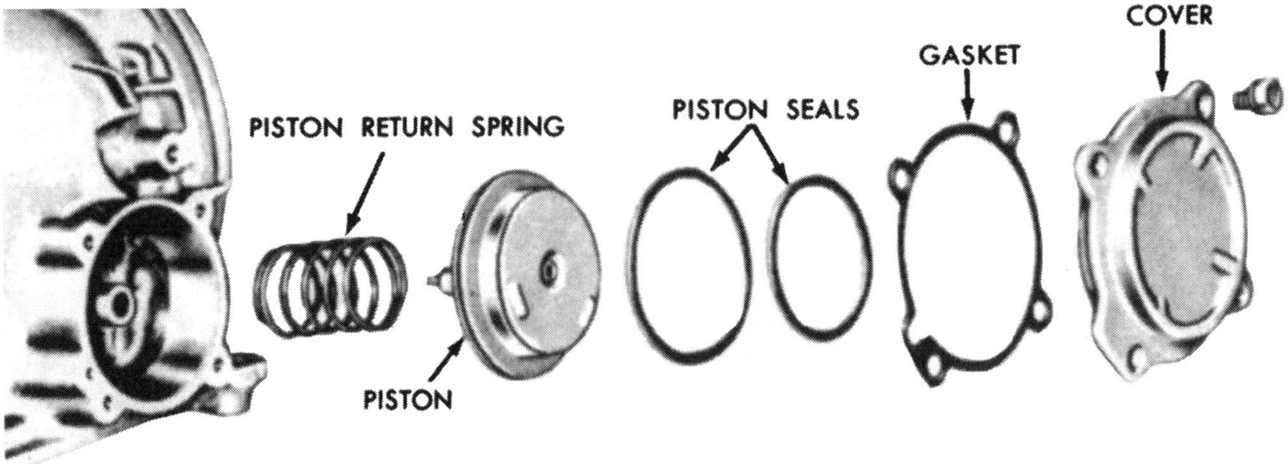

FIG 7:22 The Intermediate servo components

If dirt or sludge is thought to be in the system, drain out all the fluid from the torque converter by tilting it around with both drain plugs removed. Install the drain plugs and fill the unit with clean kerosene or suitable cleansing fluid. Preferably rotate the turbine but if this is not possible, because of lack of the correct tools, shake the converter around to swill the inside. Drain out the old kerosene and repeat the procedure as often as is required to remove all dirt from inside the unit.

7:4 Stall test

This test will give a rapid guide to the condition of the automatic transmission.

Connect an accurate tachometer to the ignition circuit and fit the instrument so that it can be seen readily from the drivers seat.

Start the engine and run it until the transmission has reached its normal operating temperature. Apply the parkbrake fully and hold the vehicle on the service brake as well. Select a drive gear and floor the throttle pedal. **Note the maximum engine speed and release the throttle pedal within five seconds.** If the engine speed is above the normal stall speed, stop the test immediately as it is an indication of slip.

The churning of the oil in the converter with the turbine stalled will produce large amounts of heat and the full throttle period must not exceed five seconds otherwise the transmission will overheat. Check each drive selection in turn and once the maximum speed of the engine has been noted in a selection, select Neutral and allow the engine to run at 1000 rev/min for at least fifteen seconds to allow the fluid to cool. When the stall speed has been checked in all four drive selections, select Neutral and leave the engine running at 1000 rev/min to allow the transmission to cool and the temperature to stabilize.

The transmission and engine are satisfactory if the stall speed lies between 1750 to 1950 rev/min for 289 cu in engines and 1780 to 2000 rev/min for 302 cu in engines.

Low stall speed:

This can be caused by lack of power from the engine, so make sure that the engine is satisfactory and correctly tuned.

The other cause of a low stall speed is a defective converter. If the stall speed is more than 300 to 400 rev/min low, take the car for a road test. If the vehicle cruises normally but acceleration (particularly from standstill) is poor and if the vehicle will only start with difficulty on gradients, then the stator one-way clutch is slipping.

If acceleration and starting on gradients is satisfactory, but the car drags at cruise, and the maximum speeds in the gears (particularly top) are limited, and the transmission overheats easily, then the stator one-way clutch is seized.

In both cases, remove the transmission so that the torque converter can be removed and checked at a service garage.

High stall speed:

A high stall speed indicates slip in the gearbox. Check each selection in turn and note the selections that give high stall speeds. Excessive speed in forward selections only indicate forward clutch slippage. Excessive speed in R selection indicates either reverse-high clutch slippage or low-reverse band slippage. Excessive speed in D1 only indicates gear train one-way clutch slippage.

7:5 Adjustments

The adjustments of the brake bands have already been dealt with in **Section 7:2.**

For full and accurate checks of the transmission, test equipment is required so that the operating pressures can be noted. The vehicle should therefore be taken to a service garage for checking and diagnosis of faults that are beyond the power of the owner to adjust or correct.

Throttle linkage:

Typical throttle linkages are shown in **FIGS 7:11** and **7:12.** Start the engine and run it until it has reached its normal operating temperature. Connect a tachometer to the ignition circuit and adjust the idling speed to the correct limits. When setting the fast-idle screw note that it must be just touching the cam at the correct mark on the cam. The large dot is for the D1 selection and the small dot for the D2 selection. Make sure that the floor covering and carpets are correctly in place, removing any matting that is not original equipment. Check the height of the accelerator pedal from the floor pan, and if it is outside the limits adjust the height at the pedal link.

Turn the engine off and press the accelerator pedal right down and block it there. On models fitted with a downshift rod, adjust the rod so that the lever on the transmission is in the kickdown detent position with approximately $\frac{1}{16}$ inch of free travel on the bellcrank assembly. On models fitted with a downshift cable, disconnect the cable at the point B. Rotate the downshift lever counterclockwise until it is against the internal stop. Hold the lever on the transmission in this position and adjust the length of the cable until its trunnion will slide easily into the lever on the throttle linkage. Turn the adjuster one additional turn clockwise and reconnect the trunnion to the lever.

Road test the car for idling and kickdown. A low idling speed will cause the engine to stall while a high idling speed will make the car creep when a drive selection is made.

Column shift:

The adjustment point is shown in **FIG 7:13.** Slacken the nut at the point A sufficiently to allow the shift rod to be moved along the column shift lever. Place the selector lever against the D stop in the D position. Move the transmission manual lever into the D position (third from rear) and check that the selector lever has not moved. Tighten the nut A so that the shift rod is secured in position. Check the selector lever in all selections.

Neutral start switch:

This is fitted to ensure that the starter motor will only operate in the Park and Neutral selections.

The attachment of the switch to the column is shown in **FIG 7:14.** Slacken the two screws that secure the switch to the column. Set the selector lever firmly against the stop of the neutral detent and rotate the switch about

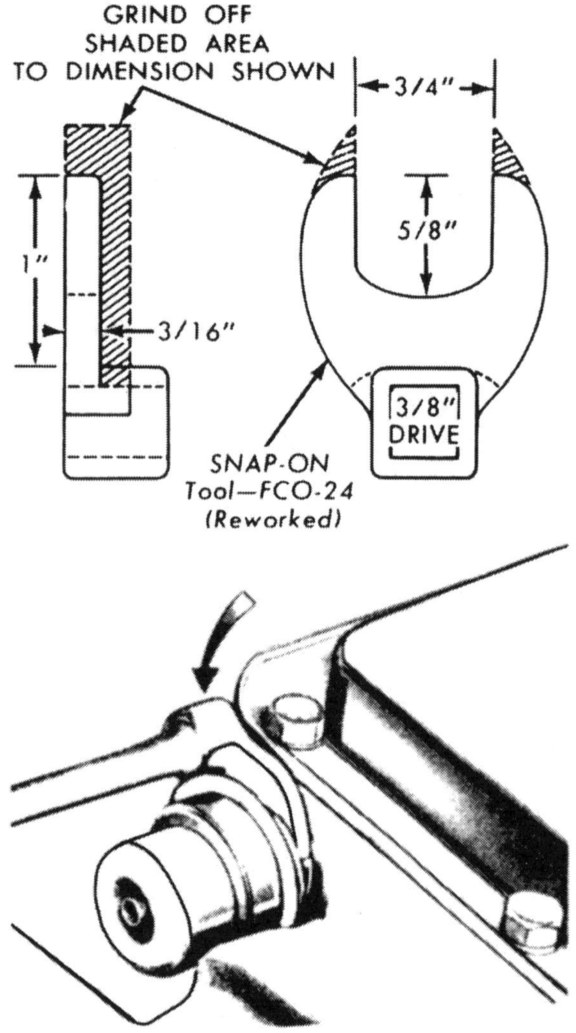

FIG 7:23 Removing/installing the vacuum unit

the column, in its elongated holes, until a No. 43 drill shank can be inserted into the gauge pin holes. Tighten the switch securing screws to hold the switch and remove the drill. Check that the starter motor only operates in the Park and Neutral selections and that it does not operate when a drive selection is made.

7:6 Governor and rear extension housing

The seal and bushing in the extension housing can be renewed without dismantling the transmission or removing it from the vehicle.

Preparation:

Raise the car on a hoist or work from a pit. Remove the drive shaft (see **Chapter 8**). Disconnect the speedometer drive cable from the extension housing if the housing is to be removed. Support the transmission on a transmission jack and free the rear mounting bolts from the extension housing.

If the extension housing is to be removed, slacken the bolts securing it so that the transmission fluid can drain out. Disconnect the muffler inlet pipes from the exhaust manifolds on the engine. Disconnect the handbrake cable from the equalizer lever.

Seal and bush assembly:

Remove the extension housing seal using the tools shown in **FIG 7:15.** Carefully use the bushing remover shown in **FIG 7:16,** taking great care not to damage the spline seal, and withdraw the bushing. The new bushing is installed using the tools shown in **FIG 7:17** and the new seal using the tools shown in **FIG 7:18.** Before installing the bushing and seal, check the counterbore of the housing for nicks or burrs and use Crocus paper to remove them. Check the yoke of the drive shaft for scores or damage and renew it if it is defective. Lubricate the inside diameter of the fibre portion of the seal and the yoke splines with B8A-19589-A lubricant.

The governor:

Remove the bolts that secure the extension housing to the main case and withdraw the extension. The attachments of the governor are shown in **FIG 7:19** and the components in **FIG 7:20**. Remove the governor retaining bolts and withdraw the governor assembly from the distributor.

Remove the primary valve securing snap ring and take out the primary valve parts. Remove the retaining clip so that the secondary valve parts can be removed.

Wash the parts in a suitable solvent. Check that the valves slide freely in their bores. If the valves stick, they may be lightly polished with Crocus paper but take care not to round off the sharp edges of the valves. Fit new parts if the old ones are scored or badly worn.

The parts are installed in the reverse order of disassembly. Lubricate them lightly with transmission fluid. When installing the secondary valve parts make sure that the retaining clip is fitted with its small concave surface downwards so that it locates the spring.

Install the extension housing using a new gasket. The remainder of the parts are then installed in the reverse order of disassembly. Fill the transmission to the correct level with fresh fluid.

7:7 Servo assemblies and vacuum unit

The components of the low-reverse servo are shown in **FIG 7:21** and those of the intermediate servo in **FIG 7:22**. The pistons and seals can be renewed but take great care to ensure scrupulous cleanliness when working on the parts. Lubricate the parts lightly with clean transmission fluid before installation.

The tool made for removing the vacuum unit and its method of use are shown in **FIG 7:23.** If the vacuum unit is thought to be faulty, check that the vacuum line is clear and either have the vacuum unit tested at a service garage or install a new unit.

7:8 Trouble shooting guide

(a) Transmission overheats

1 Excessive dirt or mud on outside of unit
2 Oil cooler or lines plugged or oil cooler defective
3 Converter stator clutch seized
4 Incorrect brake band adjustment
5 Brakes binding or other drag on car causing converter to slip

(b) Noisy operation

1 Incorrect fluid level or incorrect type of fluid
2 Incorrectly adjusted linkage
3 Oil screen to pump intake choked

(c) Incorrect shift speeds

1 Check in (b)
2 Defective vacuum unit
3 Incorrectly adjusted brake bands or defective servos
4 Defective governor assembly

(e) Poor acceleration

1 Check 3 and 4 in (a)
2 Converter stator clutch slipping

(d) Jumps in engagement or rough engagement

1 Check 4 in (a); 2 in (b) and 2 in (c)
2 Incorrect idling speed

(g) Vehicle does not hold in Park selection

1 Check 2 in (b)

(h) Fluid frothing or being forced out of filler tube

1 Too high fluid level
2 Water in system (suspect oil cooler)
3 Air leaks on suction side of pump

This guide is not fully comprehensive and only covers faults which can be cured by the owner. For other faults take the vehicle to a service garage.

Inches	Decimals	Milli-metres	Inches to Millimetres Inches	mm	Millimetres to Inches mm	Inches
1/64	.015625	.3969	.001	.0254	.01	.00039
1/32	.03125	.7937	.002	.0508	.02	.00079
3/64	.046875	1.1906	.003	.0762	.03	.00118
1/16	.0625	1.5875	.004	.1016	.04	.00157
5/64	.078125	1.9844	.005	.1270	.05	.00197
3/32	.09375	2.3812	.006	.1524	.06	.00236
7/64	.109375	2.7781	.007	.1778	.07	.00276
1/8	.125	3.1750	.008	.2032	.08	.00315
9/64	.140625	3.5719	.009	.2286	.09	.00354
5/32	.15625	3.9687	.01	.254	.1	.00394
11/64	.171875	4.3656	.02	.508	.2	.00787
3/16	.1875	4.7625	.03	.762	.3	.01181
13/64	.203125	5·1594	.04	1.016	.4	.01575
7/32	.21875	5.5562	.05	1.270	.5	.01969
15/64	.234375	5.9531	.06	1.524	.6	.02362
1/4	.25	6.3500	.07	1.778	.7	.02756
17/64	.265625	6.7469	.08	2.032	.8	.03150
9/32	.28125	7.1437	.09	2.286	.9	.03543
19/64	.296875	7.5406	.1	2.54	1	.03937
5/16	.3125	7.9375	.2	5.08	2	.07874
21/64	.328125	8.3344	.3	7.62	3	.11811
11/32	.34375	8.7312	.4	10.16	4	.15748
23/64	.359375	9.1281	.5	12.70	5	.19685
3/8	.375	9.5250	.6	15.24	6	.23622
25/64	.390625	9.9219	.7	17.78	7	.27559
13/32	.40625	10.3187	.8	20.32	8	.31496
27/64	.421875	10.7156	.9	22.86	9	.35433
7/16	.4375	11.1125	1	25.4	10	.39370
29/64	.453125	11.5094	2	50.8	11	.43307
15/32	.46875	11.9062	3	76.2	12	.47244
31/64	.484375	12.3031	4	101.6	13	.51181
1/2	.5	12.7000	5	127.0	14	.55118
33/64	.515625	13.0969	6	152.4	15	.59055
17/32	.53125	13.4937	7	177.8	16	.62992
35/64	.546875	13.8906	8	203.2	17	.66929
9/16	.5625	14.2875	9	228.6	18	.70866
37/64	.578125	14.6844	10	254.0	19	.74803
19/32	.59375	15.0812	11	279.4	20	.78740
39/64	.609375	15.4781	12	304.8	21	.82677
5/8	.625	15.8750	13	330.2	22	.86614
41/64	.640625	16.2719	14	355.6	23	.90551
21/32	.65625	16.6687	15	381.0	24	.94488
43/64	.671875	17.0656	16	406.4	25	.98425
11/16	.6875	17.4625	17	431.8	26	1.02362
45/64	.703125	17.8594	18	457.2	27	1.06299
23/32	.71875	18.2562	19	482.6	28	1.10236
47/64	.734375	18.6531	20	508.0	29	1.14173
3/4	.75	19.0500	21	533.4	30	1.18110
49/64	.765625	19.4469	22	558.8	31	1.22047
25/32	.78125	19.8437	23	584.2	32	1.25984
51/64	.796875	20.2406	24	609.6	33	1.29921
13/16	.8125	20.6375	25	635.0	34	1.33858
53/64	.828125	21.0344	26	660.4	35	1.37795
27/32	.84375	21.4312	27	685.8	36	1.41732
55/64	.859375	21.8281	28	711.2	37	1.4567
7/8	.875	22.2250	29	736.6	38	1.4961
57/64	.890625	22.6219	30	762.0	39	1.5354
29/32	.90625	23.0187	31	787.4	40	1.5748
59/64	.921875	23.4156	32	812.8	41	1.6142
15/16	.9375	23.8125	33	838.2	42	1.6535
61/64	.953125	24.2094	34	863.6	43	1.6929
31/32	.96875	24.6062	35	889.0	44	1.7323
63/64	.984375	25.0031	36	914.4	45	1.7717

UNITS	Pints to Litres	Gallons to Litres	Litres to Pints	Litres to Gallons	Miles to Kilometres	Kilometres to Miles	Lbs. per sq. In. to Kg. per sq. Cm.	Kg. per sq. Cm. to Lbs. per sq. In.
1	.57	4.55	1.76	.22	1.61	.62	.07	14.22
2	1.14	9.09	3.52	.44	3.22	1.24	.14	28.50
3	1.70	13.64	5.28	.66	4.83	1.86	.21	42.67
4	2.27	18.18	7.04	.88	6.44	2.49	.28	56.89
5	2.84	22.73	8.80	1.10	8.05	3.11	.35	71.12
6	3.41	27.28	10.56	1.32	9.66	3.73	.42	85.34
7	3.98	31.82	12.32	1.54	11.27	4.35	.49	99.56
8	4.55	36.37	14.08	1.76	12.88	4.97	.56	113.79
9		40.91	15.84	1.98	14.48	5.59	.63	128.00
10		45.46	17.60	2.20	16.09	6.21	.70	142.23
20				4.40	32.19	12.43	1.41	284.47
30				6.60	48.28	18.64	2.11	426.70
40				8.80	64.37	24.85		
50					80.47	31.07		
60					96.56	37.28		
70					112.65	43.50		
80					128.75	49.71		
90					144.84	55.92		
100					160.93	62.14		

UNITS	Lb ft to kgm	Kgm to lb ft	UNITS	Lb ft to kgm	Kgm to lb ft
1	.138	7.233	7	.967	50.631
2	.276	14.466	8	1.106	57.864
3	.414	21.699	9	1.244	65.097
4	.553	28.932	10	1.382	72.330
5	.691	36.165	20	2.765	144.660
6	.829	43.398	30	4.147	216.990

CHAPTER 8

THE DRIVE SHAFT, REAR AXLE AND REAR SUSPENSION

8:1 The drive shaft

A typical drive shaft assembly is shown in **FIG 8:1**. The pinion flange on the rear axle forms part of the rear universal joint and two of the bearings on the spider are secured to the pinion drive flange by U-bolts. The slip yoke on the front of the drive shaft slides into the extension housing of the transmission and fits onto the splines of the output shaft of the transmission. Seals are fitted to prevent lubricant from leaking out of the transmission, as shown in **FIG 8:2**.

Removal:

To preserve the balance of the drive shaft the rear yoke of the drive shaft must be kept in the same alignment with the drive pinion on the rear axle. When the parts are new they are marked with yellow aligning marks but if these marks have worn off or become covered with dirt, new ones should be made.

Remove the nuts and carefully drive out the U-bolts that secure the shaft to the pinion flange. **Wrap tape around the rear universal joint to prevent the bearings from falling off from the spider.**

Lower the rear end of the shaft and withdraw the slip yoke from the transmission. **Take care not to damage or mark the slip yoke when the shaft is free as nicks or burrs on the yoke can cause leaks from the transmission.** Light damage or burrs can be smoothed off with a stone but if the yoke is badly scored or damaged a new one will have to be fitted.

Install the drive shaft in the reverse order of removal. Lubricate the splines in the slip yoke with B8A-19589-A lubricant. Make sure that the alignment marks align after the drive shaft has been installed. Tighten the U-bolt nuts to a torque of 15 to 20 lb ft.

Maintenance:

The spider on each universal joint is fitted with a plug. Unscrew the plug and screw a grease fitting back in its place. Inject a few strokes of Ford Universal Lube through the grease fitting and then remove the fitting and install the plug.

Vibration:

The drive shaft is balanced on manufacture by the addition of balance weights. If vibration is traced to the drive shaft, first check that the balance weights are present and have not fallen off and then make sure that

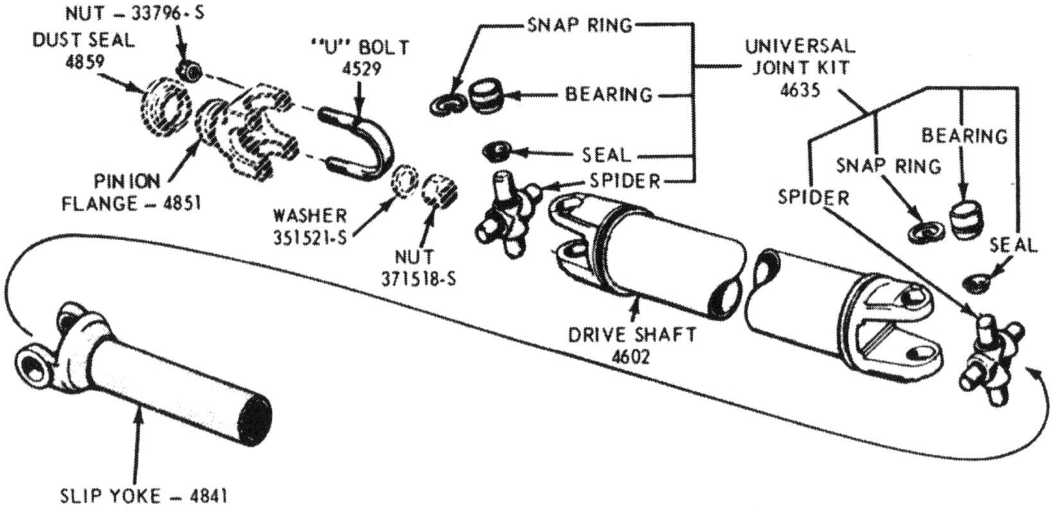

FIG 8:1 The parts of the drive shaft

mud or underseal is not clinging to the shaft and upsetting the balance. Check the universal joints for wear. The joints are checked by smartly rotating them backwards and forwards by hand and trying to lift the shaft by hand. Wear will show up as excessive play or 'clonk' as the direction of drive is reversed.

If there is no obvious cause of vibration, try rotating the shaft 180 deg. in relation to the pinion drive flange.

If rotating the shaft through 180 deg. does not cure the vibration, jack up the rear of the vehicle and place it very securely on stands. **Chock the front wheels both in front and behind.** Start the engine and select a gear. Speed the engine up until the speedometer is reading approximately 40 to 50 mile/hr. From underneath the car, bring a wax crayon slowly and carefully into light contact with the drive shaft, as shown in **FIG 8:3. Avoid the balance weights.** Stop the engine and if the mark has been made lightly enough it will only show on the heavy side of the shaft. Install two Whittek clamps to the drive shaft so that their heads are 180 deg. away from the crayon mark, as shown in **FIG 8:4.** Tighten the clamps to prevent them from moving. Start the engine and increase the speed until 65 to 70 mile/hr is shown on the speedometer. If vibration is still present, stop the engine and rotate the two clamps so that they make an angle of 45 deg. between their heads, as shown in **FIG 8:5.** If vibration is still present on test, move the heads of the clamps slightly further apart and repeat the test. Make small adjustments as required until vibration is eliminated. **Do not run the vehicle for long periods with the wheels off the ground as there is a danger of the transmission or engine overheating.**

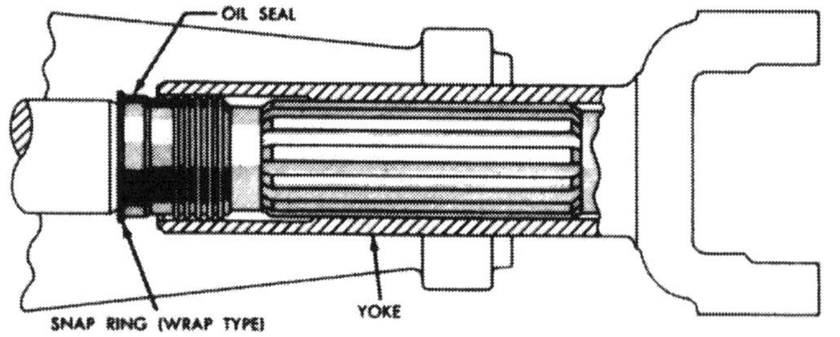

FIG 8:2 The fitting of the drive shaft slip yoke to the output shaft of the transmission

Servicing universal joints:

Kits of parts are available and these should be used. **Individual parts of the universal joint should not be renewed separately.** Before dismantling the shaft, make aligning marks on the slip yoke and shaft so that they will be reassembled and the balance of the shaft preserved.

The method of removing the bearings from a universal joint is shown in **FIG 8:6.** Remove the snap rings that secure the bearings. If the snap rings stick, clean away all rust and enamel. Mount the shaft in the padded jaws of a vice and use the tool to press out the bearing on one side as far as possible. If the bearing cannot be removed fully with the tool, grip the bearing with grips or in a vice and pull it out, using a drift if the bearing is tight as shown in **FIG 8:7.** Remove the opposite bearing in a similar manner. When all four bearings have been removed, the spider can be manoeuvred out of the yoke and shaft.

Before installing new parts, check the bores in the yoke arms. The bearings must be a light drive fit through them. If the bores are worn oval a new shaft or yoke assembly must be fitted. Check that the needle rollers in the bearings are in place and held by the grease. If a needle roller becomes displaced, and pressure is applied across it, it will snap.

Position the spider in the yoke and press a bearing into place so that it is $\frac{1}{4}$ inch below the surface. Install a new snap ring onto the bearing. The bearing is pressed back into place using the special tool as shown in **FIG 8:8.** Start a new bearing into the yoke and press it in until the original bearing contacts the snap ring. Remove the tool and secure the second bearing with a new snap ring. Install the remaining bearings using the same method.

When all the bearings are in place, check that the joint articulates freely. Binding can be relieved by tapping the ears of the yoke arms sharply. **It is essential that the joints do move freely before installing the shaft.**

8:2 Rear suspension description

A typical rear suspension is shown in **FIG 8:9** and the components are shown in **FIG 8:10.** Semi-elliptic springs and beam rear axles are used on all the models covered by this manual. Rubber insulated bushings are

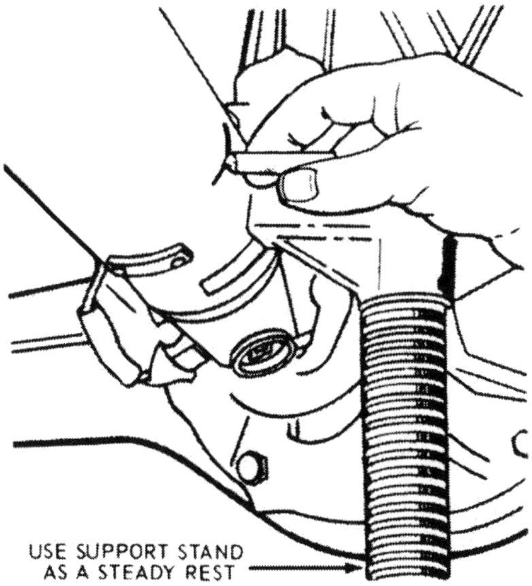

FIG 8:3 Marking the drive shaft for unbalance

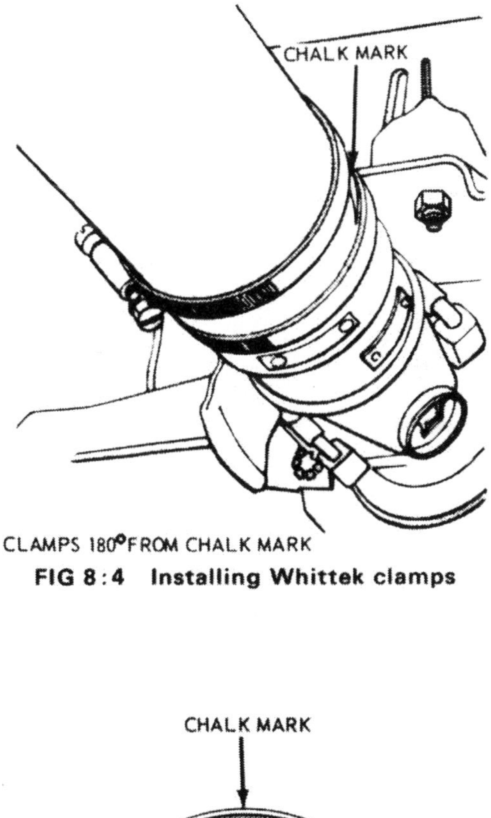

FIG 8:4 Installing Whittek clamps

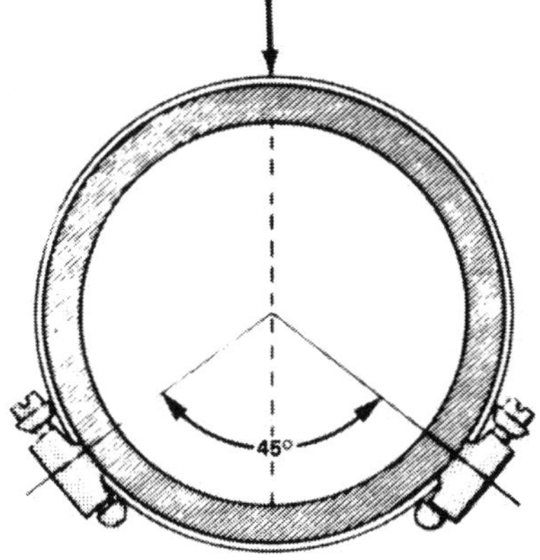

FIG 8:5 Rotating the Whittek clamps to balance the drive shaft

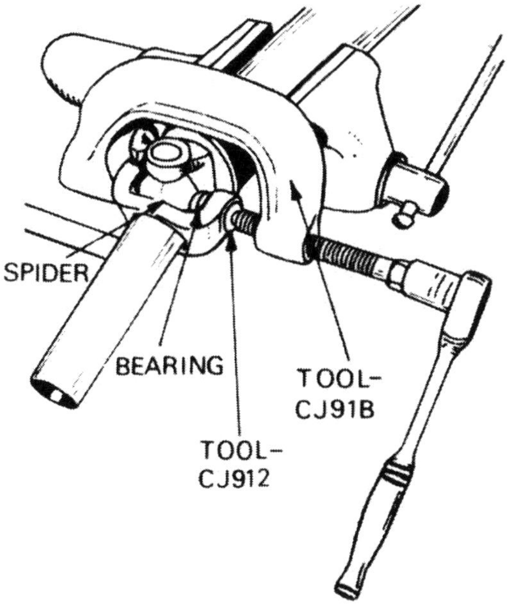

FIG 8:6 Removing a universal joint bearing

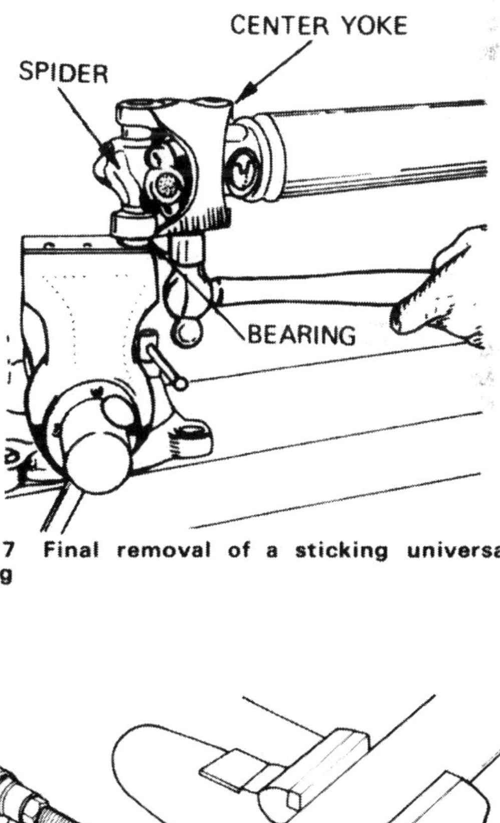

FIG 8:7 Final removal of a sticking universal joint bearing

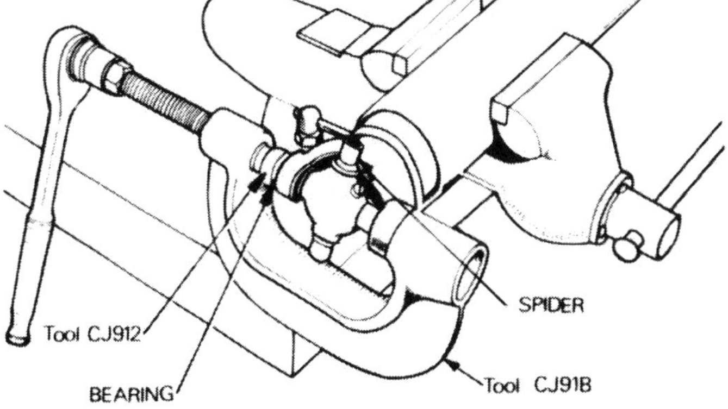

FIG 8:8 Installing a universal joint bearing

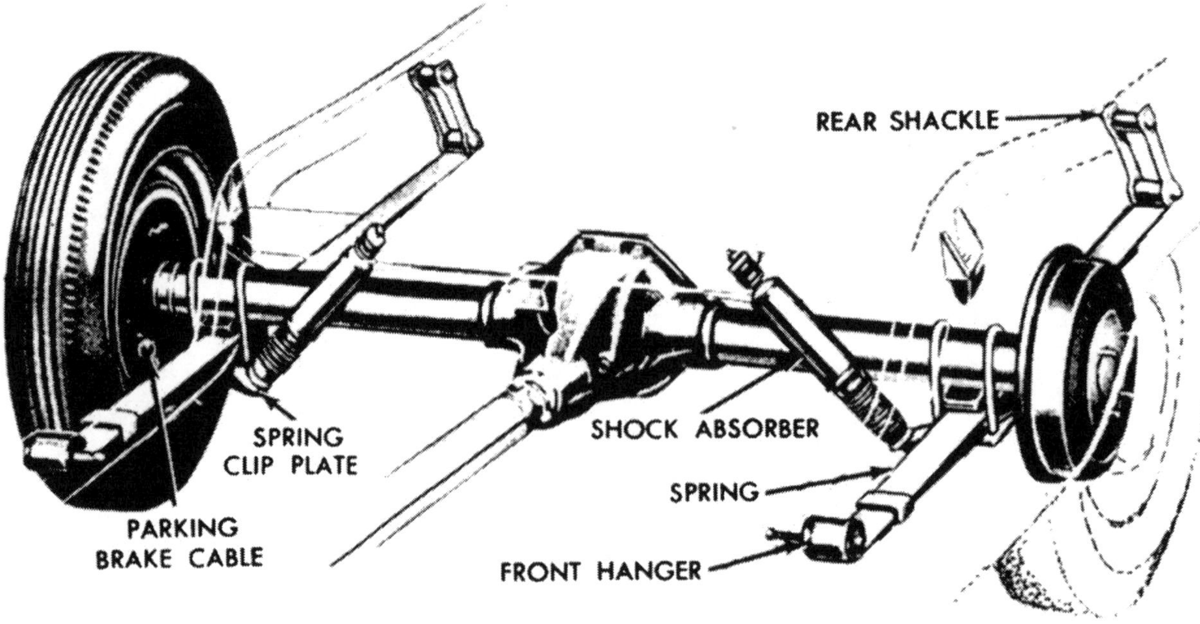

FIG 8:9 A typical rear suspension

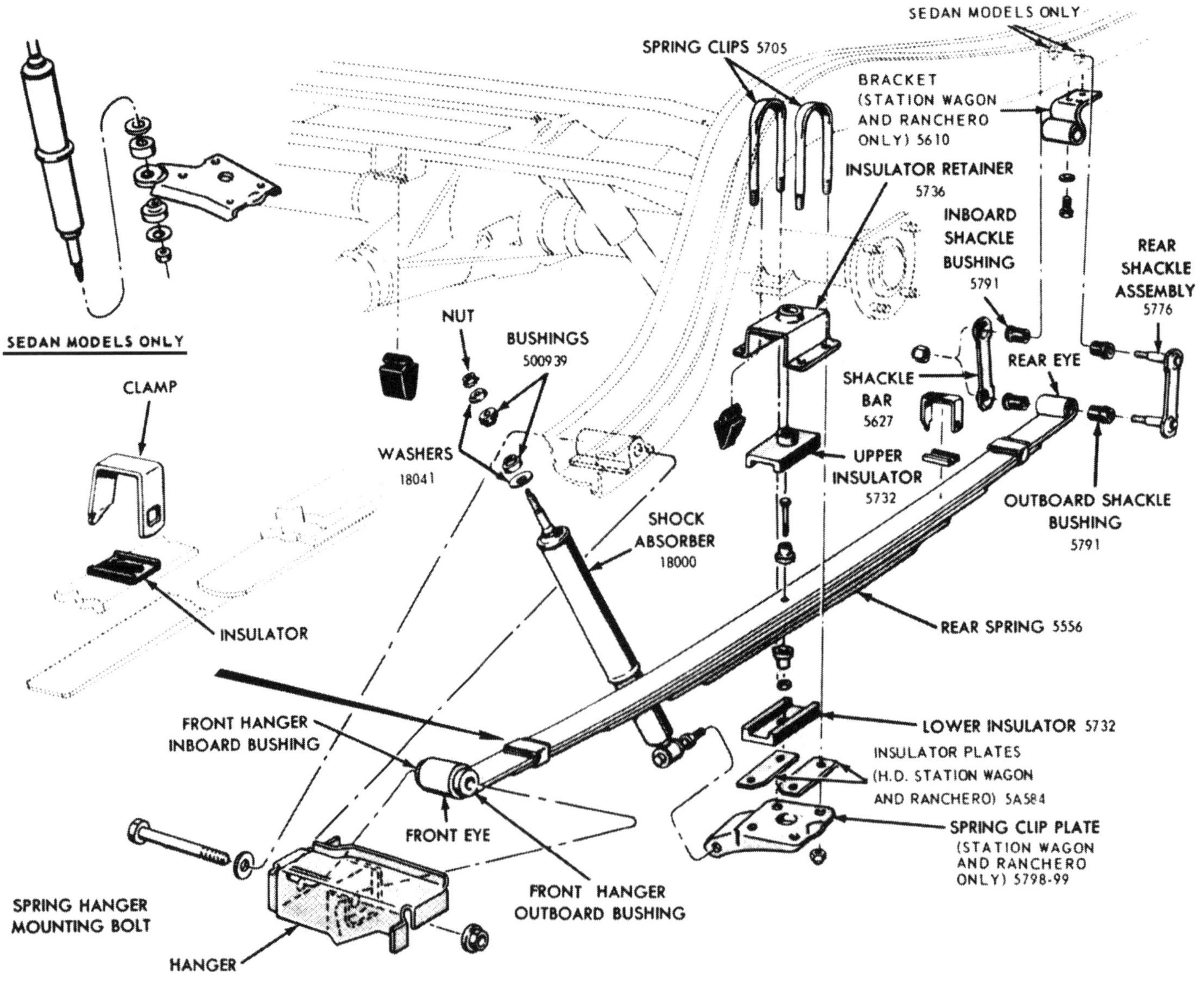

FIG 8:10 Typical rear suspensions components

fitted into the eyes at the end of the springs and insulators are fitted between the axle and spring to ensure that road shocks and noise are not transmitted to the body. The front of the spring is supported on a hanger mounting bolt but the rear end is supported on shackles to allow the spring to flex. The axle is mounted above the spring to lower the height of the suspension and is secured by U-bolts and nuts. Sealed, telescopic shock absorbers are mounted between the frame and a plate under the spring to control the movement of the suspension. The shock absorbers are inclined inwards at the top to provide lateral location for the suspension and rear axle.

Each spring is made up of several leaves. A clamp bolt secures the leaves together at their centres and the ends of the leaves are fitted with insulators between the leaves. Clamps secure the outer ends of the leaves. **The springs should be kept dry and clean and do not require lubricant to be sprayed on them.**

8:3 The shock absorbers

The owner has no means of accurately testing the shock absorbers but they should be suspected to be faulty if the vehicle continues pitching after a disturbance or rolls excessively on curves. The average life of a shock absorber is 36,000 miles so consideration should be given to renewal after this mileage. As a rough test, bounce the corner of the vehicle and if it continues oscillating without rapid damping action then the shock absorber is probably defective.

Once the shock absorber has been removed from the vehicle it can be checked for defects. Install a new damper if the old one has physical damage such as a dented body, bent ram or fluid leaks. Note that it is always best to renew shock absorbers in axle pairs.

Mount the shock absorber vertically with the lower eye in the padded jaws of a vice. Pump the shock absorber, starting off with short strokes about the mid-point and gradually increasing the length of the stroke until it reaches the full travel of the shock absorber. There should be uniform and equal resistance to motion in both directions. If there are pockets of no or weak resistance, particularly when reversing stroke, the shock absorber is defective and must be renewed. Similarly the shock absorber is defective if it operates noisily or leaks. This test will ensure that any air is bled to the top of the shock absorber, if it has been in storage for some time, and it will also detect a defective shock absorber. Even if the unit passes the test this does not guarantee that it is operating at maximum efficiency.

Removal:

The task will be easier if the rear of the car is jacked up and the road wheel removed. Open the luggage compartment lid and remove the spare wheel and tyre. On station wagons, remove the access cover from the opening in the seat riser. Fold back floor matting and pry out the access cover to expose the upper end of the shock absorber, as shown in **FIG 8:11.** Remove the nut, washer and outer bushing that secure the shock absorber to the floor pan. From underneath the car, remove the nuts and bushings that secure the shock absorber to the spring plate clip. Compress the shock absorber and remove it from the car.

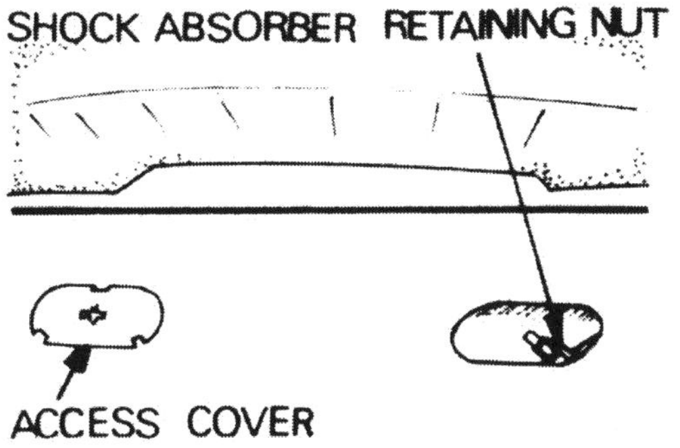

FIG 8:11 Rear shock absorber upper mounting point

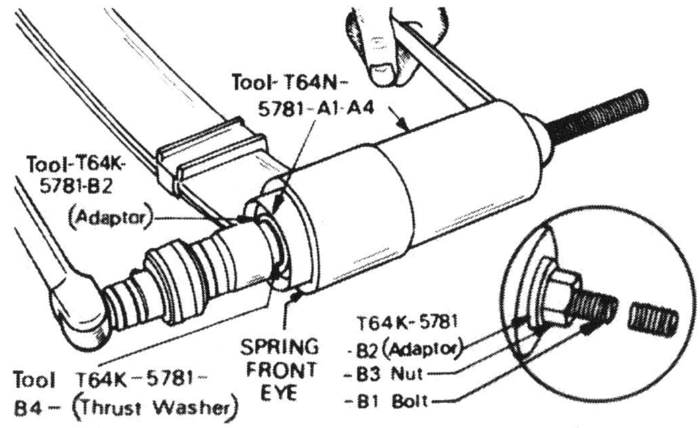

FIG 8:12 Removing the hanger bushing from a spring eye

Install the shock absorber in the reverse order of removal. **Be sure that the spring clip plate is free from damage or burrs.** Check the rubber bushings for wear or damage and renew them if they are defective.

8:4 The springs

Removal:

Raise the rear of the car and support it securely on stands beneath the underbody. It will be necessary to have jacks for supporting the rear axle.

Support the rear axle and disconnect the shock absorbers from the spring clip plates (see previous Section). Push the shock absorbers up and out of the way. Lower the jacks under the axle so that it is no longer raised. Refer to **FIG 8:10.** Take off the nuts and remove the spring clip plate from underneath the spring, collecting the insulator and any plates fitted. Raise the rear axle

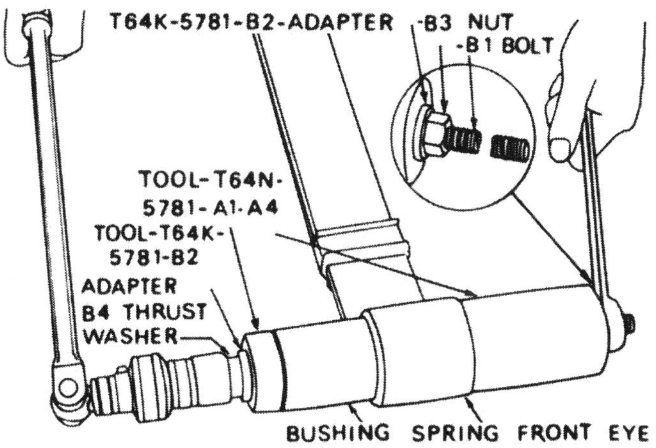

FIG 8:13 Installing the hanger bushing in a spring eye

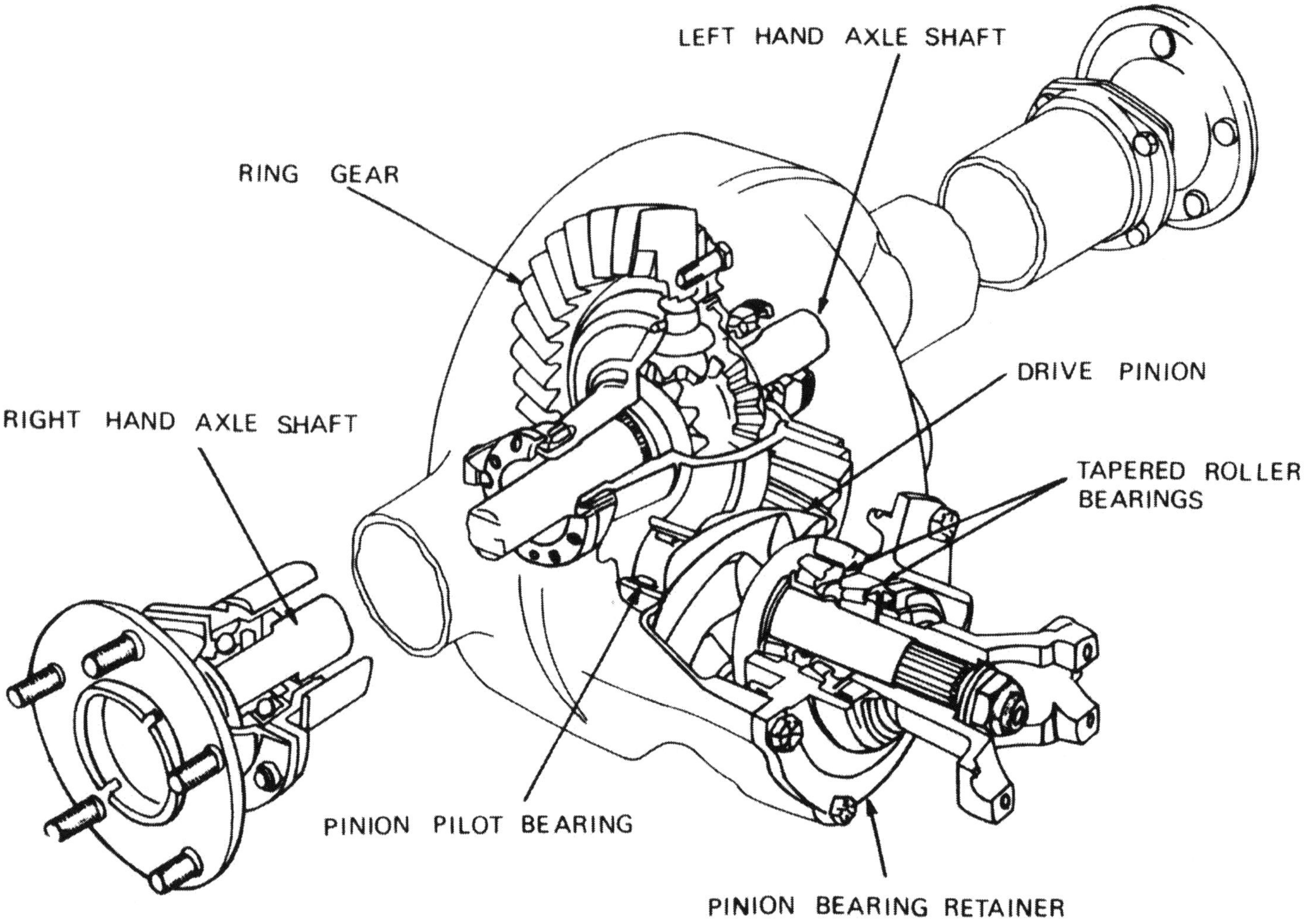

FIG 8:14 Rear axle assembly

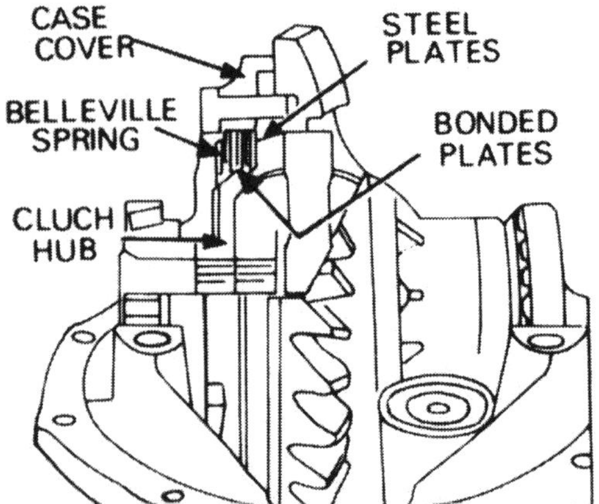

FIG 8:15 Limited-slip differential

on the jack sufficiently to take its weight off the spring. Remove the U-bolts, insulator retainer and upper insulator.

Remove the nuts from the rear shackle bar and take off the shackle bar so that the inboard bushings can be taken out. Withdraw the outer shackle assembly and outboard bushings. Support the spring by hand or on a suitable stand.

Remove the nut and washer from the front end. Withdraw the hanger bolt and remove the spring assembly from the vehicle.

Examination:

Check the leaves for cracks and inspect the anti-squeak inserts between the leaves. The spring can be disassembled by taking off the clamps and removing the center tie bolt. **When disassembling or reassembling the spring, clamp the leaves between the jaws of a strong vice to compress them together.** Slacken the vice off gradually to release the pressure once the center tie bolt has been removed.

Scrub the leaves with a wire brush to remove any dirt and corrosion before reassembling the spring. Use a long piece of rod ground to a taper at one end to align the

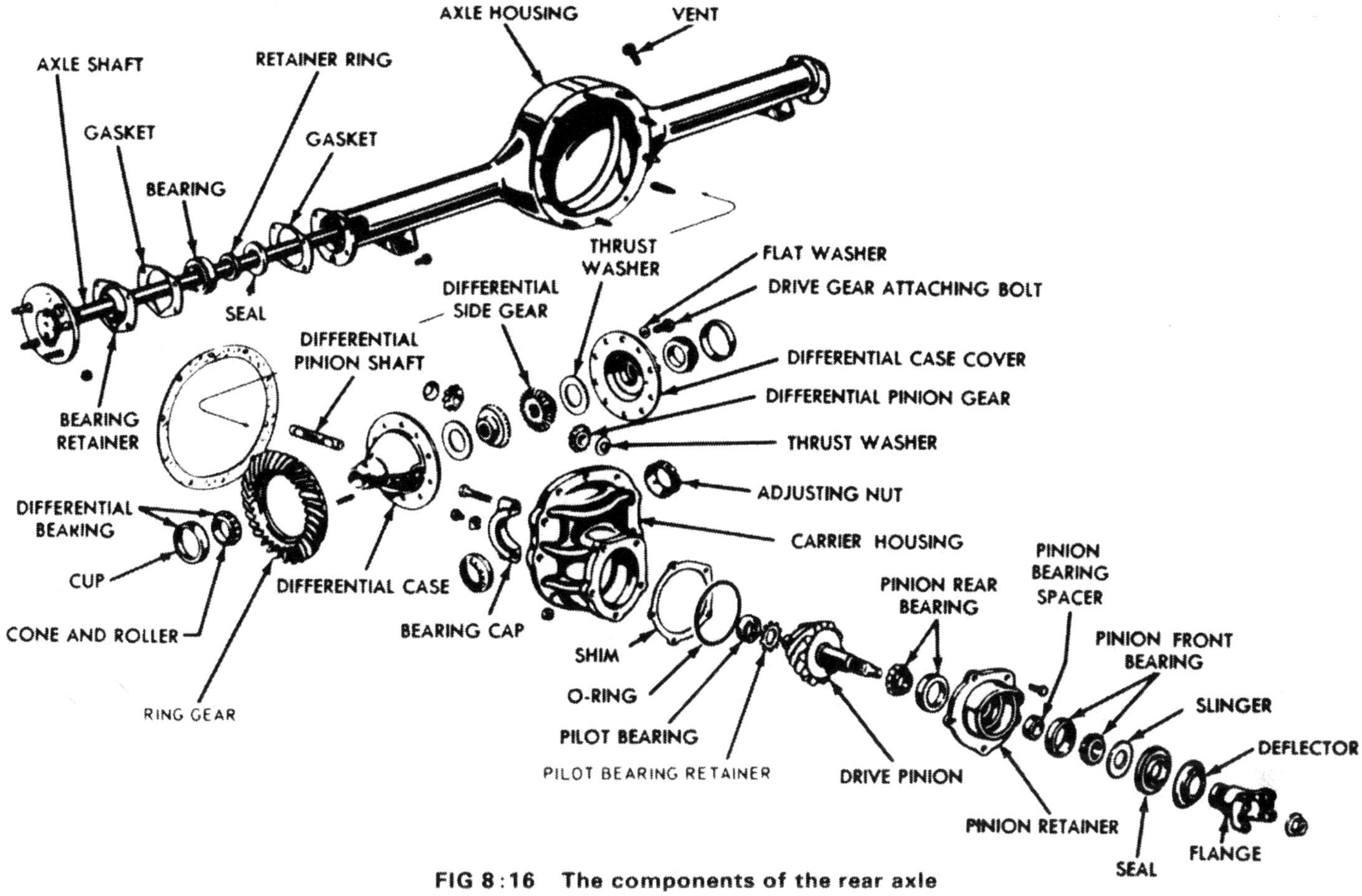

FIG 8:16 The components of the rear axle

center holes in the leaves. Align the ends of the leaves and compress them together in a vice, making sure that the anti-squeak inserts are fitted. Use a new center tie bolt and new end clamps. Be sure that the clamps are fully tightened before installing the spring.

New rear shackle bushings should be fitted if the old ones are worn, distorted or perished. Similarly the front hanger bushing should be renewed if it is defective. The special tools used for removing the old bushing are shown in **FIG 8:12** and the tools being used to install a new bushing are shown in **FIG 8:13**. If care is taken, an arbor press and suitable mandrels can be used instead of the special tools. If lubricant is required for the installation of the bushings, use water or soap and water. **Do not use grease or oil as they will perish the material of the bushings.**

Installation:

The spring is installed in the reverse order of removal. Install it first at the hanger then at the rear shackle, lower the axle back into place (with insulator and retainer in place) and secure it to the spring using the U-bolts, spring clip plate and nuts. Leave the hanger and shackle bolts finger tight. When all the parts are installed, using new insulators if the old ones are defective, place a jack under the spring and raise the spring and axle assembly up to their normal working position, **taking great care not to lift the vehicle off its stands. Torque load the nuts for the shackle and hanger with the spring held in this position as this ensures that the bushings will have the minimum strain put on them when the weight of the vehicle is resting on the suspension.**

Reconnect the shock absorber and install the road wheel. Lower the vehicle back to the ground.

8:5 The rear axle

The rear axle assembly is shown in **FIG 8:14**. A limited-slip differential can be fitted as an optional extra and a semi-sectional view of the unit is shown in **FIG 8:15.** Under cornering forces the clutch slips and allows the differential to act normally. When one wheel loses traction, such as on ice or mud, the clutch transmits torque from the wheel that has no grip to the wheel that has grip, helping to prevent wheel spin and at the same time allowing the wheel with grip to drive the vehicle.

Maintenance:

A drain plug is not fitted as seasonal lubricant changes are not required. The factory fill should remain in the housing for the life of the axle. When major repairs have to be carried out, drain the axle by slackening the rear cover bolts and allowing the lubricant to seep out.

The lubricant level should be checked at intervals of 6000 miles. Clean the area around the combined filler and level plug on the rear cover and remove the plug. Top up as required to bring the level to the bottom of the filler hole. Allow surplus lubricant to drain out before

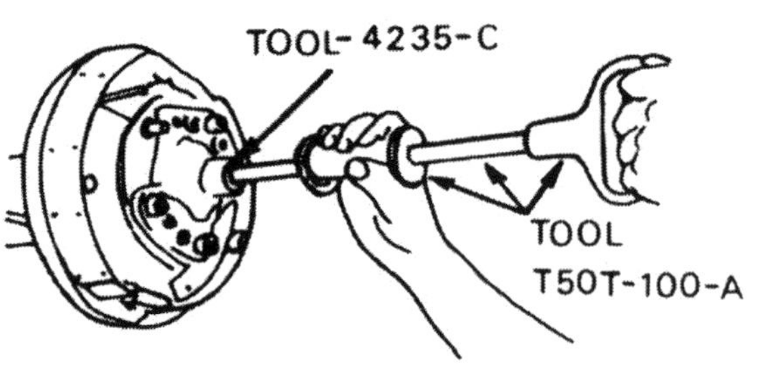

FIG 8:17 Withdrawing the axle shaft and bearing assembly

FIG 8:18 Freeing the bearing retainer ring

installing the filler plug. Normally Hypoid SAE.90 oil should be used (to factory specification) but in climates below —25°F use SAE.80 grade.

When checking the lubricant level, check the vent in the casing to be sure that it is not plugged with road dirt. A plugged vent will cause a rise in internal pressure as the unit warms up with use and lubricant will be forced past oil seals.

Hubs and axle shafts:

The components of the standard rear axle are shown in **FIG 8:16**. Synthetic material oil seals are fitted on production but leather seals only are available as replacement parts. The whole length of the axle shaft passes through the oil seal when removing or installing the shaft and great care must be taken not to damage the seal. If the seals are damaged, they must be renewed otherwise axle lubricant will leak into the brakes. **A new leather seal should be fitted in place of the original seal whenever the axle shaft is removed. Note that if the wheel bearings are removed from the axle shafts it will make the bearings unfit for further use.**

Raise the rear of the vehicle, after slackening the road wheel nuts, and place it securely onto underbody stands. Remove the road wheel. Take off the Tinnerman nuts that secure the brake drum and withdraw the drum from the assembly. Rotate the axle shafts to a convenient position for working and remove the nuts that secure the wheel bearing retainer plate, through the holes provided in the axle shaft flange. Attach the special tools shown in **FIG 8:17** to the axle flange and use the impact hammer to draw the bearing and axle shaft out of the axle. When the axle shaft has been removed, prevent the brake carrier plate from moving by securing it with one nut.

While checking and cleaning the parts, put the new leather oil seal to soak in light SAE.10 oil for at least $\frac{1}{2}$ hour.

If the bearing is worn or runs noisily, remove it from the shaft and fit a new bearing and retaining ring. Support the shaft on the padded jaws of a vice and use a cold chisel to nick it deeply at several points around the circumference, as shown in **FIG 8:18. The aim is to nick deeply but not cut through,** and once this has been done at several points the retaining ring will slide off easily. The methods of removing and installing the bearing on the shaft are shown in **FIG 8:19.** The bearing should be pressed on right up to the shoulder before attempting to press a new retainer ring into place.

Remove the old oil seal from the axle casing tube using the special tool shown in **FIG 8:20.** Install the new seal using the tool as shown in the figure. **It is essential that the correct tool is used for installation of the seal otherwise the seal may leak in service. The seal must have had $\frac{1}{2}$ hour soak in light oil before installation.** Wipe the area where the seal will fit clean of oil and smear the outside of the seal lightly with sealing compound. Do not allow sealing compound onto the lips of the seal.

Check the shaft and axle housing for irregularities, burrs, rough spots or damage. Smooth down rough spots and renew the parts if they are excessively damaged.

Place a new gasket on either side of the brake carrier plate, discarding the old gaskets. **Very gently and carefully slide the axle shaft assembly back into the housing, taking great care that the splines do not damage the new oil seal.** Turn the shaft slightly if required and start the splines into the side gear. Push the shaft in until the bearing bottoms in the housing and install the bearing retainer onto the bolts. Secure the retainer with the nuts. Install the brake drum and road wheel and check the brake adjustment. Lower the vehicle back to the ground and check the lubricant level in the rear axle.

Pinion oil seal:

This can be renewed without having to remove the axle or differential carrier from the vehicle. Disconnect the drive shaft from the drive pinion flange, making sure that the bearings on the shaft are held in place by tape and that alignment marks are made.

1 Raise the rear of the vehicle onto stands and remove both road wheels and brake drums. Attach a in-lb torque spanner to the drive flange and use the torque spanner to rotate the drive pinion for several revolutions. Note the average force required to turn the pinion.

2 Hold the flange with a suitable tool and use a socket wrench to remove the securing nut. Remove the drive pinion flange as shown in **FIG 8:21**, or with a suitable extractor. Before removing the flange, make light punch marks on both flange and pinion shaft so that the flange will be installed into its original position.

3 Remove the old seal using a similar method to the one shown in **FIG 8:20**, noting that the adaptor

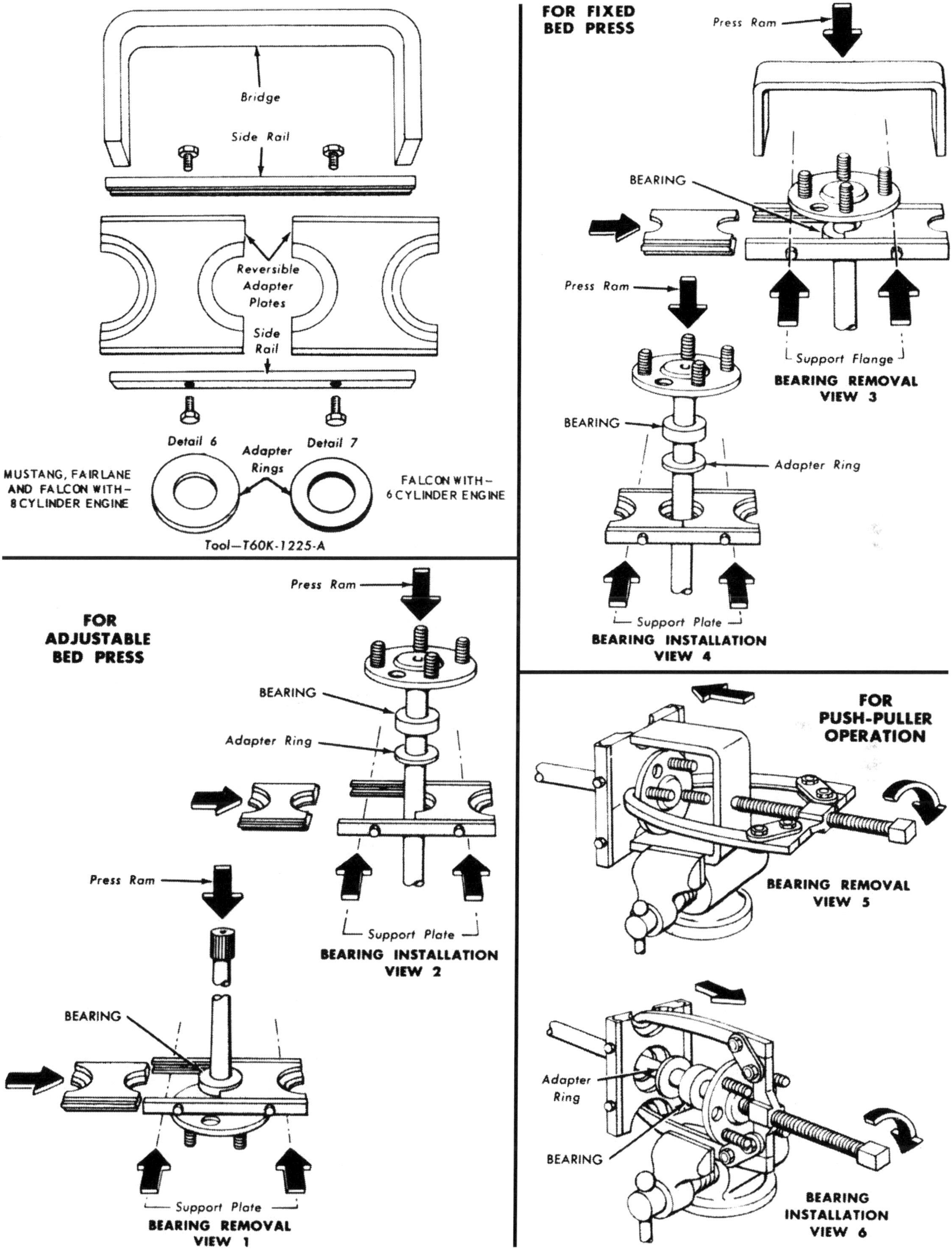

FIG 8:19 Removing and installing the bearing on the axle shaft

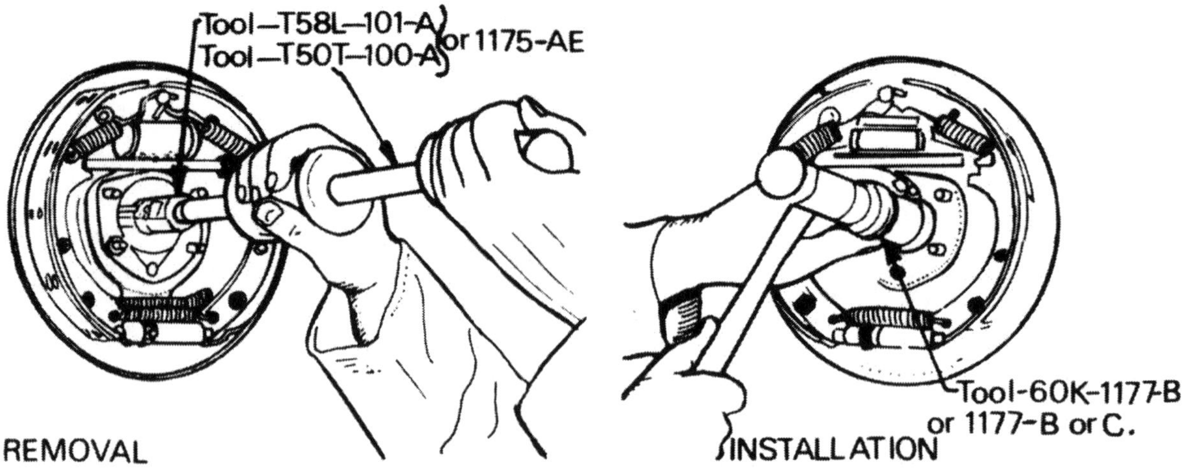

FIG 8:20 Removing and installing the axle shaft seal

1175 AB is the correct one for this seal. Lubricant will flow from the axle as the seal is removed so have a drain pan in position.

4 Check the splines on the pinion shaft for burrs or nicks. Similarly check the surface of the flange for scores or nicks. Light damage can be smoothed off using crocus paper but if the damage is deep new parts will have to be fitted.

Wipe away oil from the seal seating area, and lightly smear the external surface of the seal with sealing compound. Do not allow sealing compound onto the lips of the seal. Install the new seal by driving it in or pulling it in using suitable tools.

5 Install the drive flange and secure it using a new washer and nut. Lightly lubricate the washer side of the nut with a little oil. Tighten the nut while holding the flange and **at frequent intervals check the load required to revolve the pinion shaft,** using a torque spanner. Tighten the nut until the preload is the same as before disassembly, as found in operation 1, and then tighten the nut further until the original preload is increased by 8 to 14 inch lb. **The nut must not be slackened off to reduce the preload and if the nut is tightened too far and the preload becomes excessive the differential will have to be removed and a new spacer fitted between the pinion bearings.**

6 Install the drive shaft, brake drums and road wheels. Check the brake adjustment. Lower the vehicle back to the ground and check the lubricant level in the rear axle.

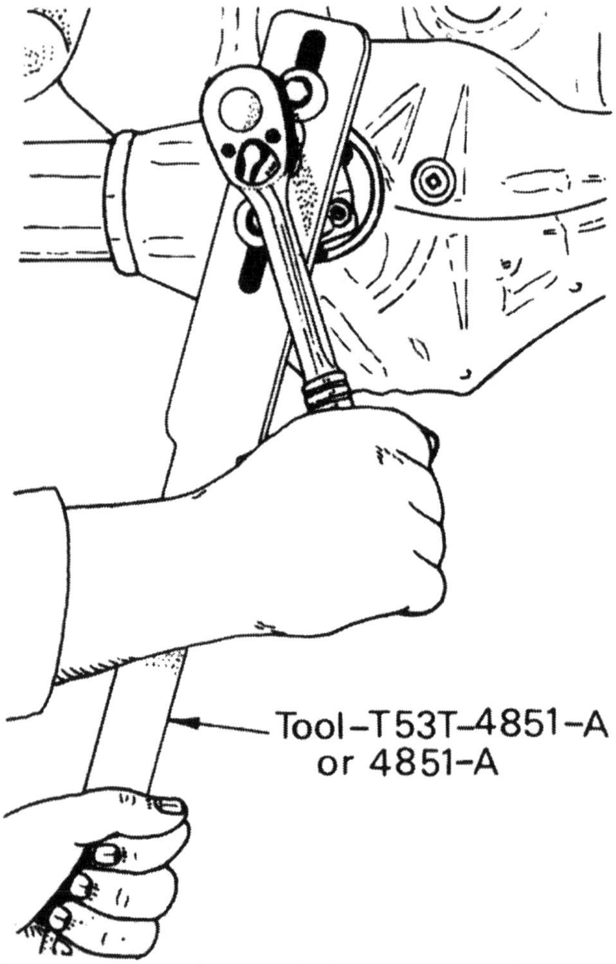

FIG 8:21 Typical method of removing the drive pinion flange

Differential:

This can be removed from the vehicle, after disconnecting the drive shaft and removing both axle shafts, leaving the axle casing fitted to the vehicle. **Work on the differential should always be carried out by specialists as the meshing of the gears and the preloads on the bearings must be set with extreme accuracy.**

Axle removal:

If the complete axle assembly is being removed, the parking brake cables must be disconnected from the brake carrier plates and the main hydraulic line disconnected at the T-piece on the axle. The hydraulic system will then have to be bled after installation of the axle.

The method given here obviates the necessity for disconnecting the braking system parts.

1 Raise the rear of the vehicle and support it on stands. Remove both road wheels, brake drums and then withdraw the axle shaft assemblies. Suspend the brake carrier plates, still attached to the brake lines and parking brake cables, safely out of the way.

2 Drain the unit by removing the rear cover. Disconnect the vent pipe from the axle casing. Free the hydraulic line and T-piece from the axle casing. Disconnect the shock absorber lower attachments from the spring clip plates.

3 Make alignment marks and disconnect the drive shaft from the drive pinion flange, carefully taping the loose bearings in place on the drive shaft.

4 Remove the nuts that secure the spring clip plates on the suspension. Remove the U-bolts, insulators and insulator retainers. Manoeuvre the axle assembly out from over the road springs.

The unit is installed in the reverse order of removal. The points to watch for when reconnecting and installing the parts have been dealt with in the relevant sections.

8:6 Trouble shooting guide

(a) Noisy suspension or axle

1 Lack of lubricant or incorrect lubricant used*
2 Defective shock absorbers
3 Defective shock absorber mounting bushings
4 Defective axle mounting insulators
5 Defective spring mounting bushings
6 Worn drive shaft universal joints
7 Defective differential unit
8 Worn bearings
9 Broken spring leaves
10 Loose spring leaf clamps

*On models fitted with limited slip differential a high pitched chattering noise on turns can be caused by lack of or incorrect lubricant. Try driving the car in five fairly tight circles clockwise and then five more circles counterclockwise. This will permit the lubricant to work into the clutch plates. Check the lubricant by draining out two tablespoonfuls of lubricant and mix it with an equal amount of white alcohol. Let the mixture stand for at least two minutes. If the sample is blue-tinted a Ford approved lubricant is used. If the sample is yellow-tinted an incorrect lubricant is being used. Drain out incorrect lubricant and refill with Ford approved lubricant. There is no need to flush the axle but the vehicle should be driven in five tight circles in both directions to distribute the lubricant onto the clutch plates.

(b) Vibration

1 Check 5 and 6 in (a)
2 Drive shaft out of balance (lost weights or dirt)
3 Drive shaft fitted 180 deg. out

(c) Lubricant leaks

1 Defective hub oil seals
2 Defective pinion oil seal
3 Defective seals on universal joint spiders
4 Plugged breather on axle
5 Too high lubricant level
6 Defective gaskets or damaged joint faces

Inches	Decimals	Milli-metres	Inches to Millimetres: Inches	Inches to Millimetres: mm	Millimetres to Inches: mm	Millimetres to Inches: Inches
1/64	.015625	.3969	.001	.0254	.01	.00039
1/32	.03125	.7937	.002	.0508	.02	.00079
3/64	.046875	1.1906	.003	.0762	.03	.00118
1/16	.0625	1.5875	.004	.1016	.04	.00157
5/64	.078125	1.9844	.005	.1270	.05	.00197
3/32	.09375	2.3812	.006	.1524	.06	.00236
7/64	.109375	2.7781	.007	.1778	.07	.00276
1/8	.125	3.1750	.008	.2032	.08	.00315
9/64	.140625	3.5719	.009	.2286	.09	.00354
5/32	.15625	3.9687	.01	.254	.1	.00394
11/64	.171875	4.3656	.02	.508	.2	.00787
3/16	.1875	4.7625	.03	.762	.3	.01181
13/64	.203125	5·1594	.04	1.016	.4	.01575
7/32	.21875	5.5562	.05	1.270	.5	.01969
15/64	.234375	5.9531	.06	1.524	.6	.02362
1/4	.25	6.3500	.07	1.778	.7	.02756
17/64	.265625	6.7469	.08	2.032	.8	.03150
9/32	.28125	7.1437	.09	2.286	.9	.03543
19/64	.296875	7.5406	.1	2.54	1	.03937
5/16	.3125	7.9375	.2	5.08	2	.07874
21/64	.328125	8.3344	.3	7.62	3	.11811
11/32	.34375	8.7312	.4	10.16	4	.15748
23/64	.359375	9.1281	.5	12.70	5	.19685
3/8	.375	9.5250	.6	15.24	6	.23622
25/64	.390625	9.9219	.7	17.78	7	.27559
13/32	.40625	10.3187	.8	20.32	8	.31496
27/64	.421875	10.7156	.9	22.86	9	.35433
7/16	.4375	11.1125	1	25.4	10	.39370
29/64	.453125	11.5094	2	50.8	11	.43307
15/32	.46875	11.9062	3	76.2	12	.47244
31/64	.484375	12.3031	4	101.6	13	.51181
1/2	.5	12.7000	5	127.0	14	.55118
33/64	.515625	13.0969	6	152.4	15	.59055
17/32	.53125	13.4937	7	177.8	16	.62992
35/64	.546875	13.8906	8	203.2	17	.66929
9/16	.5625	14.2875	9	228.6	18	.70866
37/64	.578125	14.6844	10	254.0	19	.74803
19/32	.59375	15.0812	11	279.4	20	.78740
39/64	.609375	15.4781	12	304.8	21	.82677
5/8	.625	15.8750	13	330.2	22	.86614
41/64	.640625	16.2719	14	355.6	23	.90551
21/32	.65625	16.6687	15	381.0	24	.94488
43/64	.671875	17.0656	16	406.4	25	.98425
11/16	.6875	17.4625	17	431.8	26	1.02362
45/64	.703125	17.8594	18	457.2	27	1.06299
23/32	.71875	18.2562	19	482.6	28	1.10236
47/64	.734375	18.6531	20	508.0	29	1.14173
3/4	.75	19.0500	21	533.4	30	1.18110
49/64	.765625	19.4469	22	558.8	31	1.22047
25/32	.78125	19.8437	23	584.2	32	1.25984
51/64	.796875	20.2406	24	609.6	33	1.29921
13/16	.8125	20.6375	25	635.0	34	1.33858
53/64	.828125	21.0344	26	660.4	35	1.37795
27/32	.84375	21.4312	27	685.8	36	1.41732
55/64	.859375	21.8281	28	711.2	37	1.4567
7/8	.875	22.2250	29	736.6	38	1.4961
57/64	.890625	22.6219	30	762.0	39	1.5354
29/32	.90625	23.0187	31	787.4	40	1.5748
59/64	.921875	23.4156	32	812.8	41	1.6142
15/16	.9375	23.8125	33	838.2	42	1.6535
61/64	.953125	24.2094	34	863.6	43	1.6929
31/32	.96875	24.6062	35	889.0	44	1.7323
63/64	.984375	25.0031	36	914.4	45	1.7717

UNITS	Pints to Litres	Gallons to Litres	Litres to Pints	Litres to Gallons	Miles to Kilometres	Kilometres to Miles	Lbs. per sq. In. to Kg. per sq. Cm.	Kg. per sq. Cm. to Lbs. per sq. In.
1	.57	4.55	1.76	.22	1.61	.62	.07	14.22
2	1.14	9.09	3.52	.44	3.22	1.24	.14	28.50
3	1.70	13.64	5.28	.66	4.83	1.86	.21	42.67
4	2.27	18.18	7.04	.88	6.44	2.49	.28	56.89
5	2.84	22.73	8.80	1.10	8.05	3.11	.35	71.12
6	3.41	27.28	10.56	1.32	9.66	3.73	.42	85.34
7	3.98	31.82	12.32	1.54	11.27	4.35	.49	99.56
8	4.55	36.37	14.08	1.76	12.88	4.97	.56	113.79
9		40.91	15.84	1.98	14.48	5.59	.63	128.00
10		45.46	17.60	2.20	16.09	6.21	.70	142.23
20				4.40	32.19	12.43	1.41	284.47
30				6.60	48.28	18.64	2.11	426.70
40				8.80	64.37	24.85		
50					80.47	31.07		
60					96.56	37.28		
70					112.65	43.50		
80					128.75	49.71		
90					144.84	55.92		
100					160.93	62.14		

UNITS	Lb ft to kgm	Kgm to lb ft	UNITS	Lb ft to kgm	Kgm to lb ft
1	.138	7.233	7	.967	50.631
2	.276	14.466	8	1.106	57.864
3	.414	21.699	9	1.244	65.097
4	.553	28.932	10	1.382	72.330
5	.691	36.165	20	2.765	144.660
6	.829	43.398	30	4.147	216.990

CHAPTER 9

THE FRONT SUSPENSION AND HUBS

9:1 Description

A typical front suspension assembly is shown in **FIG 9:1**. The 1965 models differ in the inboard attachments of the lower and upper arms and this affects the method of setting the suspension geometry.

The upper and lower arms are attached by pivot bolts to the frame and can therefore move vertically. The outboard end of the lower arm is located by a strut to prevent it from moving in the fore and aft direction. A spindle and stub axle assembly swivels in the outboard ends of the suspension arms in ball joints. The ball joints allow the suspension to move, as well as allowing the spindle to rotate about its vertical axis for the steering movement of the front wheels.

A coil spring and concentric shock absorber are installed between the upper suspension arm and a tower on the frame. The spring takes the weight of the vehicle while the shock absorber controls the movement of the suspension and damps out oscillations.

The road wheel is attached to a hub assembly which rotates about the stub axle on two opposed taper bearings.

A stabilizer bar interconnects the two lower suspension arms. When one suspension is at a different height from the other, the stabilizer bar is twisted and exerts a restoring force between the two suspensions to bring them back to the same height.

9:2 Maintenance

The suspension ball joints should be greased at intervals of 12,000 miles. Remove the plug from the ball joint and install a grease fitting. Inject grease through the fitting until the dust cover just starts to swell. Do not inject so much grease that it exudes past the dust cover, otherwise the weatherseal will be damaged. Check that the dust covers are in good condition and not starting to crack or split. If a dust cover has split, it must be renewed to prevent dirt from entering the ball joint. If dirt has enter the ball joint, try to remove it by pumping through plenty of fresh grease with the dust cover off. If the dirt cannot be flushed out, fit a new ball joint.

When the ball joints have been lubricated, remove the grease fittings and install the plugs.

At approximately 30,000 miles the wheel hub should be removed, old grease washed out and the hub installed packed with fresh grease after checking the condition of the wheel bearings. This operation is dealt with in the next section.

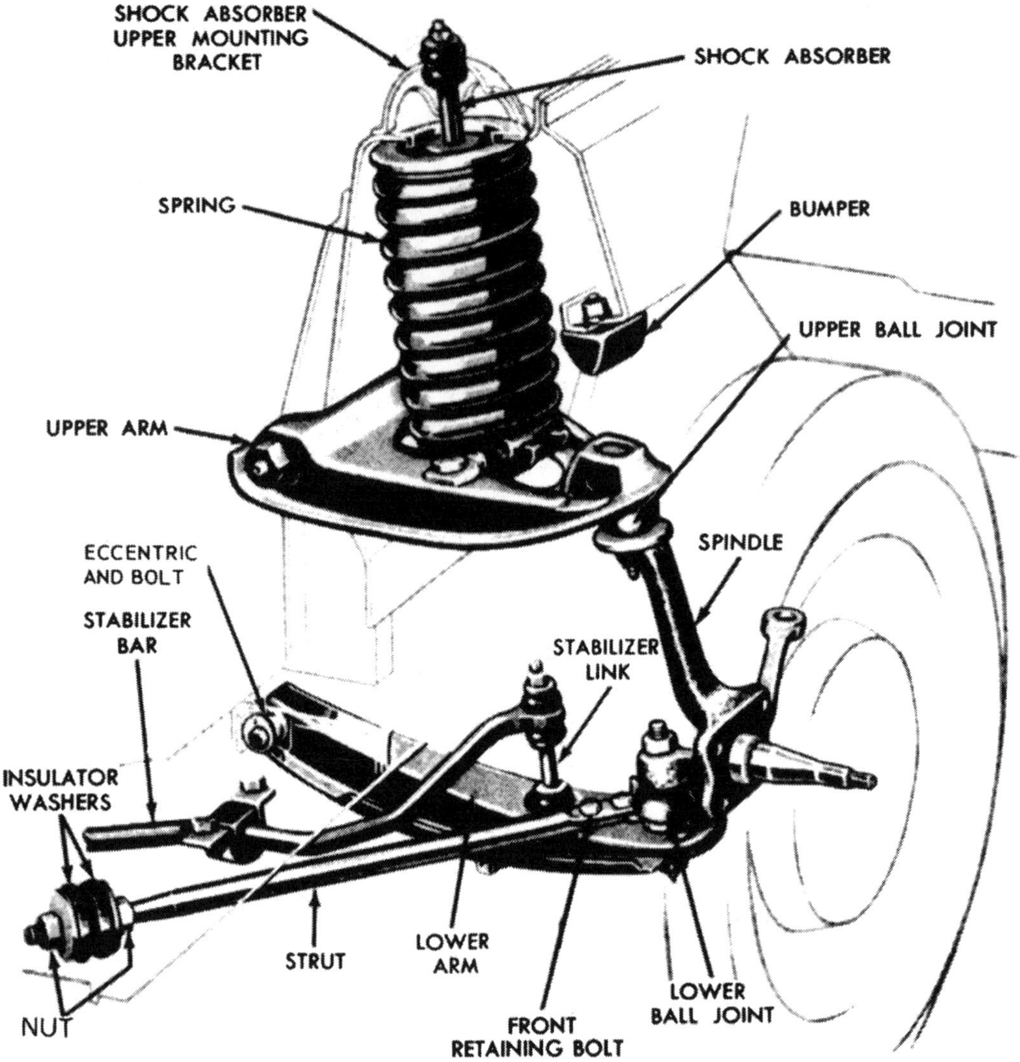

FIG 9:1 Typical front suspension assembly

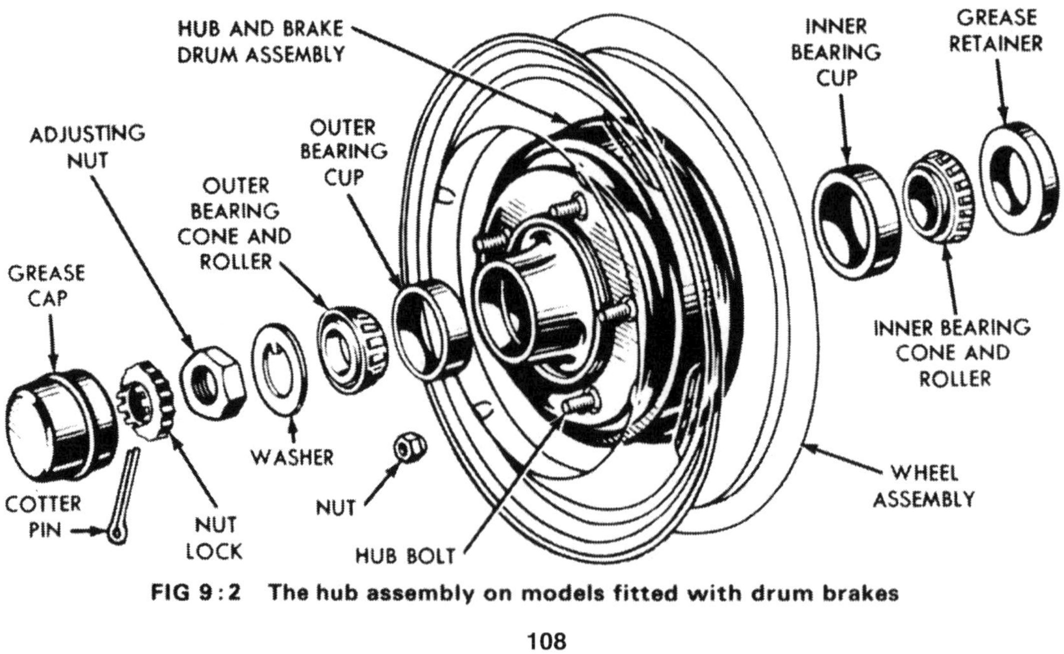

FIG 9:2 The hub assembly on models fitted with drum brakes

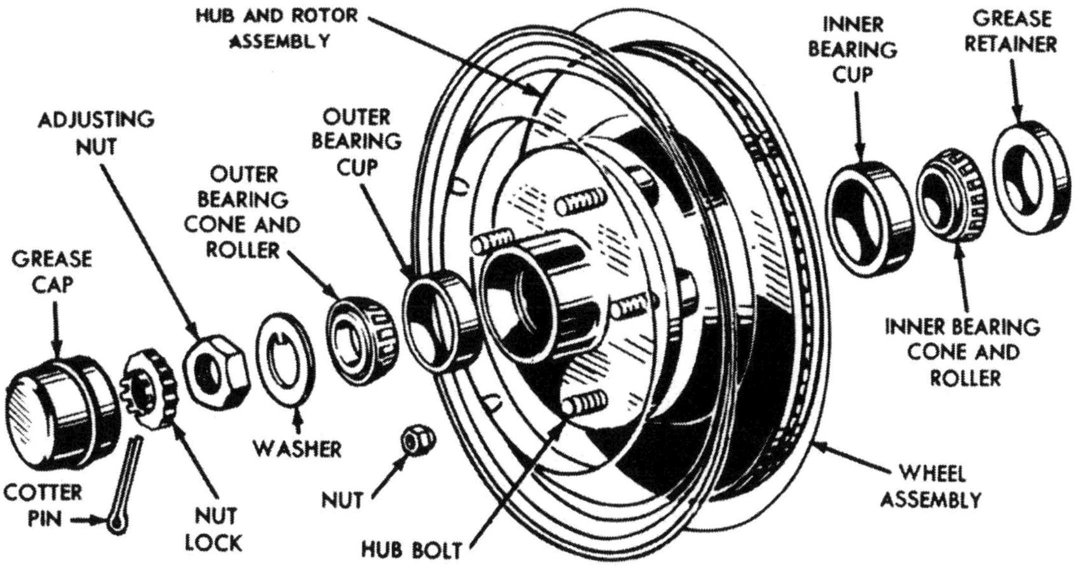

FIG 9:3 The hub assembly on models fitted with disc brakes

9:3 Wheel hubs

The components of a wheel hub fitted with drum brakes are shown in **FIG 9:2** and those fitted with disc brakes in **FIG 9:3.** Whenever the front of the vehicle is raised off the ground, briefly check the wheel bearings. Grip the tire at the nine and three o'clock positions and rock it as though trying to make it steer. If there is excessive play then the wheel bearings are likely to need adjustment. Do not confuse steering movement with play. Spin the wheel and check that it rotates freely without grinding noises. Do not confuse brake noise with that from a defective bearing. **If the bearings are noisy, the wheel hub should be removed so that they can be visually examined.** Packing in more grease on the outer bearing and tightening up the adjustment alone can be dangerous, as the inner bearing may be on the point of failure.

Removal:

Raise the front of the car and place it securely onto stands. On models fitted with drum brakes, the hub assembly can be removed with the road wheel and tire but on models fitted with disc brakes the road wheel must be taken off in order to remove the brake caliper. Slacken the road wheel attachment nuts slightly before raising the car, otherwise difficulty will be found in slackening them once the wheel can spin freely.

Remove the grease cap from the hub and extract the cotterpin that secures the nut lock. On models fitted with disc brakes, slacken back the nut by three turns and rock the road wheel to press the friction pads and pistons back into the brake caliper. On models fitted with drum brakes, slacken the brake adjustors if necessary. Remove the road wheel. This is advisable on all models, though not necessary on models with drum brakes, as it will be found easier to handle the parts and to clean them without getting solvent or grease onto the tire.

On models fitted with disc brakes, remove the caliper from the suspension and hang it safely out of the way so that the flexible hose is not strained (see **Chapter 11**).

Remove the locknut, nut, and special washer. Pull the hub off from the spindle, carefully collecting the inner race of the outer bearing as it comes free.

Pry out the grease retainer so that the inner race assembly of the inner bearing can be removed from the hub. The bearing cups can be left installed in the hub, unless the bearings are defective. If the bearing cups must be removed, drive them out evenly and squarely with a suitable drift.

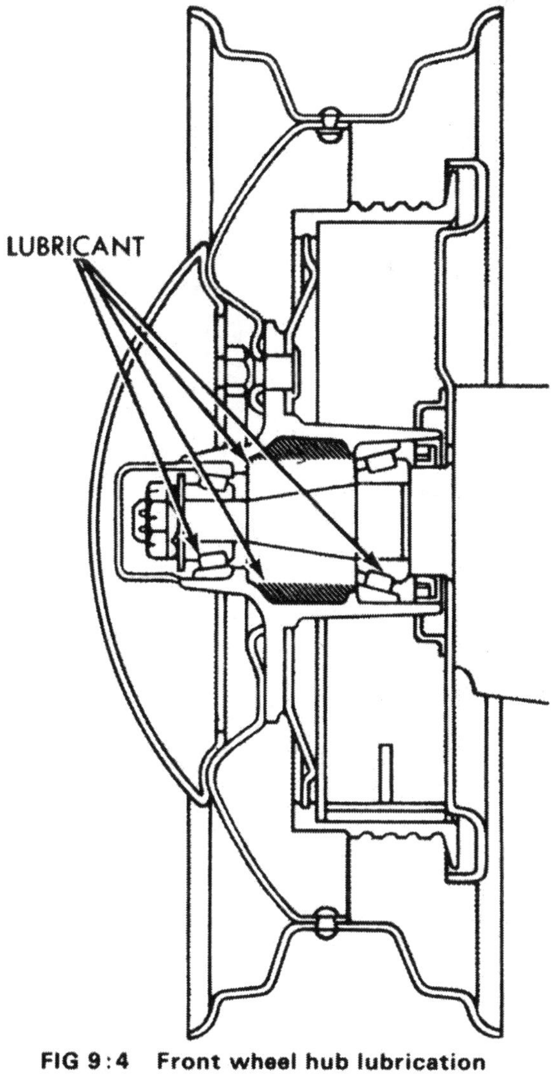

FIG 9:4 Front wheel hub lubrication

FIG 9:5 Front wheel bearing adjustment

Cleaning and inspection:

Wipe out the surplus old grease using rags. Wash the hub assembly in a suitable solvent (such as gasoline or kerosene) to remove the remainder of the old grease. Rotate the inner races of the bearings under clean solvent until all old lubricant and dirt have been washed out. Dry the parts with an air-line. **When drying the bearings, do not allow them to rotate under the air pressure or they will become chipped and damaged.**

Brush away all dust and dirt from the brake carrier plate, or disc brake dust shield. Wipe down the spindle with a piece of cloth wetted with solvent.

Lubricate the bearings with thin oil and insert each one in turn into its bearing cup. Press the race firmly into the cup and while applying pressure oscillate the bearing backwards and forwards. If the bearing feels at all rough clean it again, as specks of dirt may be the cause. If cleaning cannot cure the roughness, drive out both bearing cups and drive new ones back into place, using the new inner races on installation of the hub. Similarly both bearings complete must be renewed if either one of them shows any signs of wear, pitting, cracking or fretting.

Check the spindle for cracks or corrosion. Give the spindle a wipe over with a clean oily or greasy cloth to prevent corrosion.

Installation and lubrication:

The points to apply grease are shown in **FIG 9:4**. Pack the inside of the hub to the depth shown with fresh grease. Work ample grease into the inner races of both bearings and apply a light smear of grease to the bearing cup. Install the inner race of the inner bearing and hold it in place by installing a new grease retainer. The new grease retainer should have been soaked in light oil for 30 minutes before installation.

Before installing the hub assembly, use a harsh solvent (such as carbon tetrachloride) to clean the operating face of the disc or drum free from all lubricant or finger marks.

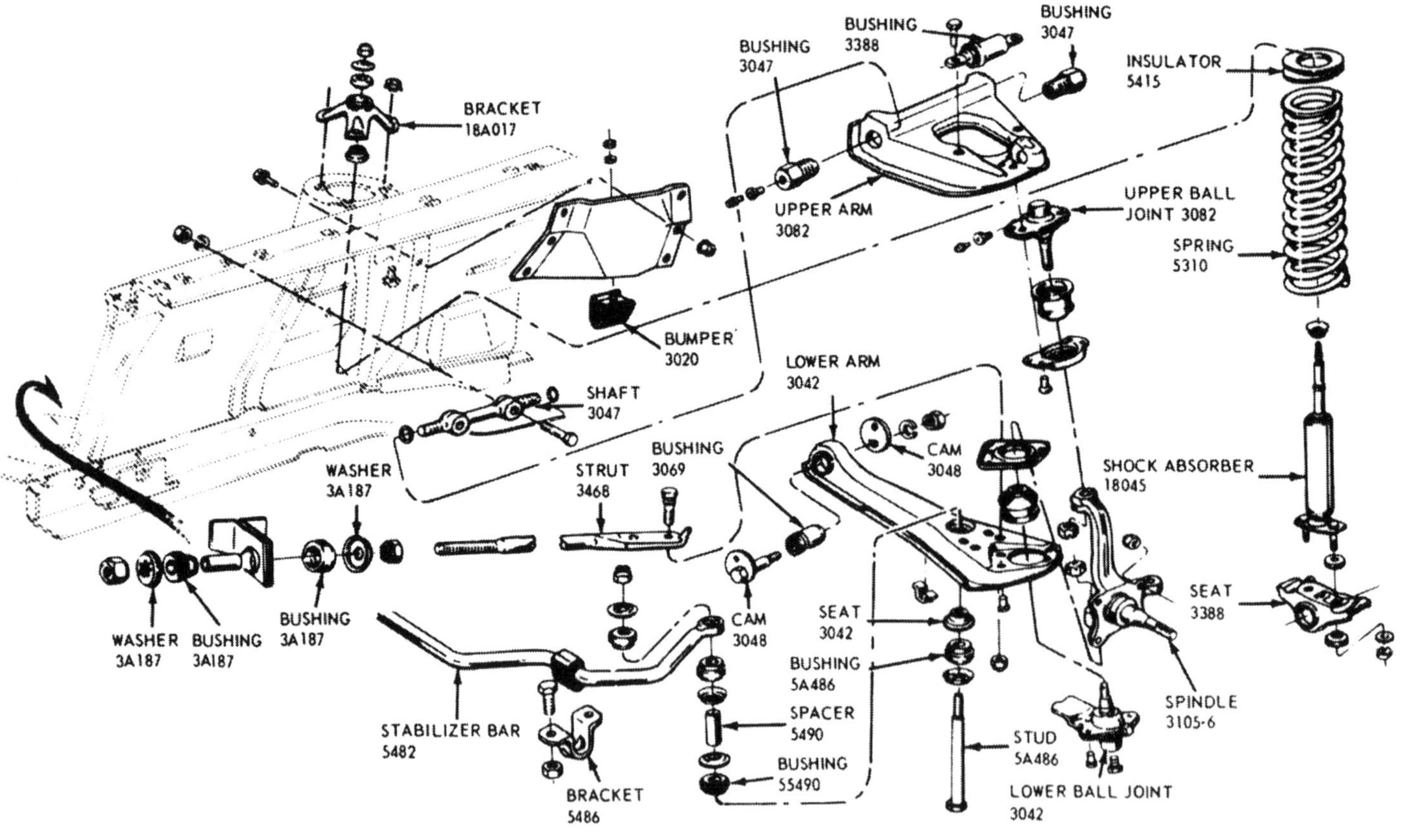

FIG 9:6 Typical suspension components

Slide the hub assembly back onto the spindle, taking great care to keep it central and avoid damaging the grease retainer.

Press the inner race of the outer bearing back into position. Install the special washer and nut followed by the locknut left loosely in place. Once the adjustment of the bearings has been completed, install the brake caliper, hub grease cap and road wheel. Check that the road wheel spins freely before lowering the vehicle back to the ground. **On models fitted with disc brakes, pump the brake pedal hard several times to take up the adjustment in the brakes before using the vehicle.**

Adjustment:

The method of setting the correct adjustment is shown in **FIG 9:5.** Spin the hub and tighten the nut to a torque of 17 to 25 lb ft. If a torque spanner is not available, tighten the nut until there is a definite but slight resistance to spinning the hub (showing that the bearings are under slight preload and have been pulled fully back into place) and slacken the nut again. Tighten the nut using moderate pressure until all end play is taken up. Install the locknut so that the castellations are aligned with a cotterpin hole. Back of the nut and locknut together by one castellation and then lock them using a new cotterpin. Note that the ends of the cotterpin must be opened circumferentially as shown otherwise the ends may damage the static pick-up point in the grease cap.

9:4 The shock absorbers

Testing of the shock absorbers for defects is the same as for the shock absorbers on the rear suspension (see **Chapter 8, Section 8:3**).

The components of the later suspension are shown in **FIG 9:6.** On some of the earlier as well as some of the the latest, the shock absorber may have the eye at the bottom and be secured to the upper attachment bracket by nuts and bushing. A typical shock absorber and its attachments are shown in **FIG 9:7.**

Removal:

Raise the front of the car onto stands and support the suspension with a jack under the lower arm. Keep the suspension low enough in relation to the vehicle to be able to work on the lower attachments.

Remove the nuts that secure the lower attachment and remove the nuts that secure the upper mounting bracket to the frame. Lift the shock absorber and its bracket out of the vehicle and separate the shock absorber from the bracket.

Install the shock absorber in the reverse order of removal, making sure that all bushings are in good condition.

9:5 The stabilizer bar

The attachments of the stabilizer bar are shown in **FIG 9:6** and the figure is typical for all model stabilizer bars.

Raise the front of the vehicle and place it securely on stands. Access will be easier if the road wheels are removed. Support each suspension with a jack or small stand under the lower arm, ensuring that both suspensions are set to the same height.

Take off the nuts and withdraw the stabilizer link bolts from each lower arm, carefully collecting the spacer, bushings and washers as they come free.

Remove the bolts that secure the stabilizer brackets to the frame and withdraw the stabilizer from under the vehicle.

Before installing the stabilizer bar, check all the bushings and insulators for wear, perishing or deformation. Renew all defective bushings.

Lubricate the bushings with RUGLYDE or comparable lubricant and install the parts in the reverse order of disassembly. Tighten the bolts and nuts to the correct torque loads before fitting the road wheels and lowering the vehicle back to the ground.

FIG 9:7 Typical front shock absorber and its attachments

9:6 Suspension ball joints

Typical suspension components are shown in **FIG 9:6.**

Upper ball joint inspection:

Raise the front of the vehicle and support it on underbody stands so that both suspensions hang freely at the end of their range of movement. This ensures that the upper ball joints are unloaded. Be sure that the wheel bearings are correctly adjusted (see **Section 9:3**).

Mount a dial indicator on the upper arm and position it so that its plunger rests vertically on the underside of the spindle at the upper ball joint. Lift and lower the

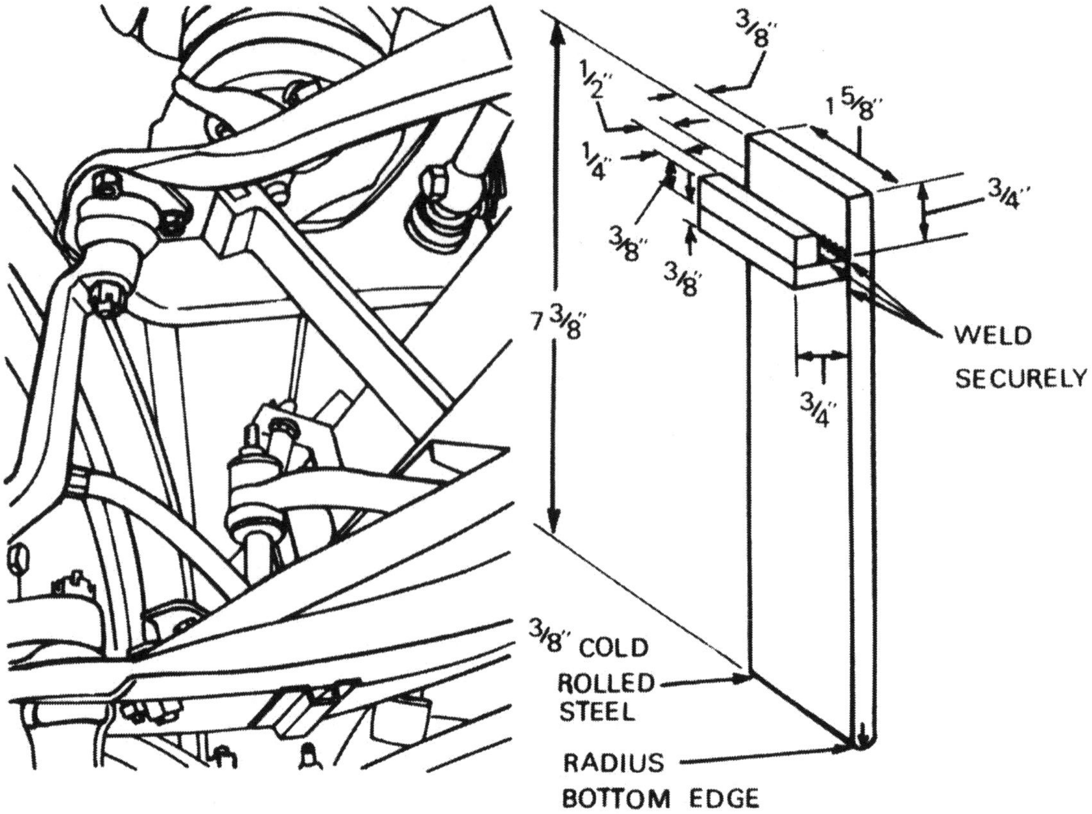

FIG 9:8 Upper arm support for 1965 models

spindle and suspension assembly, noting the play on the dial indicator. The play should not exceed .200 inch. Mount the dial indicator on the upper arm so that the plunger now rests vertically against the inside wheel rim. Grasp the top of the tire and rock it in and out so that the radial play on the ball joint can be measured on the dial indicator. The radial play must not exceed .250 inch. If the play is excessive in either direction, a new ball joint must be fitted.

Lower ball joint inspection:

With the front of the vehicle raised off the ground on underbody stands, have an assistant rock the lower edge of the tire in and out. Any movement between the lower end of the spindle and the lower arm indicates that the ball joint is worn and the lower arm must be renewed. While carrying out this check there will be movement on the upper ball joint, but this movement may be safely ignored—unless it is found to exceed the limits given.

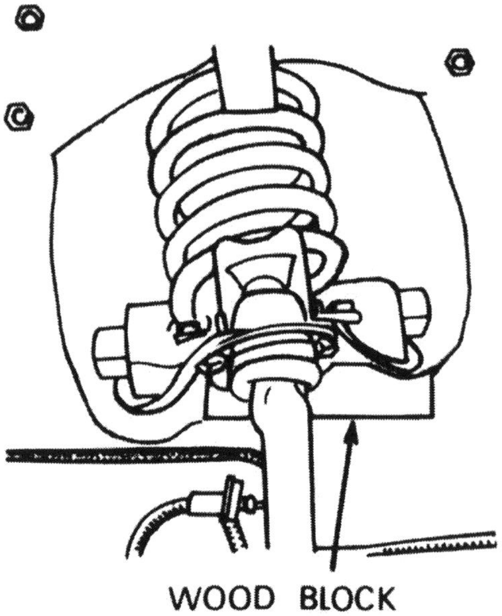

FIG 9:9 Upper arm support for all later models

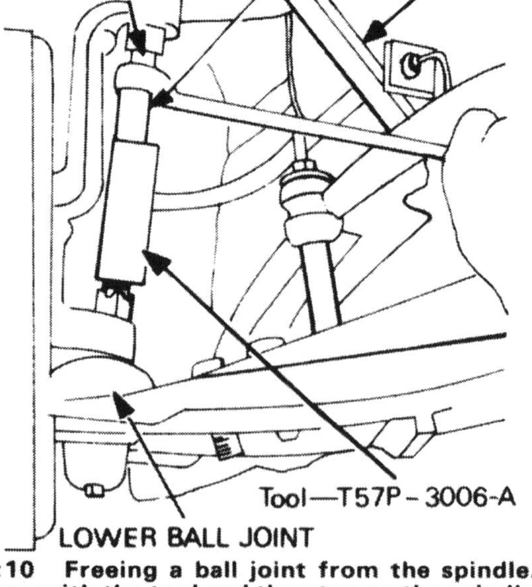

FIG 9:10 Freeing a ball joint from the spindle. Apply pressure with the tool and then tap on the spindle

Renewing upper ball joint:

On the 1964 and 1965 models, have a support made to the dimensions shown in **FIG 9:8,** and mount it as shown in the figure. On later models a block of wood fitted between upper arm and frame, as shown in **FIG 9:9,** will take the pressure of the road spring. Raise the front of the vehicle and **make sure that the support fits securely and safely into place.** Stand the vehicle on underbody supports.

Remove the cotterpin from the ball joint nut and slacken the nut by several turns. Install the tool as shown in **FIG 9:10.** The ball joint can be freed from the spindle by pressing out the threaded pin with the tool and tapping on the spindle while pressure is applied. Under no circumstances hit on the end of the threaded portion, as even with a slave nut fitted to protect the threads internal damage will be caused to the ball joint. If the special tool is not available, pull firmly down on the spindle, lay a block of metal on one side of the spindle eye and hit on the opposite side of the eye with a copper mallet. The tapers will quickly free.

Partially drill through the heads of the rivets that secure the old ball joint, **ensuring that the upper arm itself is not marked.** Knock off the heads with a cold chisel and punch out the stems so that the ball joint is freed from the upper arm.

Check the end of the arm for cracks or damage. Clean the end of the arm and remove any burrs from the edges of the mounting holes.

Secure the new ball joint in place with the special nuts and bolts supplied. **Do not use ordinary bolts and do not attempt to rivet the new ball joint into place.** Install the spindle onto the ball joint pin and tighten the nut to the specified torque. Lock the nut with a new cotterpin, tightening the nut if required to align the cotterpin hole and castellations.

Lubricate the ball joint until the dust cover just starts to swell. Install the road wheel and tire, lower the vehicle back to the ground and remove the block or support that held the upper arm.

Renewal of lower arm:

This is best left to a service garage, not because the work is beyond the owners capabilities but because the positioning of the shims and bushes on the lower arm inboard pivot controls the camber of the suspension, and the position of the strut in its frame bracket controls the castor of the suspension. **If the owner does carry out the work, the car must be taken to a service garage for checking and adjusting the suspension geometry.**

If the arm is to be renewed, raise the car with the supports holding the upper arm, as described for renewing upper ball joint. Disconnect the stabilizer bar link. Remove the nuts and bolts that secure the strut to the lower arm. Note that the strut bushings must be renewed if they are worn or perished and the strut cup washers must be renewed if they are distorted or damaged.

Disconnect the lower ball joint, using the tool shown in **FIG 9:10.** It may be necessary to remove the cotterpin from the upper ball joint nut in order to position the tool satisfactorily on the upper ball joint. Remove the pivot bolt and take off the lower arm.

Install the arm in the reverse order of disassembly.

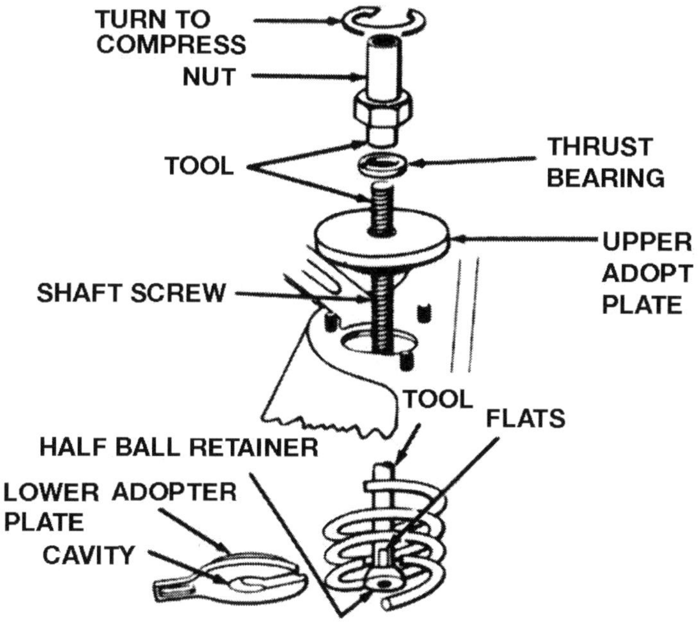

FIG 9:11 Typical spring compresser for removing or installing the spring

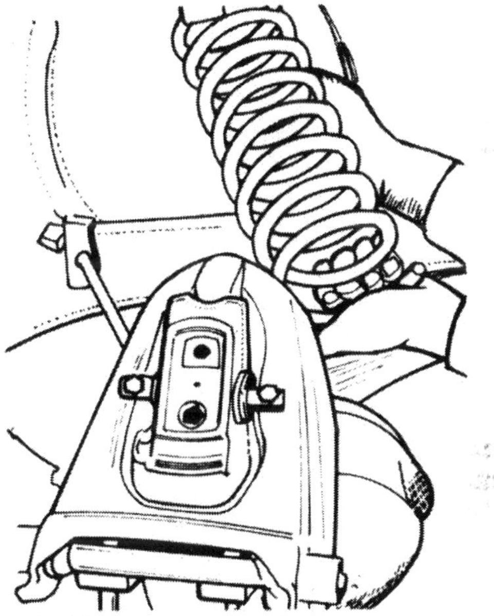

FIG 9:12 Removing or installing the spring

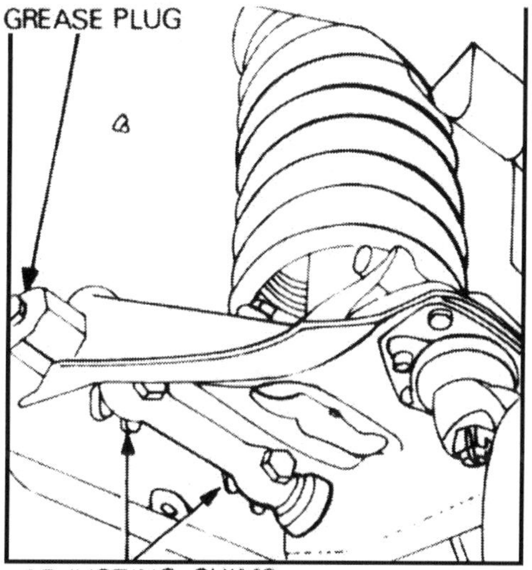

FIG 9:13 Setting the castor and camber on a 1965 model

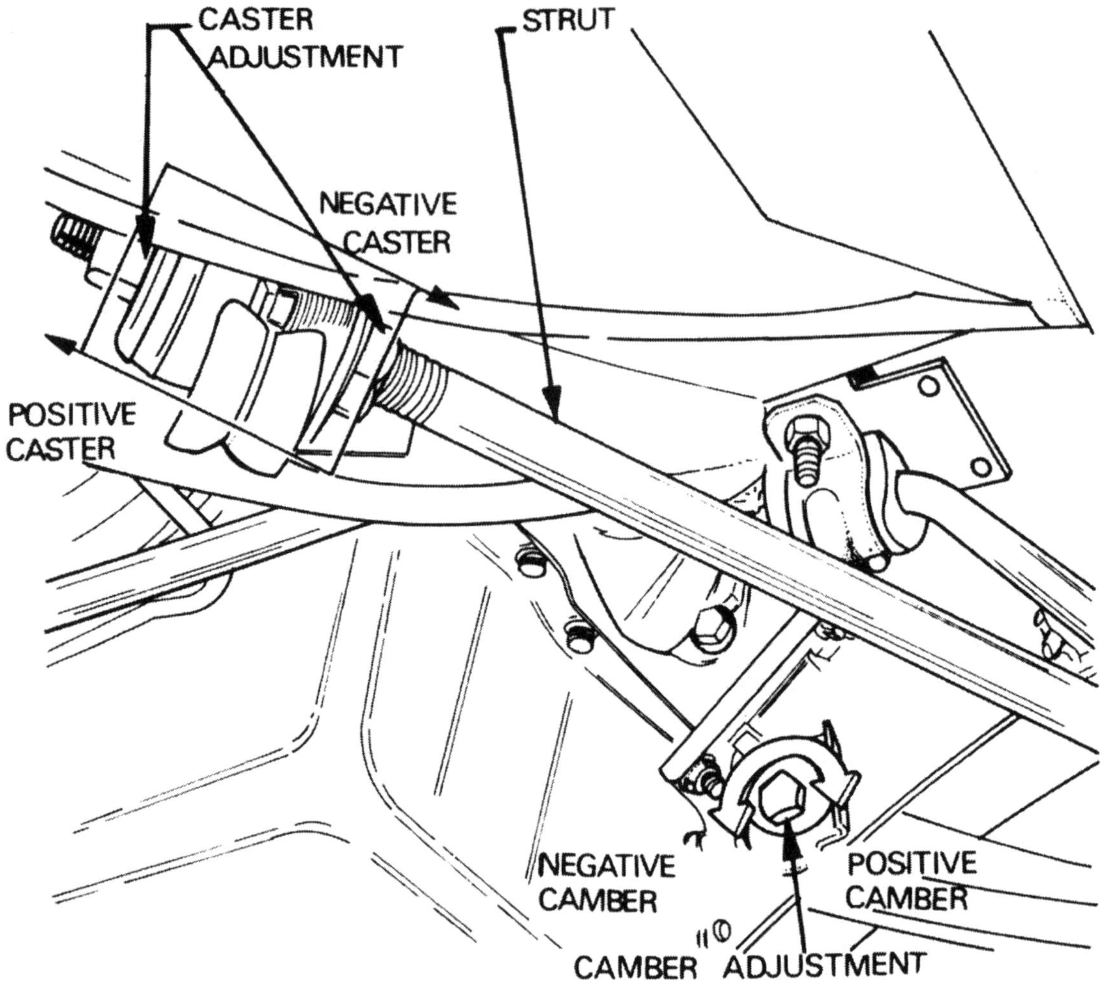

FIG 9:14 Setting the caster and camber on all later models

9:7 The springs

Removal:

1 Remove the air cleaner from the engine to gain access room. Raise the front of the car and place it securely onto underbody stands. Remove the road wheel.

2 Remove the shock absorber as described in **Section 9:4.** Install the parts of the special tool shown in **FIG 9:11.** The actual part numbers of the tool will vary, so check with a dealer for the correct parts. The lower plate is installed under the third coil from the bottom and locked into the half ball retainer. The upper adaptor plate is held by the three studs of the damper upper attachment. Use a suitable socket and ratchet drive to compress the spring until it is clear of the upper arm.

3 Disconnect the upper ball joint from the spindle. Remove the nuts that secure the upper arm pivot to the frame and take out the upper arm assembly, carefully collecting the shims fitted.

4 Slowly and carefully release the spring compressing tool and when the pressure is released, remove the tool parts and remove the spring as shown in **FIG 9:12.**

Check the spring for cracks, damage, or weakening and renew it if it is defective.

The parts are installed in the reverse order of disassembly. Insert the spring into position and compress it with the special tools. Install the upper arm. **When installing the upper arm use the specified type keystone lockwashers and fit the shims back into their original positions.**

9:8 Suspension geometry

On the 1964 and 1965 models, shims are fitted between the lower arm pivot and frame to control both the castor and camber angles of the suspension, as shown in **FIG 9:13.** On all the later models the castor is set by the positioning of the support strut in the frame bracket and the camber is set by the position of the eccentric on the lower inboard pivot, as shown in **FIG 9:14.** The suspension angles are illustrated in **FIG 9:15.**

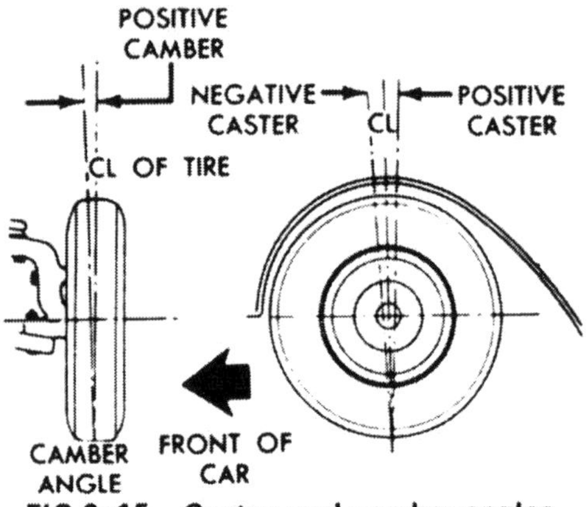

FIG 9:15 Castor and camber angles

Incorrect suspension geometry will cause poor handling and excessive tyre wear. Accurate gauges and instruments are required to check the angles and for this reason the vehicle should be taken to a service garage if it is suspected that the geometry is incorrect. The cost of a check will be less than the cost of new tyres, so always have the geometry checked after major repairs or 'curbing' a wheel.

9:9 Trouble shooting guide

(a) Wheel wobble or tramp

1 Unbalanced front wheels and tires
2 Broken or weak road springs
3 Defective suspension arm inboard pivots
4 Worn or loose hub bearings
5 Worn or defective suspension ball joints
6 Loose road wheel attachments
7 Uneven tire wear

(b) Vehicle bottoms on suspension

1 Check 2 in (a)
2 Bumpers missing or worn
3 Defective shock absorbers

(c) Heavy steering

1 Neglected lubrication
2 Incorrect suspension geometry

(d) Excessive tire wear

1 Check 2, 3, 4 and 5 in (a); 2 in (b)

(e) Rattles

1 Check 2, 3 and 5 in (a) and 3 in (b)
2 Stabilizer bar broken or bushings worn

(e) Excessive rolling

1 Check 2 in (a); 3 in (b) and 2 in (e)

Inches	Decimals	Milli-metres	Inches to Millimetres		Millimetres to Inches	
			Inches	mm	mm	Inches
1/64	.015625	.3969	.001	.0254	.01	.00039
1/32	.03125	.7937	.002	.0508	.02	.00079
3/64	.046875	1.1906	.003	.0762	.03	.00118
1/16	.0625	1.5875	.004	.1016	.04	.00157
5/64	.078125	1.9844	.005	.1270	.05	.00197
3/32	.09375	2.3812	.006	.1524	.06	.00236
7/64	.109375	2.7781	.007	.1778	.07	.00276
1/8	.125	3.1750	.008	.2032	.08	.00315
9/64	.140625	3.5719	.009	.2286	.09	.00354
5/32	.15625	3.9687	.01	.254	.1	.00394
11/64	.171875	4.3656	.02	.508	.2	.00787
3/16	.1875	4.7625	.03	.762	.3	.01181
13/64	.203125	5·1594	.04	1.016	.4	.01575
7/32	.21875	5.5562	.05	1.270	.5	.01969
15/64	.234375	5.9531	.06	1.524	.6	.02362
1/4	.25	6.3500	.07	1.778	.7	.02756
17/64	.265625	6.7469	.08	2.032	.8	.03150
9/32	.28125	7.1437	.09	2.286	.9	.03543
19/64	.296875	7.5406	.1	2.54	1	.03937
5/16	.3125	7.9375	.2	5.08	2	.07874
21/64	.328125	8.3344	.3	7.62	3	.11811
11/32	.34375	8.7312	.4	10.16	4	.15748
23/64	.359375	9.1281	.5	12.70	5	.19685
3/8	.375	9.5250	.6	15.24	6	.23622
25/64	.390625	9.9219	.7	17.78	7	.27559
13/32	.40625	10.3187	.8	20.32	8	.31496
27/64	.421875	10.7156	.9	22.86	9	.35433
7/16	.4375	11.1125	1	25.4	10	.39370
29/64	.453125	11.5094	2	50.8	11	.43307
15/32	.46875	11.9062	3	76.2	12	.47244
31/64	.484375	12.3031	4	101.6	13	.51181
1/2	.5	12.7000	5	127.0	14	.55118
33/64	.515625	13.0969	6	152.4	15	.59055
17/32	.53125	13.4937	7	177.8	16	.62992
35/64	.546875	13.8906	8	203.2	17	.66929
9/16	.5625	14.2875	9	228.6	18	.70866
37/64	.578125	14.6844	10	254.0	19	.74803
19/32	.59375	15.0812	11	279.4	20	.78740
39/64	.609375	15.4781	12	304.8	21	.82677
5/8	.625	15.8750	13	330.2	22	.86614
41/64	.640625	16.2719	14	355.6	23	.90551
21/32	.65625	16.6687	15	381.0	24	.94488
43/64	.671875	17.0656	16	406.4	25	.98425
11/16	.6875	17.4625	17	431.8	26	1.02362
45/64	.703125	17.8594	18	457.2	27	1.06299
23/32	.71875	18.2562	19	482.6	28	1.10236
47/64	.734375	18.6531	20	508.0	29	1.14173
3/4	.75	19.0500	21	533.4	30	1.18110
49/64	.765625	19.4469	22	558.8	31	1.22047
25/32	.78125	19.8437	23	584.2	32	1.25984
51/64	.796875	20.2406	24	609.6	33	1.29921
13/16	.8125	20.6375	25	635.0	34	1.33858
53/64	.828125	21.0344	26	660.4	35	1.37795
27/32	.84375	21.4312	27	685.8	36	1.41732
55/64	.859375	21.8281	28	711.2	37	1.4567
7/8	.875	22.2250	29	736.6	38	1.4961
57/64	.890625	22.6219	30	762.0	39	1.5354
29/32	.90625	23.0187	31	787.4	40	1.5748
59/64	.921875	23.4156	32	812.8	41	1.6142
15/16	.9375	23.8125	33	838.2	42	1.6535
61/64	.953125	24.2094	34	863.6	43	1.6929
31/32	.96875	24.6062	35	889.0	44	1.7323
63/64	.984375	25.0031	36	914.4	45	1.7717

UNITS	Pints to Litres	Gallons to Litres	Litres to Pints	Litres to Gallons	Miles to Kilometres	Kilometres to Miles	Lbs. per sq. In. to Kg. per sq. Cm.	Kg. per sq. Cm. to Lbs. per sq. In.
1	.57	4.55	1.76	.22	1.61	.62	.07	14.22
2	1.14	9.09	3.52	.44	3.22	1.24	.14	28.50
3	1.70	13.64	5.28	.66	4.83	1.86	.21	42.67
4	2.27	18.18	7.04	.88	6.44	2.49	.28	56.89
5	2.84	22.73	8.80	1.10	8.05	3.11	.35	71.12
6	3.41	27.28	10.56	1.32	9.66	3.73	.42	85.34
7	3.98	31.82	12.32	1.54	11.27	4.35	.49	99.56
8	4.55	36.37	14.08	1.76	12.88	4.97	.56	113.79
9		40.91	15.84	1.98	14.48	5.59	.63	128.00
10		45.46	17.60	2.20	16.09	6.21	.70	142.23
20				4.40	32.19	12.43	1.41	284.47
30				6.60	48.28	18.64	2.11	426.70
40				8.80	64.37	24.85		
50					80.47	31.07		
60					96.56	37.28		
70					112.65	43.50		
80					128.75	49.71		
90					144.84	55.92		
100					160.93	62.14		

UNITS	Lb ft to kgm	Kgm to lb ft	UNITS	Lb ft to kgm	Kgm to lb ft
1	.138	7.233	7	.967	50.631
2	.276	14.466	8	1.106	57.864
3	.414	21.699	9	1.244	65.097
4	.553	28.932	10	1.382	72.330
5	.691	36.165	20	2.765	144.660
6	.829	43.398	30	4.147	216.990

CHAPTER 10

THE STEERING GEAR

10:1 Description

A recirculating ball steering unit is fitted to all models covered by this manual. A sectional view of a typical unit is shown in **FIG 10:1**. The steering shaft is connected to the steering wheel, through the steering column, and therefore revolves as the wheel is turned. The ball nut moves up or down the spiral of the steering shaft, with the balls reducing friction, and as it moves it rotates the sector shaft by the sector gear teeth.

The sector shaft is linked to the steering linkage, shown in **FIG 10:2**, by the Pitman arm. An idler arm is mounted on the opposite side of the frame to the steering unit and a rod interconnects the Pitman arm to the idler arm so that the center rod will always be at right angles to the fore and aft axis of the vehicle. Spindle connecting rods transmit the motion of the tie rod to the spindles on the suspension so that the front wheels move for steering.

Power steering can be fitted as an optional extra. The parts are fitted as an addition to the normal system, as shown in **FIG 10:3**.

The pump is integral with the reservoir and mounted inside it. The pump and reservoir assembly is mounted on the engine and the pump is driven by a belt from the engine. The control valve is operated when the steering is turned, and returns fluid from the appropriate side of the power cylinder to assist in turning the steering. The valve is designed to ensure that the fluid circulates through the pump and valve when the steering is in the straight-ahead position. The pump is of the slipper type. Eight slippers are fitted and as the pump revolves they move in and out in the track, altering the volume of the space between them. The slippers are so arranged that, as they part, the fluid is drawn in through the inlet, and as the track forces them together, the fluid is pumped out through the outlet port. A sectional view of the pump and reservoir assembly is shown in **FIG 10:4**.

10:2 Maintenance

On models equipped with manual steering, check that the steering operates fully and freely without excessive play or stiffness. Check the dust covers on the spindle connecting rod ends to make sure that they are in good condition and are not allowing dirt to enter the ball joint.

On models fitted with power steering, check the fluid level in the reservoir, check the drive belt tension, and lubricate.

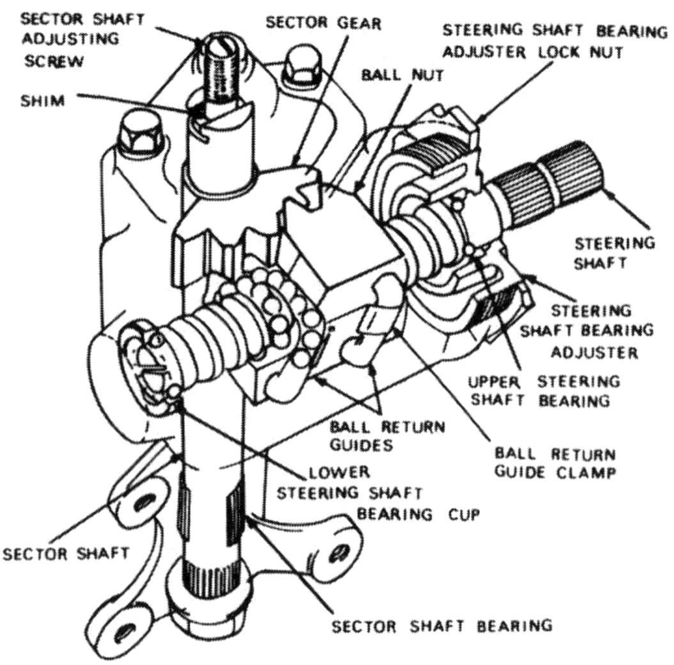

FIG 10:1 Typical steering gear unit

Start the engine and operate the steering from full lock to full lock several times and then shut off the engine. On later models the engine should be run until the power steering has reached its normal operating temperature. Withdraw the dipstick and check the level. If the level is low, top up with automatic transmission fluid to the correct mark. If the level has been allowed to fall so low that air has been drawn into the pump, and the fluid appears milky or frothing, turn the steering from full lock to lock several times while topping up.

The method of setting the drive belt tension is given in **Chapter 4, Section 4:4. Do not lever against the reservoir when setting the belt tension otherwise the reservoir will distort and cause fluid leaks.** A boss is cast into the face of the pump and a wrench should be applied to this boss when moving the pump into the position for the correct tension.

When lubricating the actuator ball stud, wipe the grease fitting clean before applying a grease gun. Inject grease until the dust cover just starts to swell or expand.

Care must be taken on all models, whether power steering is fitted or not, when turning the steering from full lock to full lock with the front wheels clear of the ground. Do not turn the steering vigorously as the steering unit will be damaged when it comes to a sudden halt at the end of its travel. Where power steering is fitted never apply heavy pressure to the steering wheel when the steering is on full lock for the same reason.

10:3 The steering linkage

A typical steering linkage is shown in **FIG 10:2**. The spindle connecting rods are made up of the ball joint ends and an internally threaded sleeve. One end of the sleeve has a lefthand thread while the other end has a righthand thread. Clamps lock the sleeve to the ends to prevent the adjustment from altering in service. The effective length of the spindle connecting rods is altered to set the correct front wheel alignment and to set the steering wheel spokes (see **Section 10:8**).

The ball joints on the ends of the spindle connecting rods are non-adjustable and non-repairable and the complete ends must be renewed if there is excessive

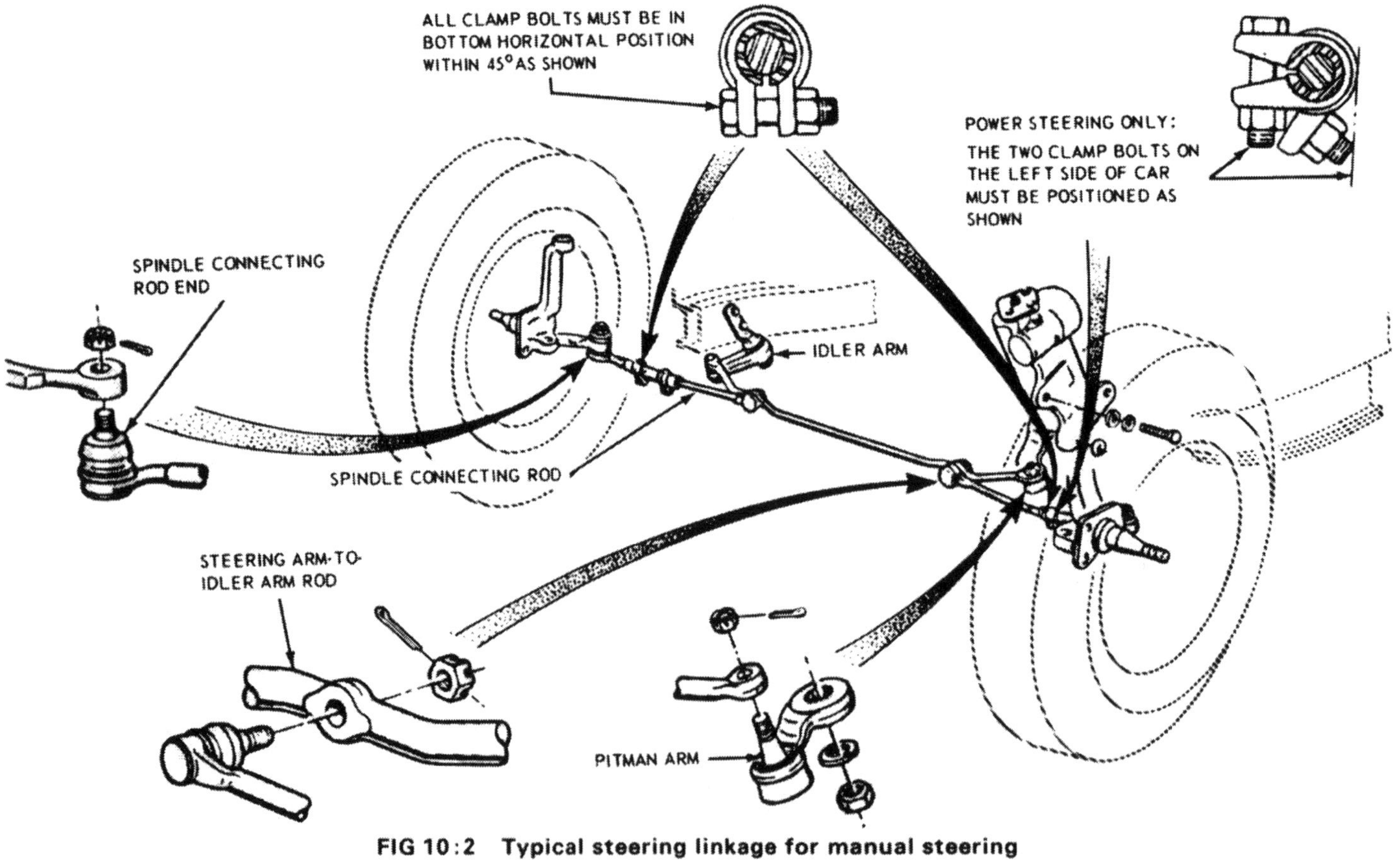

FIG 10:2 Typical steering linkage for manual steering

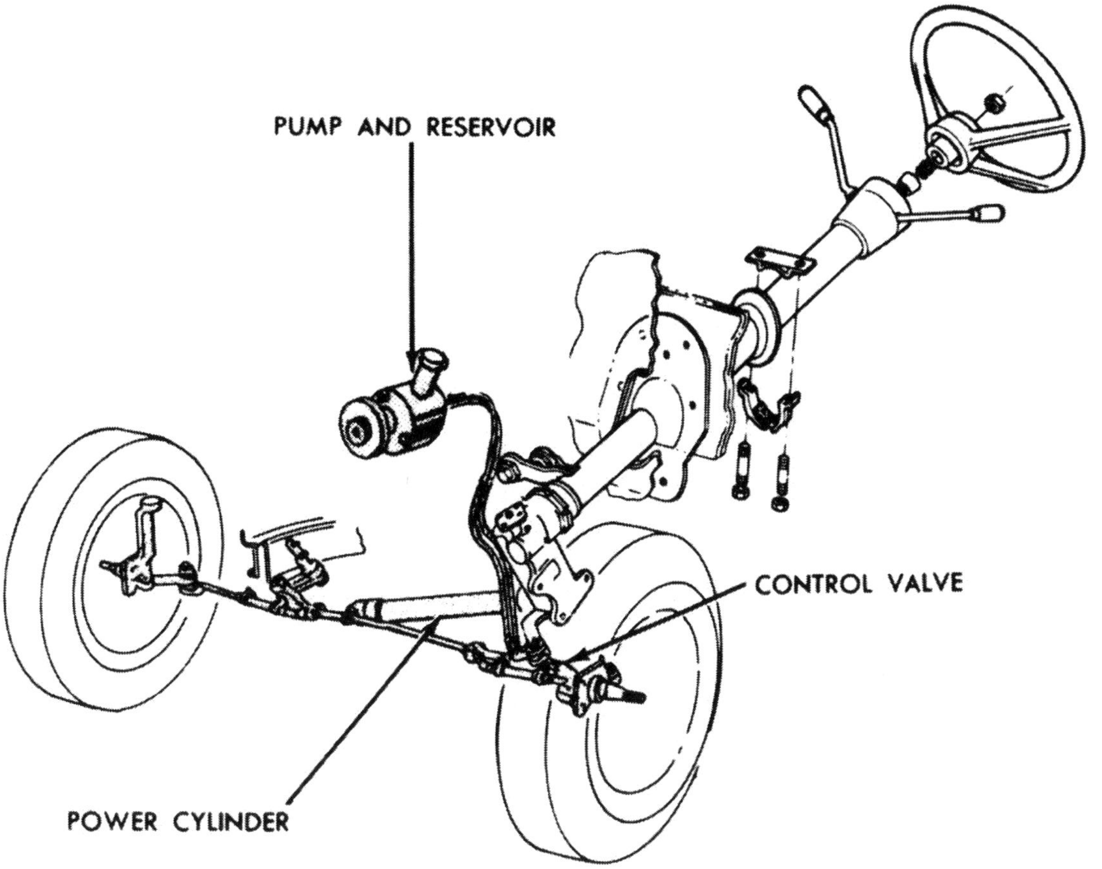

FIG 10 : 3 Power steering installed

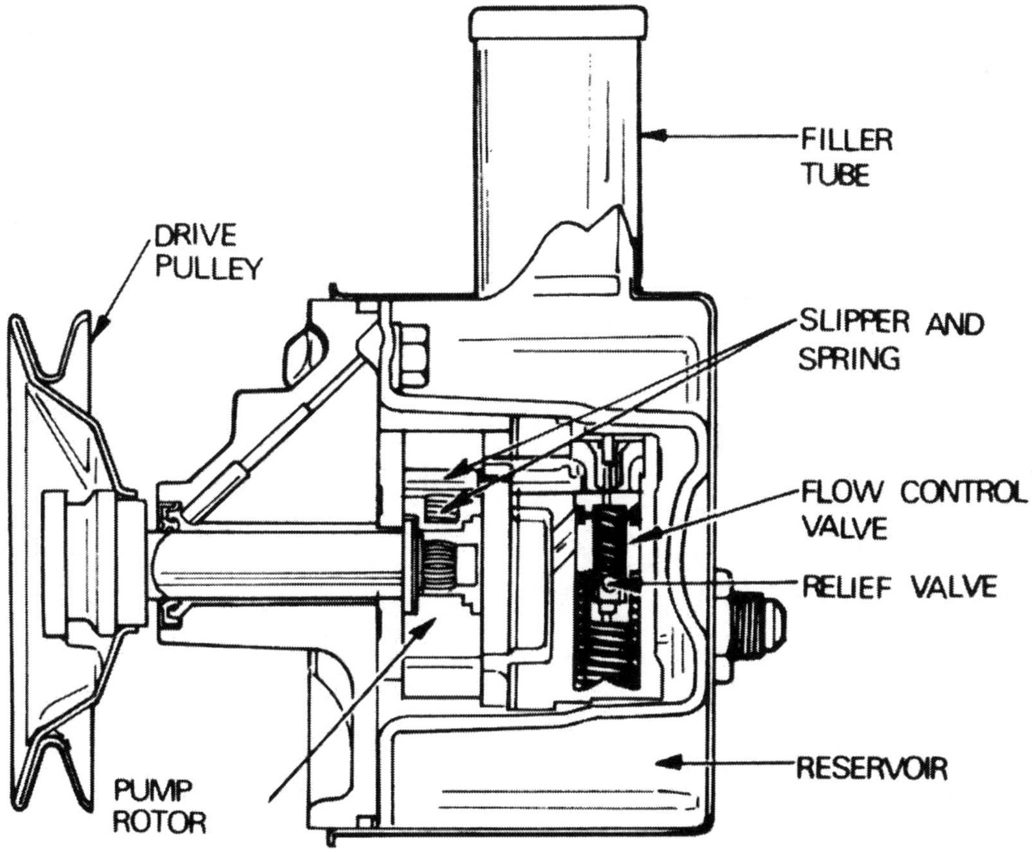

FIG 10 : 4 Sectional view of steering pump and reservoir

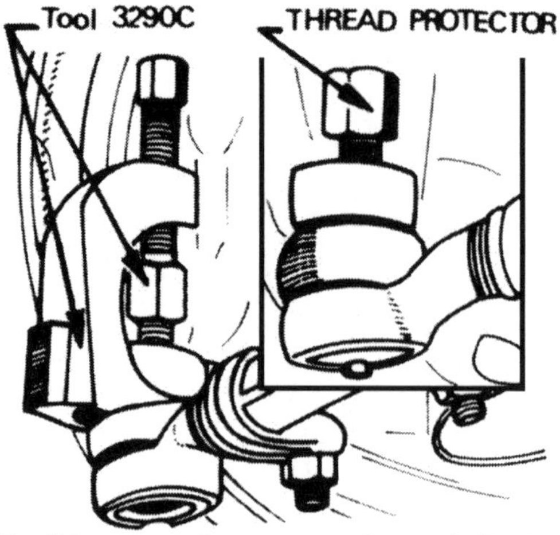

FIG 10:5 Disconnecting connecting spindle ball joint

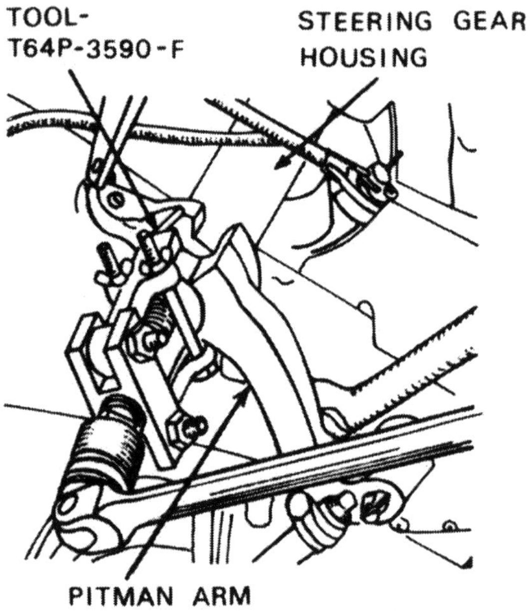

FIG 10:6 Removing the Pitman arm from the steering gear

play in a ball joint. Check for play by having an assistant gently turn the steering backwards and forwards, with the weight of the vehicle on the road wheels, and watch the linkage as it moves.

Freeing ball joints:

The safest and best method of disconnecting a ball joint from its attachment is to use the special tool, as shown in **FIG 10:5.** Extract the cotterpin and remove the nut that secures the pin of the ball joint. Fit the tool as shown and by tightening the bolt of the tool the pin will be pressed out.

If the special tool is not available the ball joint can still be freed. **Never hammer on the end of the threaded portion of the pin, even with a slave nut fitted.** Undo the nut that secures the ball joint to the last few threads. Lay a block of metal on one side of the tapered eye and pull the end with the ball joint firmly away from the other member. Hit firmly with a copper mallet on the side of the tapered eye opposite to the block of metal and the tapers will quickly free.

The ball joint is reconnected by pressing the pin into the tapered eye, by hand, and tightening the castellated nut to the correct torque before locking it with a new cotterpin.

If a new spindle connecting rod end is being fitted, the alignment of the front wheels and the setting of the steering wheel will not be unduly altered if the number of threads required to undo the old end are counted and the new end threaded back by the same number of turns. However, it is advisable to have the front wheel alignment checked by a service garage.

Note that the sleeve itself must be renewed if it is damaged or bent. **Do not attempt to straighten a bent sleeve.**

Pitman arm:

The ball joint on the Pitman arm is disconnected as described previously. On the models where the ball joint is integral with the Pitman arm a new arm must be fitted if the ball joint is defective or worn.

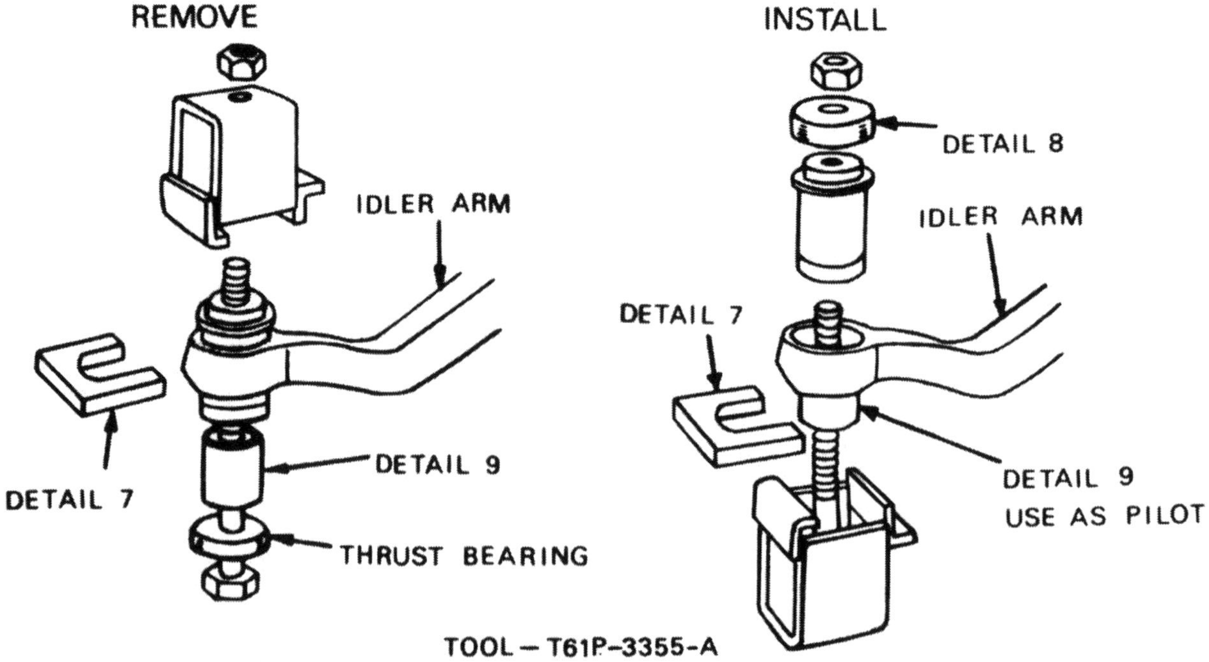

FIG 10:7 Renewing the bush on an early idler arm assembly

When removing the Pitman arm, take off the nut and washer that secure the arm to the sector shaft of the steering unit. A suitable extractor, as shown in **FIG 10:6**, must be used. **Hammering on the sector shaft or Pitman arm is liable to cause internal damage to the steering unit.**

Before the Pitman arm is fully removed from the sector shaft, make light alignment marks on the arm and shaft so that the arm will be installed into its original position.

Install the Pitman arm in the reverse order of disassembly.

Idler arm:

On earlier models the idler arm is fitted with a renewable bush but on later models the complete idler arm must be renewed if there is excessive play.

Free the idler arm from the central rod. On the earlier models, remove the nut and cotterpin that secure the arm to the bracket on the frame. The idler arm bushing can then be renewed using the special tool as shown in **FIG 10:7.**

On the later models the idler arm complete with bracket is removed from the frame and a new assembly installed.

10:4 Steering gear adjustments

There are only two adjustments that need to be made on the steering gear unit and they are both carried out, **in the correct sequence**, with the unit fitted to the vehicle.

Lubrication:

Before carrying out adjustments check the lubricant level in the unit. The filler plug and adjustment points are shown in **FIG 10:8**. Center the steering and remove the filler plug. Remove the lower cover-to-housing attachment bolt. Use a piece of clean rod or similar tool to push all loose lubricant in the filler hole back into the unit so that the hole is clear. Slowly turn the steering to full left lock and lubricant should rise within the lower cover bolt hole. Slowly turn the steering to the full opposite lock and the lubricant should rise in the filler hole. If the lubricant does not rise in the holes correctly, further lubricant should be added through the filler hole until the level is correct.

Steering column:

Remove the crash pad and sufficient hardware to expose the steering shaft nut that secures the steering wheel. Check the alignment of the shaft in the bearing. If the shaft is off center in the bearing, fit a spring scale (capacity up to 30 or 50 lbs) to the end of the shaft. Center the shaft by pulling on the spring scale and note the reading on the scale. If the pull required is less than 20 lbs no action need be taken. If the pull required exceeds 20 lbs then the steering column must be realigned. A steering column out of alignment will cause the steering to be heavy.

Steering shaft in steering gear:

Disconnect the Pitman arm from the sector shaft. Slacken the locknut on the sector adjusting screw (see **FIG 10:8**), and turn the adjusting screw counterclockwise by at least two full turns.

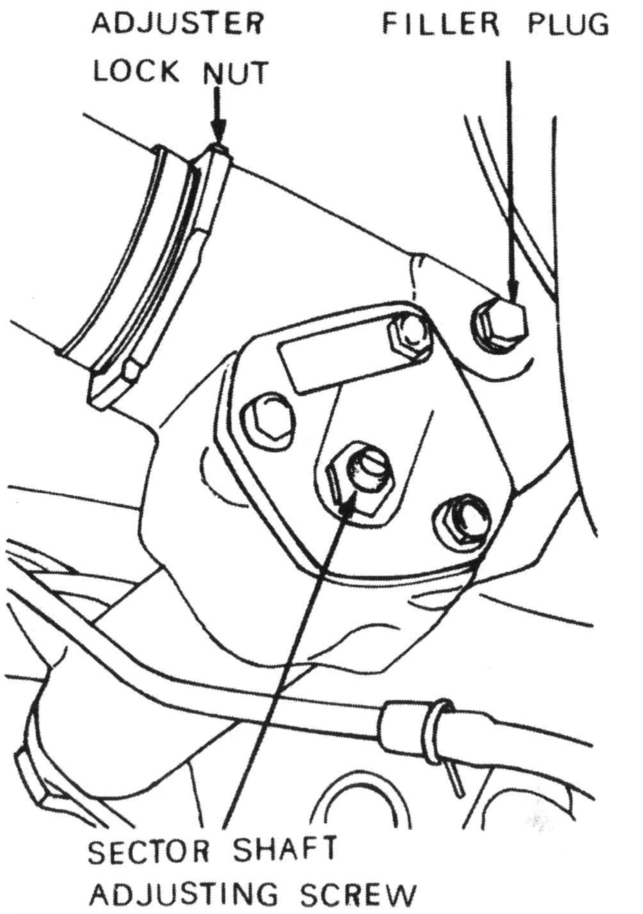

FIG 10:8 Lubrication and adjustment points on the steering gear

Attach a torque wrench (measuring in inch lb) to the steering wheel nut. Use the torque spanner to turn the steering approximately $1\frac{1}{2}$ turns either side of the center position. If the torque required is not within limits (see **Technical Data**), slacken the adjustor to bring the torque within limits. When the bearing preload is correctly set, hold the adjustor and tighten the adjustor locknut. If the bearings feel rough or noisy, the steering unit must be removed from the vehicle for dismantling and further examination.

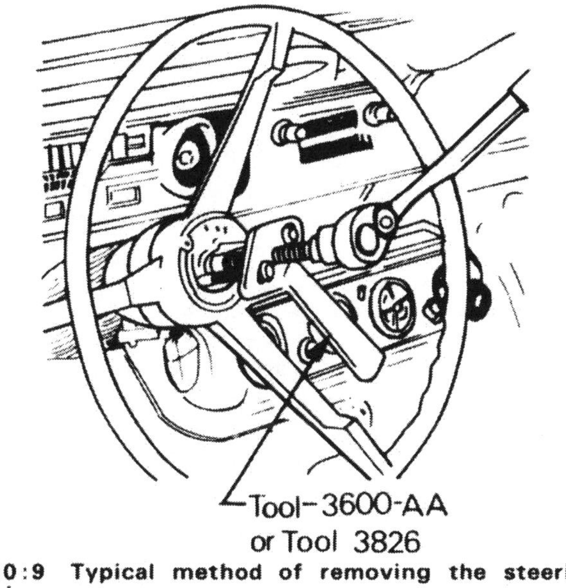

FIG 10:9 Typical method of removing the steering wheel

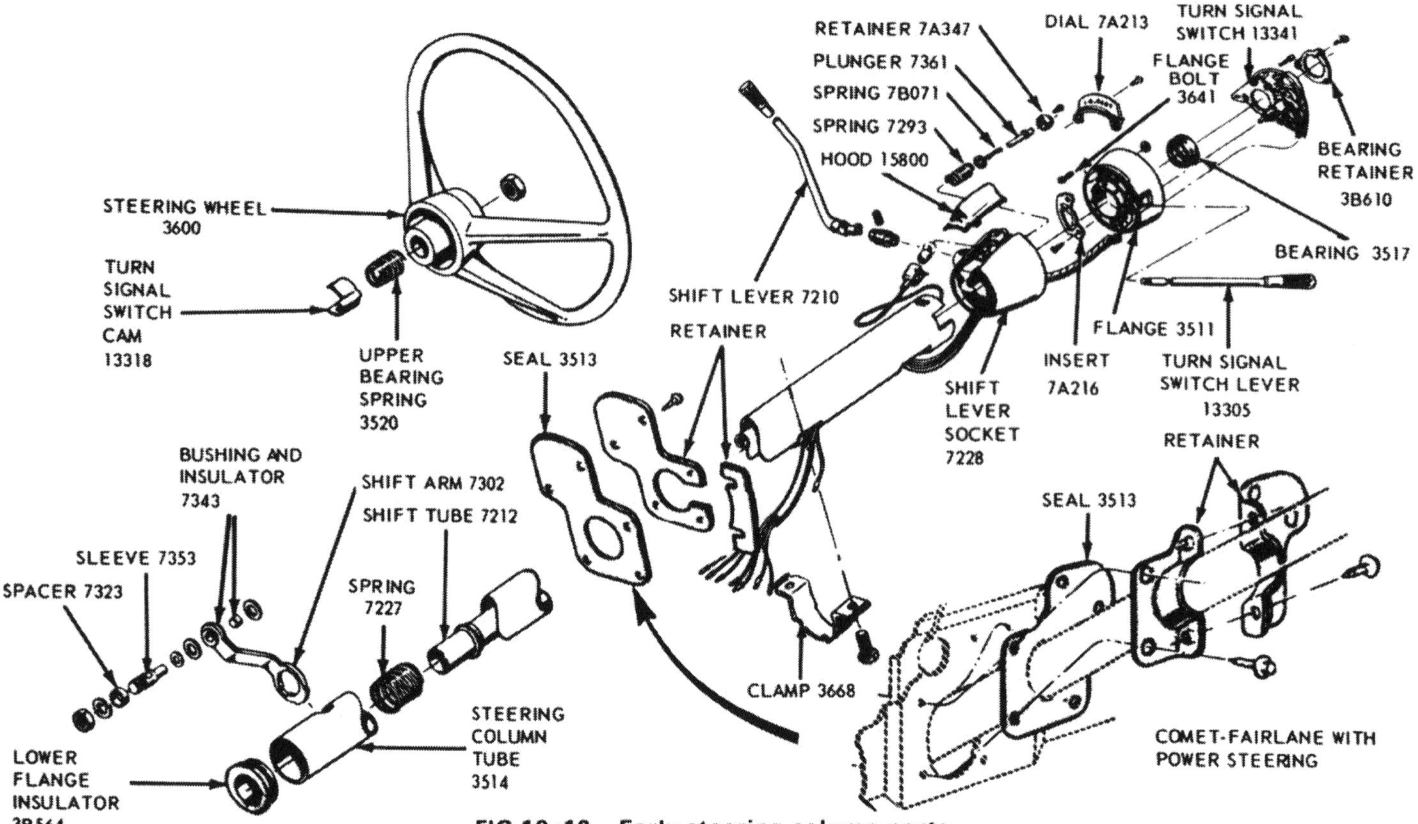

FIG 10:10 Early steering column parts

Sector shaft adjustment:

When the bearings have been correctly set, set the sector shaft adjusting screw. Turn the steering gently from full lock to full lock, counting the turns required. **Do not slam the steering against the stops as internal damage will be caused.** Turn the steering back by half the number of turns required from lock to lock so that it is now in the central position.

Use the torque spanner on the steering wheel nut to turn the steering, and turn the sector shaft adjusting screw clockwise until the torque is within the correct limits as the steering passes through the center point. There should be no perceptible backlash with the steering at 30 deg. either side of center. Hold the adjuster screw in position and tighten the locknut.

Install the hardware and crashpad to the steering wheel. Install the Pitman arm back onto the sector shaft.

10:5 The steering column

On later models a safety collapsible steering column is fitted. Under heavy impact the column collapses approximately 6 inches in a controlled manner to prevent injury to the driver. **On this safety column care must be taken when removing or installing either the steering wheel or column itself not to damage the collapsible portion inside the bellows.** This is particularly important when removing the steering wheel, after its attachment nut has been taken off, and an extractor (as shown in **FIG 10:9**) should always be used to free the wheel from its splines.

Removal:

Before attempting to remove any part of the steering column or disassemble the parts, disconnect the battery.

On the very early models, shown in **FIG 10:10**, steering column and steering gear are removed as a unit from the engine compartment. Disconnect the steering column attachments, including the wires, transmission shift linkage, brackets and Pitman arm. Remove the steering wheel and wrap the end of the steering column with rags to prevent it from damaging the fabric of the seats. It will be necessary to disconnect the wires to the left bank of spark plugs to prevent them from becoming damaged. It will also be necessary at least to loosen the air cleaner so that access room can be made, as well as remove the exhaust manifold on the lefthand side. On models fitted with power brakes it will also be necessary to remove the power brake unit.

On later models the steering column is removed from inside the car, and the steering unit removed separately.

On all models slotted holes are provided to ensure that the steering column is correctly aligned. When the steering column is in position, check the position of the inner column in the upper bearing and make small adjustments to centralize it. If the slotted holes do not allow sufficient movement for adjustment, shims will have to be fitted. The installation of the steering column on a late model is shown in **FIG 10:11**, and its components in **FIG 10:12**.

10:6 Steering gear servicing

The components of the later steering gear unit are shown in **FIG 10:13** and **FIG 10:14**.

Strict cleanliness is essential when reassembling the parts and all the parts should be washed in a suitable solvent and lubricated with clean lubricant before installation and assembly.

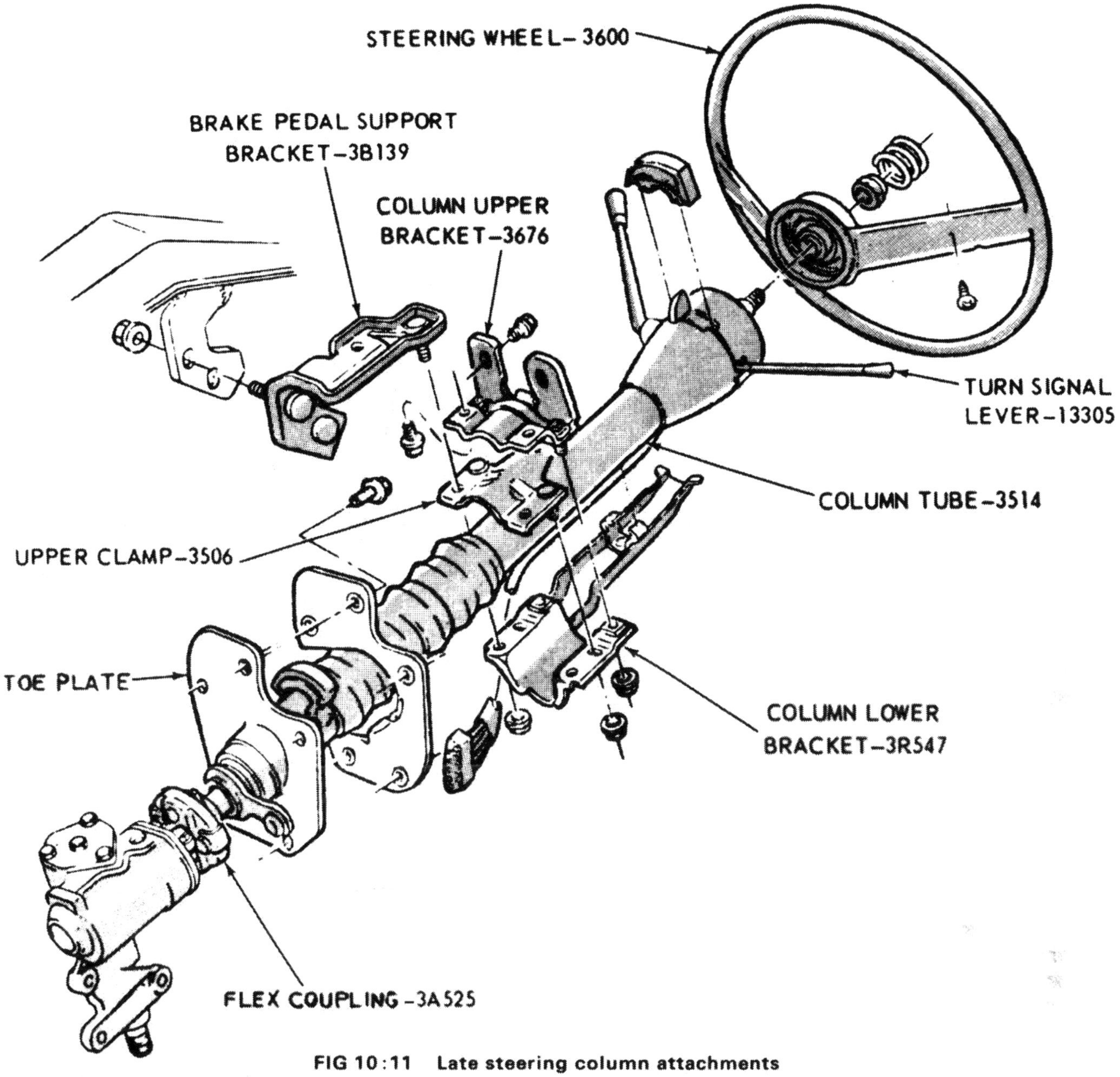

FIG 10:11 Late steering column attachments

Disassembly:

1 Turn the sector shaft gently from stop to stop, counting the turns, and then rotate it back by half the number of turns to set it in the midway position. Slacken the locknut on the sector shaft adjustor and remove the three bolts that secure the cover. Keep turning the adjustor screw clockwise until the cover comes free. Remove the adjusting screw from the sector shaft, but keep the shim with it. Withdraw the sector shaft from the housing.

2 Slacken the locknut for the ballbearing adjustor and unscrew the adjustor nut. Remove the steering shaft upper bearing assembly. Carefully withdraw the steering shaft and ball nut assembly from the housing. **Do not allow the ball nut to run down the spiral as the ball guides will be damaged when the ball nut reaches the end of its travel.**

3 Clean the ball nut and steering shaft assembly by washing the parts in solvent. Rotate the steering shaft in the ball nut but do not force it to the end of its travel. Dry the parts with compressed air and lubricate them with light oil. Rotate the shaft in the nut and check for stiffness or binding. If the shaft rotates freely then there is no need to disassemble the parts. **Only if there is stiffness or binding should the parts be dismantled.**

4 If the sector shaft bearing and oil seal show damage or wear, they should be pressed out of the housing. If the bearing has rotated in the housing, a new housing must be fitted. If the bearing and oil seal are satisfactory, leave them in place but avoid washing the oil seal with solvent.

Ball nut assembly:

If there is stiffness in the assembly, dismantle the parts and check for wear or damage. Hold the nut so that the ball guides are uppermost and keep the assembly over a container with a piece of cloth laid in the bottom. The cloth is to prevent the balls from bouncing as they fall and possible becoming lost. Remove the clamp and

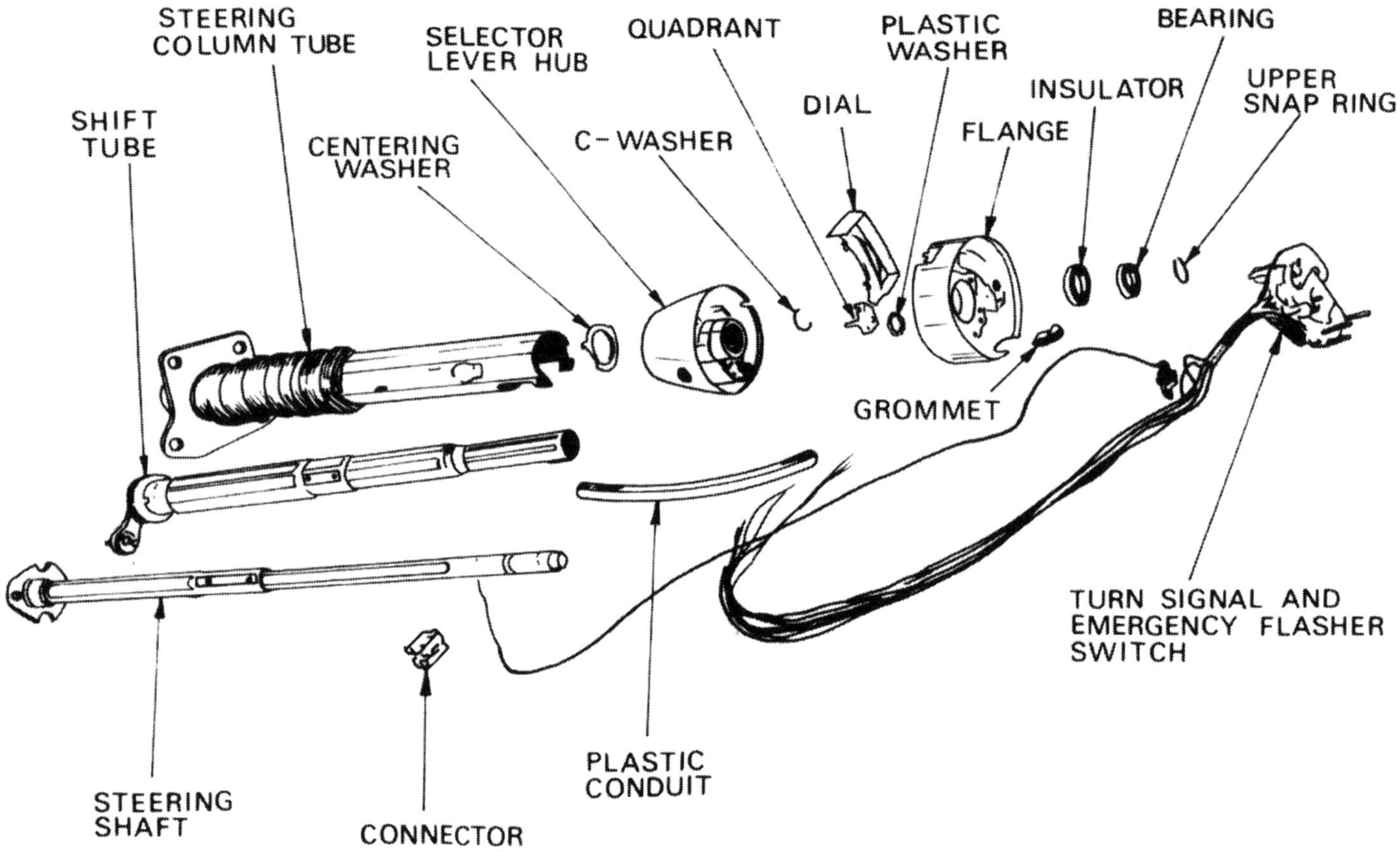

FIG 10:12 Late steering column parts

withdraw the two ball guides, taking care not to lose the balls in the guides. Turn the nut over and rotate the steering shaft backwards and forwards until all the balls are out of the assembly. The number of balls varies but it is always an even number.

Fit the nut back onto the shaft, as shown in **FIG 10:15**. Install the ball guides, tapping them back into place with a wooden handle if required. Divide the balls into two equal amounts, noting that each half of the set is fitted to one ball guide and channel. Hold the nut with the ball guides uppermost and feed the balls back through the hole in the top of each ball guide. Rotate the steering shaft backwards and forwards to distribute the balls in the circuit. When all the balls are fitted, secure the guides in place with the clamp, which also covers the holes in the ball guides. Lubricate the nuts with the correct lubricant (C3AZ-19578-A).

Assembly:

Check all the parts for wear and renew any that are defective. If there is difficulty in removing the cup for the lower bearing of the steering shaft, tap the housing onto a clean piece of wood to dislodge the cup. Renew both bearings if they are worn or show signs of fretting or breaking up.

1 If the sector shaft bearing and oil seal have been removed, press new parts back into place. Lubricate the oil seal lightly. Install a new oil seal in the bearing adjustor nut if the old seal is damaged or worn.

2 Install the lower bearing cup back into the housing. Lubricate all the bearings and install both bearings onto the steering shaft. Carefully slide the steering shaft and nut assembly back into the housing. Lightly smear the threads of the bearing adjustor with sealing compound. **Do not allow sealing compound onto the shaft bearings and do not apply it to the female threads.** Clamp the housing gently in a vice so that the sector shaft axis is horizontal and the splined end of the steering shaft is vertically upwards. Install the bearing cup, adjustor with oil seal, and finally the locknut with the marking S outwards. Leave the locknut slack. Fit a torque spanner that measures in inch lb to the splined end of the steering shaft and tighten down the adjustor nut until the correct torque is required to turn the shaft. Hold the adjustor and tighten the locknut to keep it in this position.

3 Use feeler gauges to check the clearance between the adjustor and the sector shaft, with the shim in position. If the clearance exceeds .002 inch then a new shim must be selectively fitted to reduce the clearance to a figure below this limit. Lightly smear the adjustor threads with sealing compound and partially screw it back into the cover. Install the sector shaft and shim onto the adjustor. Fill the unit with the correct quantity of lubricant. Rotate the steering shaft until the ball nut is correctly positioned for meshing with the sector teeth. Lower the sector shaft down into the housing. Lightly smear the threads of the cover

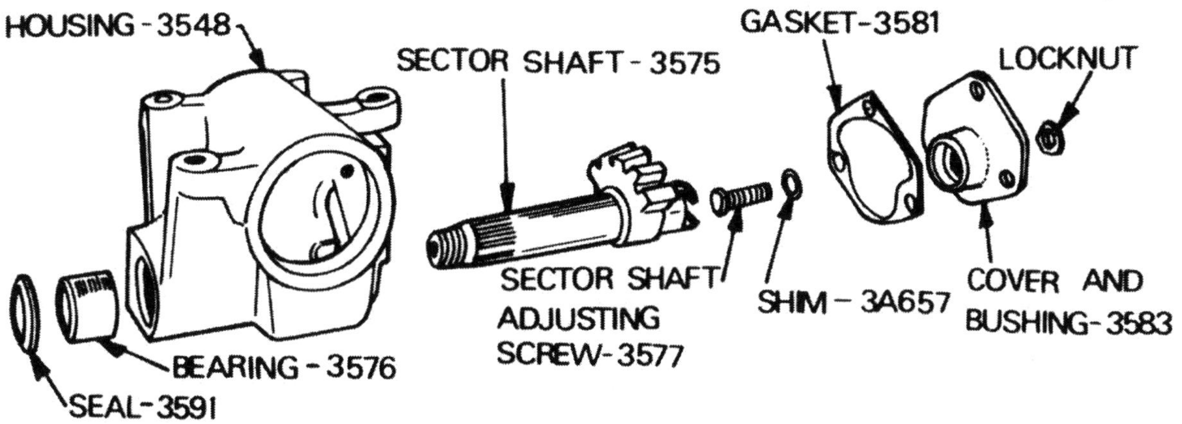

FIG 10:13 **Steering gear sector shaft and parts**

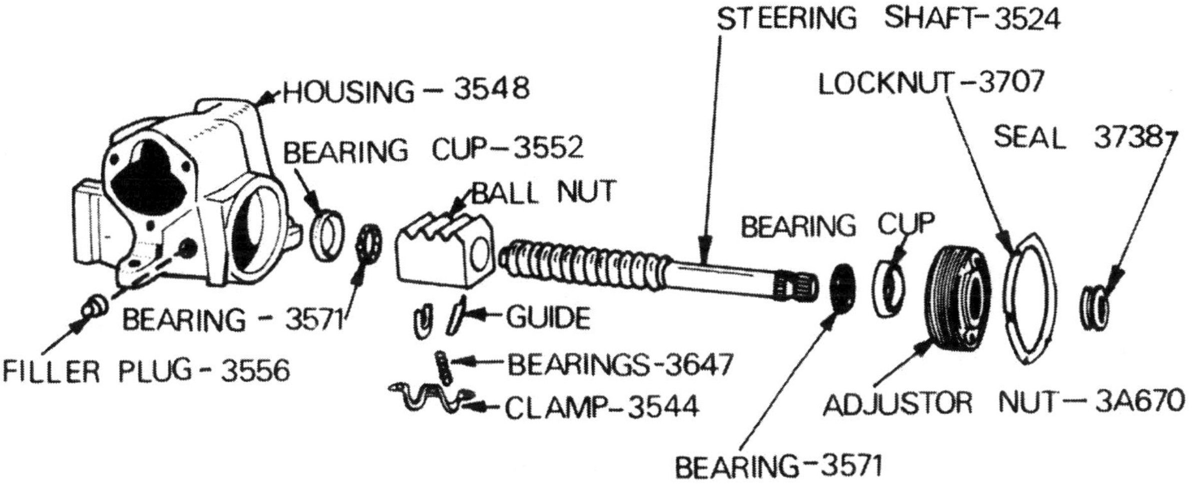

FIG 10:14 **Steering gear steering shaft and parts**

attachment bolts with sealing compound and use them to secure the cover but do not tighten the bolts at this stage. Turn the sector adjustor counterclockwise until there is end clearance and then fully tighten the cover bolts. Set the mesh adjustor until the correct torque is required to turn the steering shaft and then lock the adjustor with the locknut.

10:7 The power steering

Operation:

The operation of the system is best shown in **FIG 10:16**. When the steering wheel is rotated and the force exerted on it exceeds 4 lbs approximately, the valve spool overcomes the pressure of the centering spring and the spool moves to allow pressure to be fed to the appropriate side of the power cylinder and fluid to return from the other side, so that the pressure differential moves the power cylinder piston, and therefore body, so as to assist the steering wheel effort.

A check valve is fitted between the pressure and return line so that when excessive pressure is reached (such as when the steering is at its stops) the check valve opens and limits the maximum pressure in the system.

Adjusting centering spring:

An incorrectly adjusted centering spring can cause stiff steering. The method of adjustment, after removing the cover from the valve assembly is shown in **FIG 10:17.** Tighten the nut until there is a snug fit and then slacken it off by $\frac{1}{4}$ turn. **Do not overtighten the nut otherwise the stop pin may break at the travel regulator stop.** Install the cover and apply a little silicone grease at the sealing areas.

Pump:

A defective pump cannot be repaired and a new one must be fitted in its place. On the earlier models only the

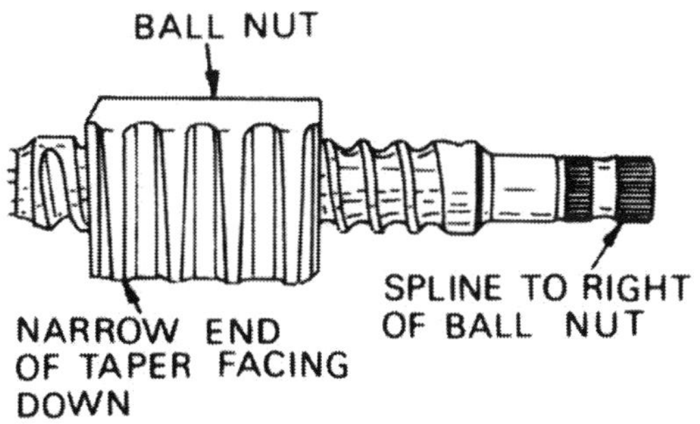

FIG 10:15 **Positioning the ball nut on the shaft**

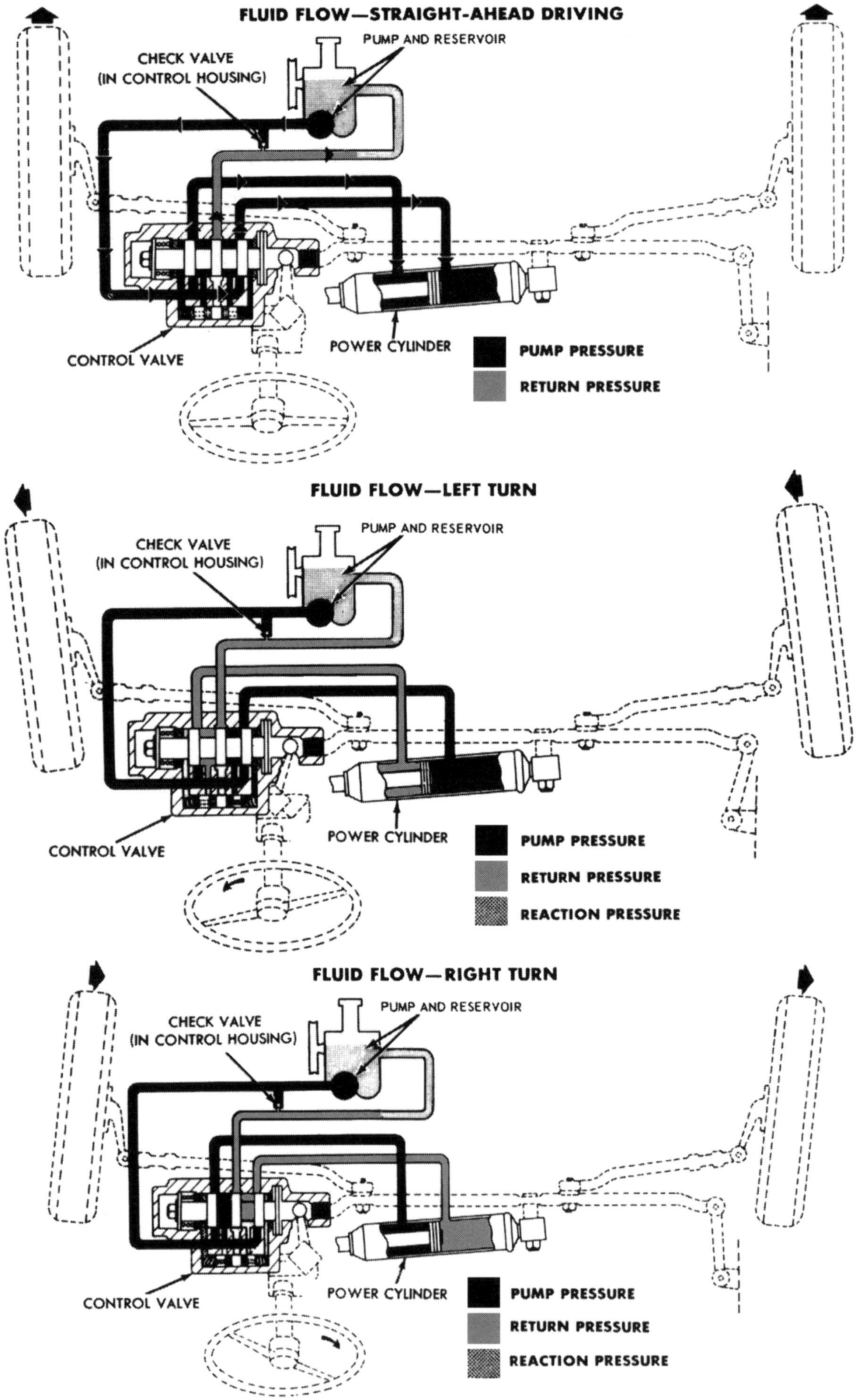

FIG 10:16 Power steering flow diagram

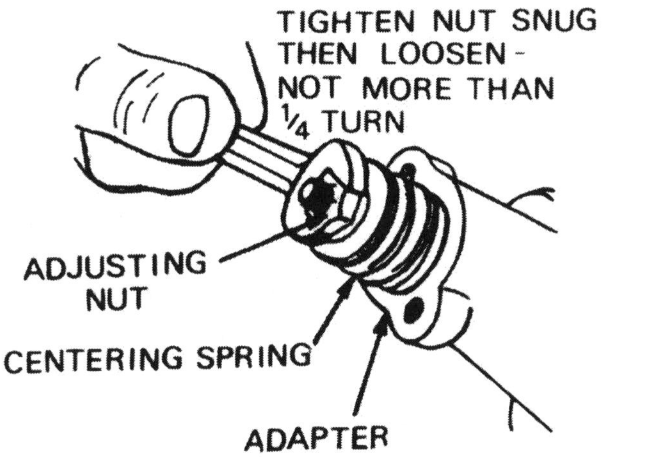

FIG 10:17 Adjusting the power steering valve centring spring

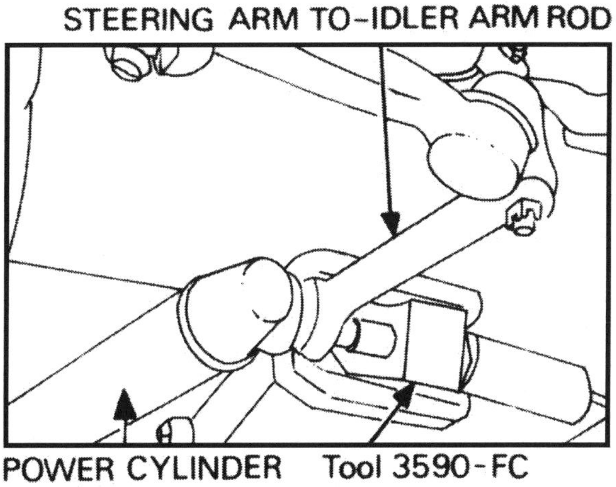

FIG 10:19 Disconnecting the power cylinder

pulley can be renewed if it is damaged but on later models the reservoir can be renewed as well.

A pressure gauge, T-piece and shut-off valve are required to check the operation of the pump, as well as a tachometer. Disconnect the pressure line at the pump and reconnect it using the T-piece and shut-off valve with the gauge on the third arm of the T-piece. Make sure that the shut-off valve is open. Start the engine and top up to replace any fluid loss. Allow the engine to run at idle for at least two minutes to warm up the fluid in the system. Cycle the steering from stop to stop several times and switch off the engine. Top up with more fluid to replace any air that was in the system. Restart the engine, leave the steering untouched and the engine at idle. The pressure in the system should be approximately 50 lb/sq inch.

Speed the engine up to 1000 rev/min and carefully close the shut-off valve. Note the maximum pressure reading and open the shut-off valve. **The shut-off valve should not be held closed for more than 5 seconds otherwise the fluid in the pump will overheat and cause undue wear in the parts.** The pressure should be a minimum of 750 lb/sq inch. A low pressure will be caused by a defective pump, but make sure that belt slip is not the cause of poor pump operation. If the pressure is above normal but the steering is stiff, it is most likely that the valve assembly is at fault.

The pump is removed by freeing the drive belt tension, disconnecting the hoses (and draining the fluid into a suitable container) and then removing the two pivot bolts and two adjustor bolts.

When the pump has been installed and the correct belt tension set, fill the reservoir to the correct level with automatic transmission fluid and start the engine. Cycle the steering several times to expel any air in the system and top up the reservoir to the correct level.

Power cylinder:

The components of a power cylinder sealing are shown in **FIG 10:18.** New seals are available in kits of service spares. When installing new seals, use a deep socket or suitable piece of tubing to press the seals back into position.

To remove the power cylinder, first disconnect the hoses from it and allow the fluid to drain into a suitable container. It is advisable to mark the hoses so that they are correctly reconnected. **If the hoses are incorrectly connected the steering wheel is liable to kick back with considerable force if it is turned.** Remove the Pal nut and nut which secure the cylinder rod to the frame. Take off the washer and insulator to free the rod. Remove the castellated nut (after extracting the cotter-pin) that secures the cylinder to the center link and then

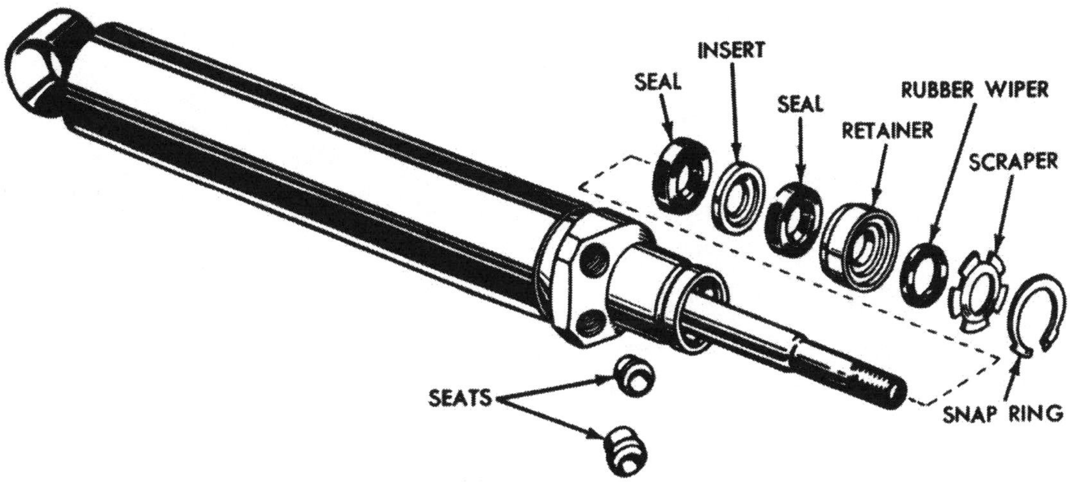

FIG 10:18 The power cylinder seals

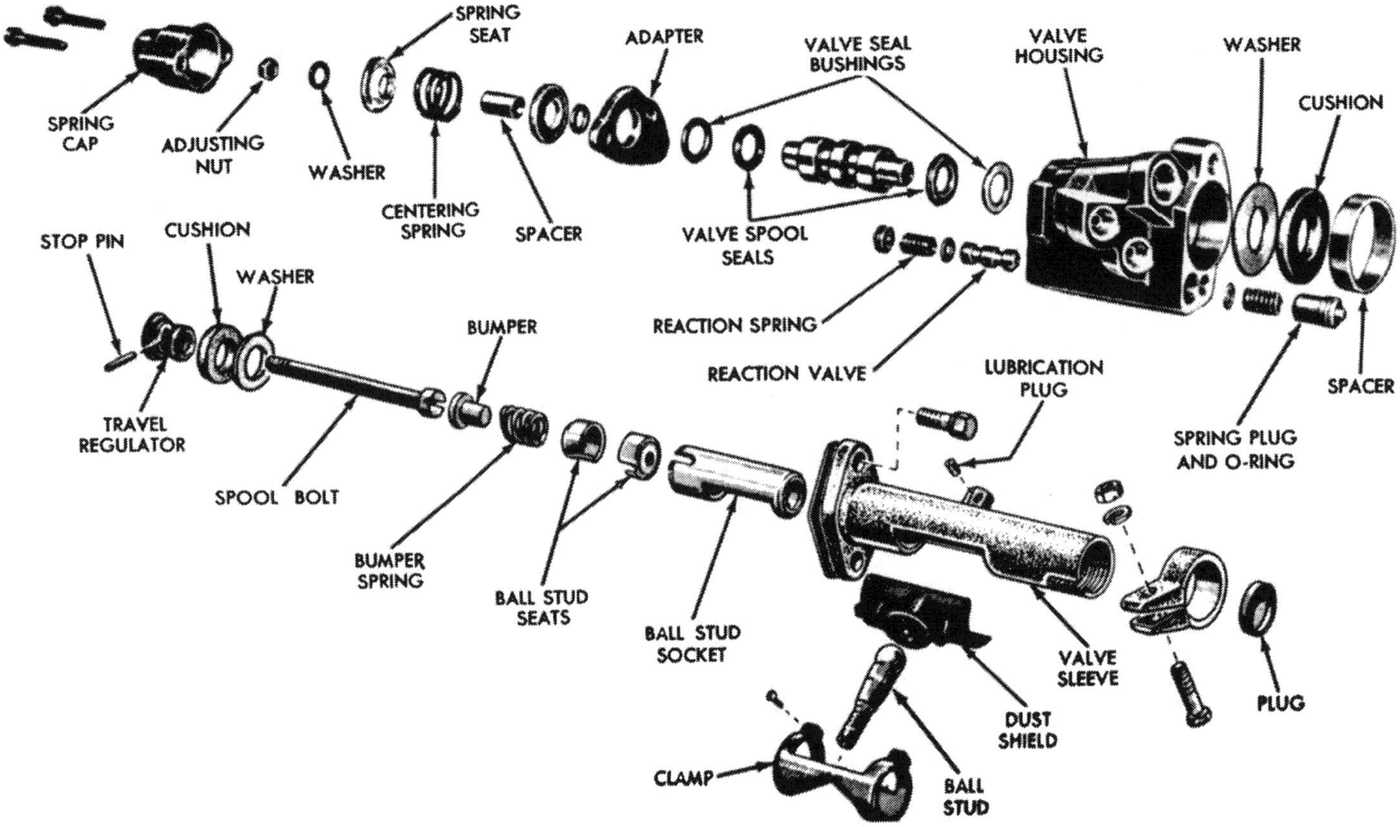

FIG 10:20 The control valve parts

disconnect the cylinder from the link using the tool shown in **FIG 10:19.**

Install the cylinder in the reverse order of removal. Fill the fluid reservoir and top it up while cycling the steering to expel air from the system. Top up to the correct level in the reservoir.

Control valve:

The components of the control valve are shown in **FIG 10:20.** The valve can be disassembled and new seals fitted as well as slight burrs removed from the valve spool, but if there is a major defect then a new valve assembly should be fitted. If crocus paper is used to remove burrs, **take great care not to round off the sharp edges on the spool.**

To remove the unit, first disconnect all four hoses from it and allow the fluid to drain into a container. Label the hoses as they are disconnected to ensure that they will be correctly installed. Turn the steering gently fully left and right to force out fluid. Loosen the clamping nut and bolt at the righthand end of the sleeve. Remove the roll-pin that secures the valve through the slot in the sleeve. Remove the nut and then free the ball stud as shown in **FIG 10:21** from the sector shaft arm. Turn the steering fully to the left and unscrew the valve assembly.

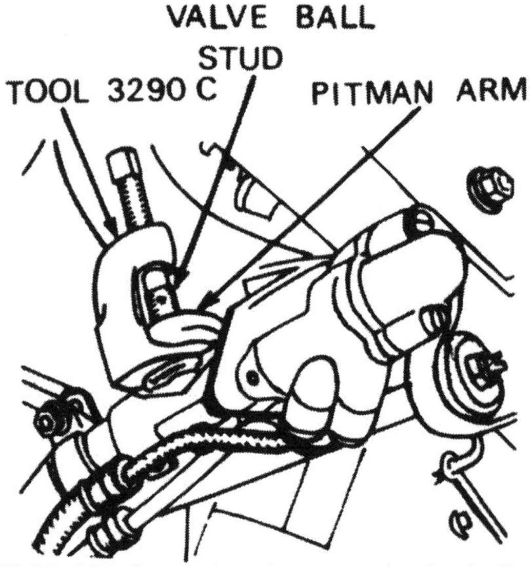

FIG 10:21 Removing the control valve ball stud

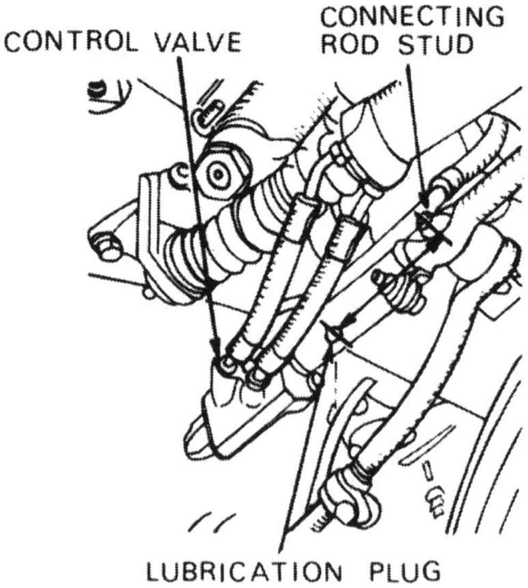

FIG 10:22 Installing the control valve

Screw the valve back into position until about four threads are left visible, and position the ball stud in the sector shaft arm. Measure the distance between the center of the grease plug and center of the stud, as shown in **FIG 10:22.** On all models, except 1964 and 1965, the valve should be screwed in or out until the dimension shown is correct at 5⅝ inch. On 1964 and 1965 models the dimension should be set between 4¼ and 4⅜ inch. When the distance is correct, align the hole in the steering arm to the idler arm rod with the slot in the sleeve and install a new roll-pin to lock the valve in position. Tighten the clamp and install the remainder of the parts in the reverse order of disassembly. Fill the reservoir to the correct level and top up as required after cycling the system and checking for leaks.

10:8 Steering geometry

The front wheel alignment should be checked by a service garage using accurate gauges. Incorrect alignment will cause excessive tire wear, leaving the tread of the tire worn with a characteristic feathered edge on one side.

The toe-in of the wheels and the setting of the steering wheel spokes is carried out by altering the effective lengths of the spindle connecting rods, as shown in **FIG 10:23.** The various adjustments that can be carried out are shown in **FIG 10:24.**

10:9 Trouble shooting guide

(a) Wheel tramp or shimmy

1 Worn hub bearings
2 Defective or worn suspension
3 Unbalanced wheels and tires
4 Slack or worn steering connections
5 Incorrect suspension geometry

(b) Wander

1 Check 2, 4 and 5 in (a)
2 Front and rear suspensions out of alignment (frame damage)
3 Uneven tire pressures or wear

(c) Hard steering

1 Check 5 in (a)
2 Neglected lubrication
3 Steering gear incorrectly adjusted
4 Steering column bent or misaligned
5 Defective bearings in steering column
*6 Defective pump
*7 Defective valve assembly
*8 Slipping drive belt
*9 Control valve spool centering not accurately set
*10 Low fluid level

(d) Lost motion

1 Check 1 and 4 in (a) and 3 and 9 in (c)
2 Defective column coupling
3 Loose steering wheel or badly worn splines

(e) Noisy operation*

1 Check 6, 8, 9 and 10 in (c)
2 Incorrect hoses used or hoses incorrectly routed

**Power steering only*

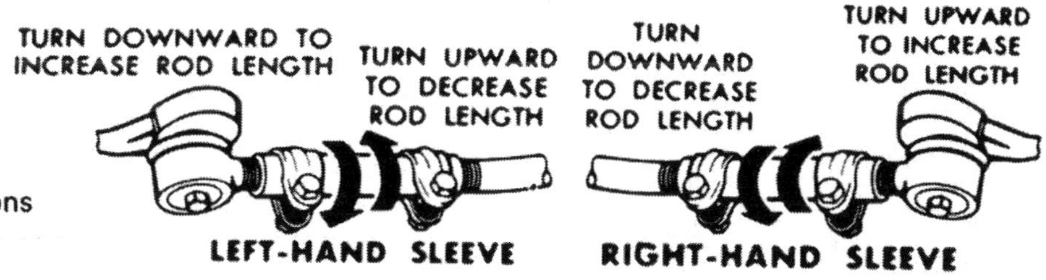

FIG 10:23 Adjusting the connecting spindles

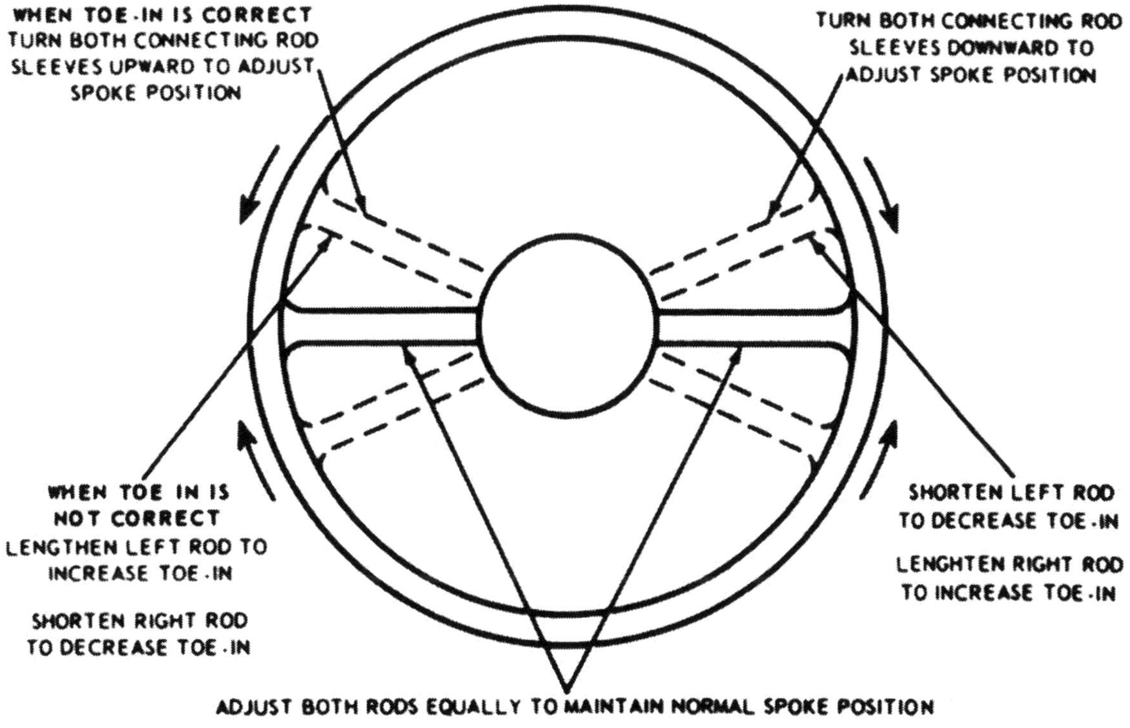

FIG 10:24 Adjusting the toe-in or spoke alignment

Inches	Decimals	Milli-metres	Inches to Millimetres		Millimetres to Inches	
			Inches	mm	mm	Inches
1/64	.015625	.3969	.001	.0254	.01	.00039
1/32	.03125	.7937	.002	.0508	.02	.00079
3/64	.046875	1.1906	.003	.0762	.03	.00118
1/16	.0625	1.5875	.004	.1016	.04	.00157
5/64	.078125	1.9844	.005	.1270	.05	.00197
3/32	.09375	2.3812	.006	.1524	.06	.00236
7/64	.109375	2.7781	.007	.1778	.07	.00276
1/8	.125	3.1750	.008	.2032	.08	.00315
9/64	.140625	3.5719	.009	.2286	.09	.00354
5/32	.15625	3.9687	.01	.254	.1	.00394
11/64	.171875	4.3656	.02	.508	.2	.00787
3/16	.1875	4.7625	.03	.762	.3	.01181
13/64	.203125	5·1594	.04	1.016	.4	.01575
7/32	.21875	5.5562	.05	1.270	.5	.01969
15/64	.234375	5.9531	.06	1.524	.6	.02362
1/4	.25	6.3500	.07	1.778	.7	.02756
17/64	.265625	6.7469	.08	2.032	.8	.03150
9/32	.28125	7.1437	.09	2.286	.9	.03543
19/64	.296875	7.5406	.1	2.54	1	.03937
5/16	.3125	7.9375	.2	5.08	2	.07874
21/64	.328125	8.3344	.3	7.62	3	.11811
11/32	.34375	8.7312	.4	10.16	4	.15748
23/64	.359375	9.1281	.5	12.70	5	.19685
3/8	.375	9.5250	.6	15.24	6	.23622
25/64	.390625	9.9219	.7	17.78	7	.27559
13/32	.40625	10.3187	.8	20.32	8	.31496
27/64	.421875	10.7156	.9	22.86	9	.35433
7/16	.4375	11.1125	1	25.4	10	.39370
29/64	.453125	11.5094	2	50.8	11	.43307
15/32	.46875	11.9062	3	76.2	12	.47244
31/64	.484375	12.3031	4	101.6	13	.51181
1/2	.5	12.7000	5	127.0	14	.55118
33/64	.515625	13.0969	6	152.4	15	.59055
17/32	.53125	13.4937	7	177.8	16	.62992
35/64	.546875	13.8906	8	203.2	17	.66929
9/16	.5625	14.2875	9	228.6	18	.70866
37/64	.578125	14.6844	10	254.0	19	.74803
19/32	.59375	15.0812	11	279.4	20	.78740
39/64	.609375	15.4781	12	304.8	21	.82677
5/8	.625	15.8750	13	330.2	22	.86614
41/64	.640625	16.2719	14	355.6	23	.90551
21/32	.65625	16.6687	15	381.0	24	.94488
43/64	.671875	17.0656	16	406.4	25	.98425
11/16	.6875	17.4625	17	431.8	26	1.02362
45/64	.703125	17.8594	18	457.2	27	1.06299
23/32	.71875	18.2562	19	482.6	28	1.10236
47/64	.734375	18.6531	20	508.0	29	1.14173
3/4	.75	19.0500	21	533.4	30	1.18110
49/64	.765625	19.4469	22	558.8	31	1.22047
25/32	.78125	19.8437	23	584.2	32	1.25984
51/64	.796875	20.2406	24	609.6	33	1.29921
13/16	.8125	20.6375	25	635.0	34	1.33858
53/64	.828125	21.0344	26	660.4	35	1.37795
27/32	.84375	21.4312	27	685.8	36	1.41732
55/64	.859375	21.8281	28	711.2	37	1.4567
7/8	.875	22.2250	29	736.6	38	1.4961
57/64	.890625	22.6219	30	762.0	39	1.5354
29/32	.90625	23.0187	31	787.4	40	1.5748
59/64	.921875	23.4156	32	812.8	41	1.6142
15/16	.9375	23.8125	33	838.2	42	1.6535
61/64	.953125	24.2094	34	863.6	43	1.6929
31/32	.96875	24.6062	35	889.0	44	1.7323
63/64	.984375	25.0031	36	914.4	45	1.7717

UNITS	Pints to Litres	Gallons to Litres	Litres to Pints	Litres to Gallons	Miles to Kilometres	Kilometres to Miles	Lbs. per sq. In. to Kg. per sq. Cm.	Kg. per sq. Cm. to Lbs. per sq. In.
1	.57	4.55	1.76	.22	1.61	.62	.07	14.22
2	1.14	9.09	3.52	.44	3.22	1.24	.14	28.50
3	1.70	13.64	5.28	.66	4.83	1.86	.21	42.67
4	2.27	18.18	7.04	.88	6.44	2.49	.28	56.89
5	2.84	22.73	8.80	1.10	8.05	3.11	.35	71.12
6	3.41	27.28	10.56	1.32	9.66	3.73	.42	85.34
7	3.98	31.82	12.32	1.54	11.27	4.35	.49	99.56
8	4.55	36.37	14.08	1.76	12.88	4.97	.56	113.79
9		40.91	15.84	1.98	14.48	5.59	.63	128.00
10		45.46	17.60	2.20	16.09	6.21	.70	142.23
20				4.40	32.19	12.43	1.41	284.47
30				6.60	48.28	18.64	2.11	426.70
40				8.80	64.37	24.85		
50					80.47	31.07		
60					96.56	37.28		
70					112.65	43.50		
80					128.75	49.71		
90					144.84	55.92		
100					160.93	62.14		

UNITS	Lb ft to kgm	Kgm to lb ft	UNITS	Lb ft to kgm	Kgm to lb ft
1	.138	7.233	7	.967	50.631
2	.276	14.466	8	1.106	57.864
3	.414	21.699	9	1.244	65.097
4	.553	28.932	10	1.382	72.330
5	.691	36.165	20	2.765	144.660
6	.829	43.398	30	4.147	216.990

CHAPTER 11

THE BRAKING SYSTEM

11 :1 Description

Hydraulically operated brakes are fitted to all four wheels. Drum brakes are fitted on all rear wheels, but on later models disc brakes can be fitted as an option to the front wheels instead of the conventional drum brakes which are standard.

The rear brakes are mechanically operated by the park brake mechanism, and a system of cables transmits the pull to the rear brakes. A system of metal lines and flexible hoses transmits the hydraulic pressure generated by the master cylinder to the brakes, and a typical system is shown in **FIG 11 :1**.

On the early models a single master cylinder is fitted and all four brakes are hydraulically interconnected. On later models a dual master cylinder is fitted. The later master cylinder contains two pistons in series and the front and rear brakes are hydraulically independent from one another so that a failure in one half of the braking system will still allow the other half to operate at full efficiency, though pedal travel will be longer and the overall stopping distance of the car will be increased.

A power unit can be fitted as an optional extra between the brake pedal and master cylinder. The unit is vacuum operated, from the intake manifold, and proportionately increases the load applied to the master cylinder. For a given retardation, less pressure needs to be applied to the brake pedal. The unit is designed so that the brakes will still operate if the unit fails but there will be no vacuum assistance. **This point must be born in mind when towing the vehicle as the reservoir suction will soon be used up and a heavier brake pedal pressure will be required.**

11 :2 Maintenance

Fluid level:

At regular intervals check the fluid level in the master cylinder reservoir. **Wipe the filler cap clean** and remove it from the master cylinder. On the dual master cylinder, the diaphragm should come away with the filler cap. Top up, if required, to the correct level using Rotunda Super Duty Brake fluid (Specification SAE.70. R3).

On models fitted with disc brakes the level will steadily drop as the disc brake pads wear and this is normal. Any undue drop requires immediate investigation

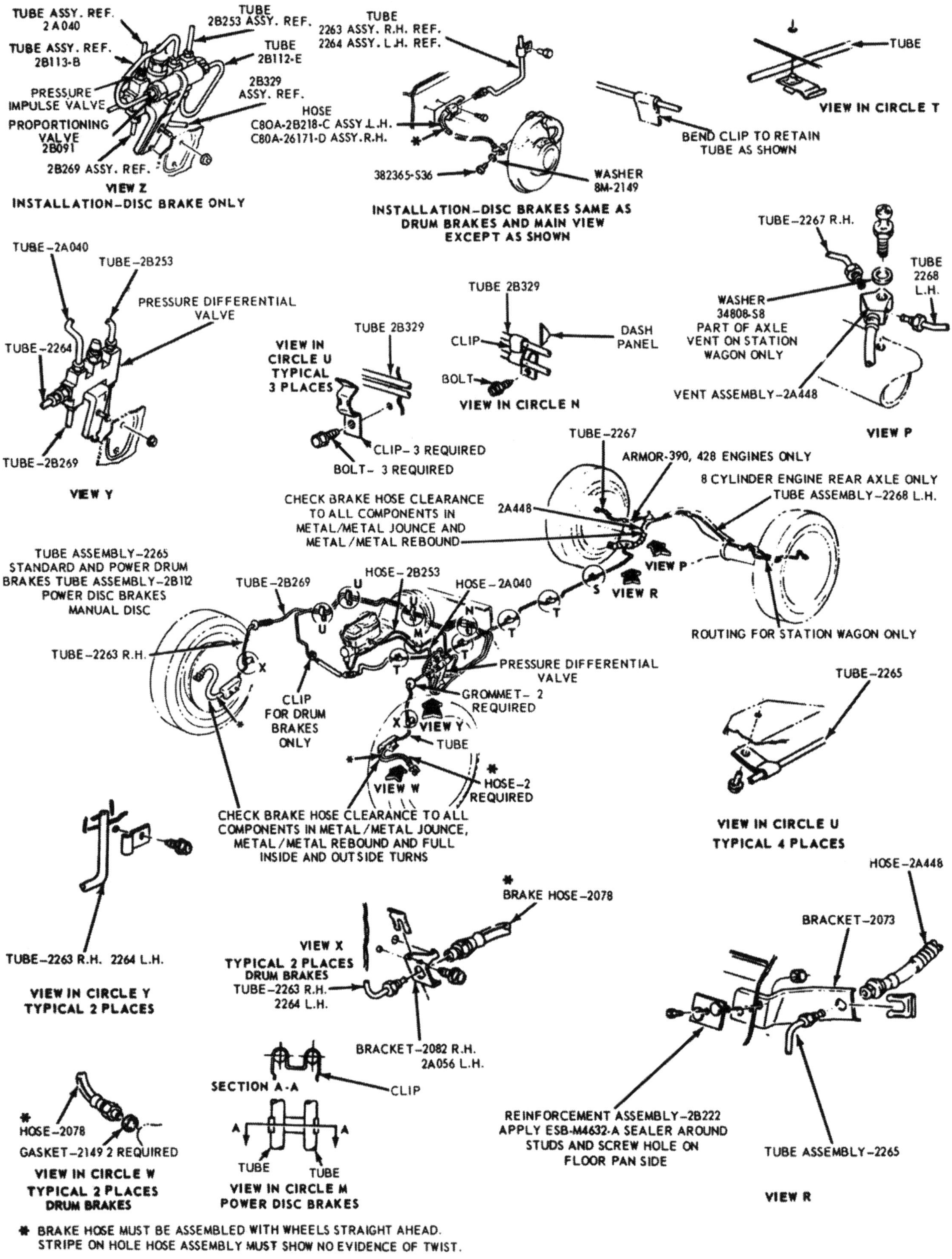

FIG 11:1 Typical service brake system, 1969 model

of the braking system as it can be caused by a fluid leak in the lines. Leaks must be repaired as soon as possible as they can lead to brake failure.

Adjustments:

All brakes are automatically adjusting with use. Disc brakes have a seal around the piston which automatically sets the clearance between pad and rotor. Drum brakes have an internal mechanism that operates an adjusting lever when the clearance between the lining of the shoe and drum becomes excessive. At regular intervals the brake drums should be removed, loose dust and dirt blown out, and the linings checked for wear. Drum brake linings must be renewed when there is less than $\frac{1}{32}$ inch of lining material left above the rivet heads.

The adjustment of the parking brake is dealt with in **Section 11:11.**

Preventative maintenance:

Regularly examine the friction linings on the brakes as well as all lines and flexible hoses. Lines which are corroding or dented by stones should be renewed as soon as possible as the damage will cause a weak spot which may fail under the stress of heavy braking.

At intervals of 18 months, drain the system and fill and bleed it using fresh hydraulic fluid. Alternatively bleed the system using sufficient fresh fluid to ensure that all the old fluid is bled out and replaced by fresh fluid. The fluid gradually accumulates moisture from the atmosphere and, coupled with the heat from the brakes, this makes the fluid lose efficiency. It is for these reasons that the fluid should be regularly changed.

At intervals of three years, drain out the fluid and disassemble all the parts of the braking system. Discard all the old seals, fitting all new ones on reassembly, and check all components for wear or damage, renewing those that are defective. Hydraulic lines may be flushed through with denatured alcohol.

11:3 Flexible hoses

When removing or installing a flexible hose, the flexible portion must not be twisted. When the hose is installed the line on the hose must be straight.

Hoses which show signs of defects, such as cracking, perishing, or chafing, must be renewed. If a hose is blocked, try clearing it with gentle air pressure and if this does not remove the obstruction fit a new hose. **Do not attempt to clear a blocked hose by poking wire through it.**

Removal:

Hold the hexagons on the adjacent end of the hose with a wrench to prevent the hose from turning. Undo the union nut that secures the brake line to the hose, using another wrench. Remove the clip or locknut that secures the hose to the bracket, still holding the hose with a wrench, and then free the hose from the bracket. The other end of the hose can then be unscrewed, allowing the flexible portion to rotate freely.

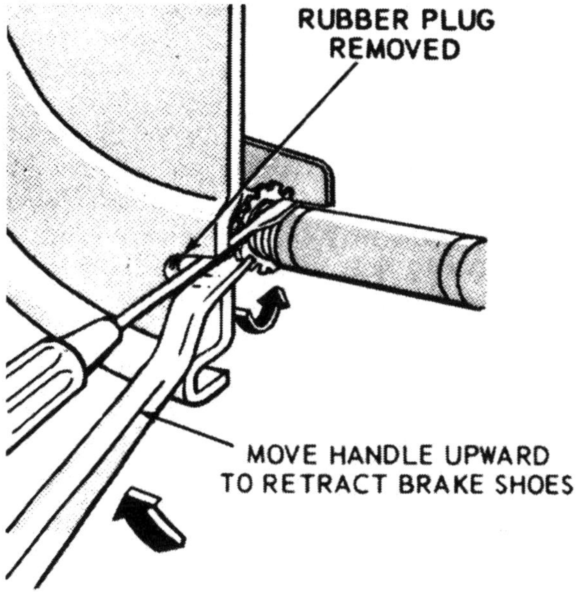

FIG 11:2 Backing off the brake adjustment

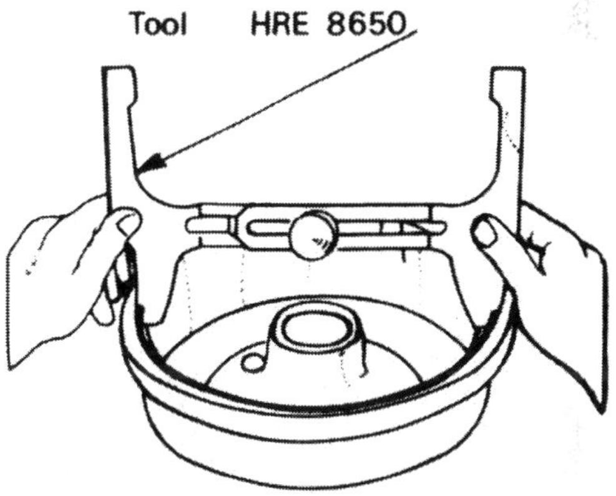

FIG 11:3 Measuring the drum diameter

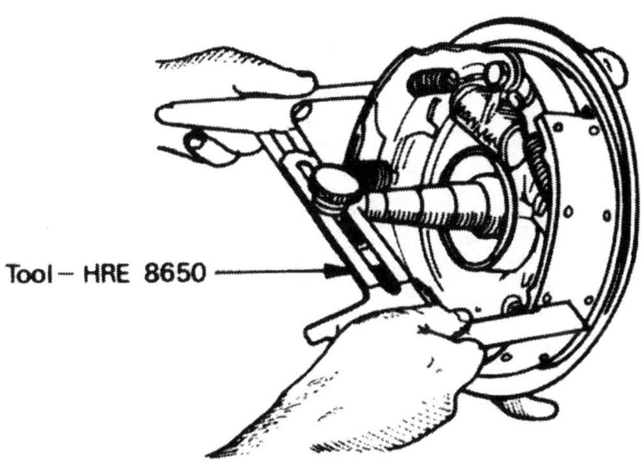

FIG 11:4 Setting the brake shoes to the correct diameter

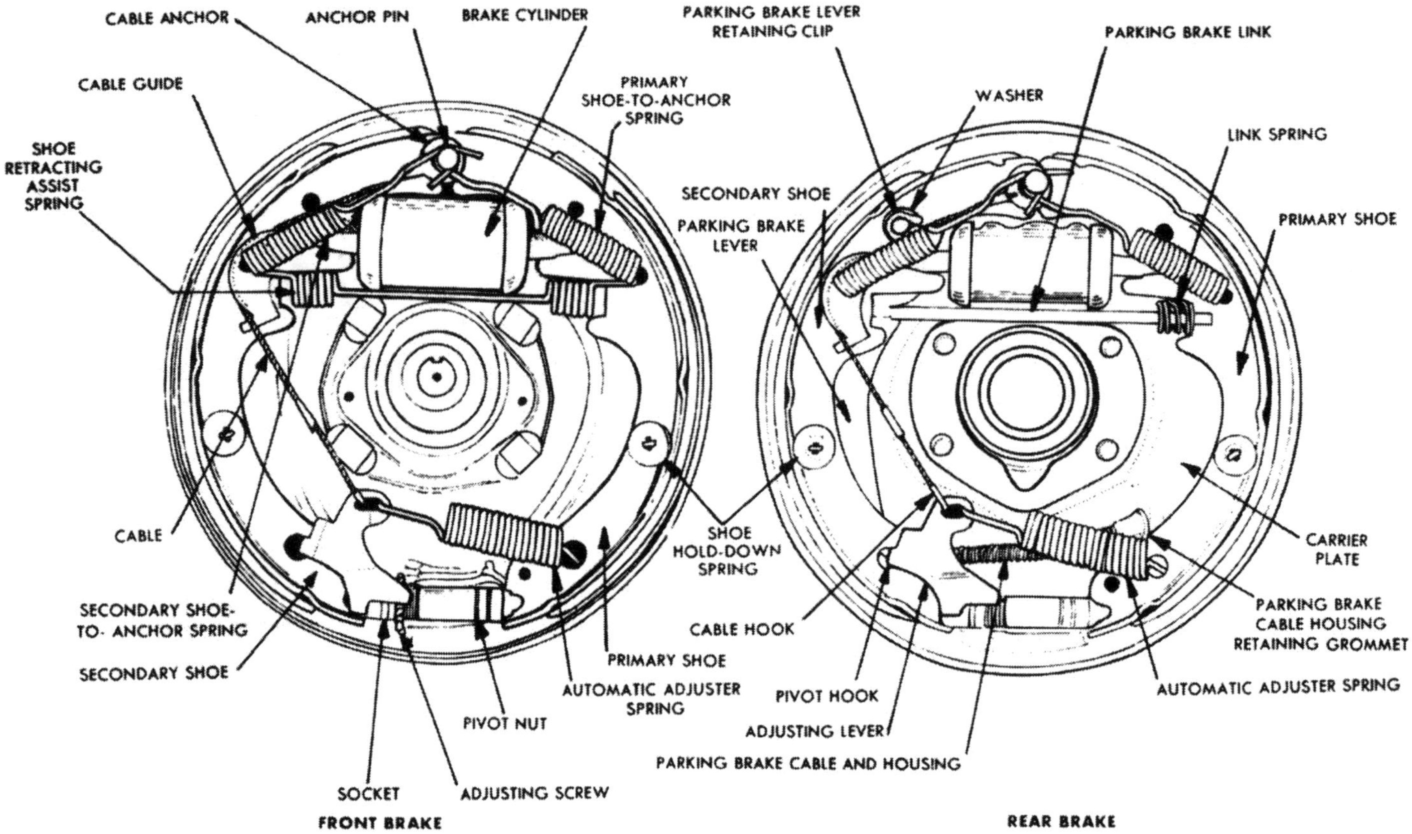

FIG 11:5 The brake assemblies with the drums removed

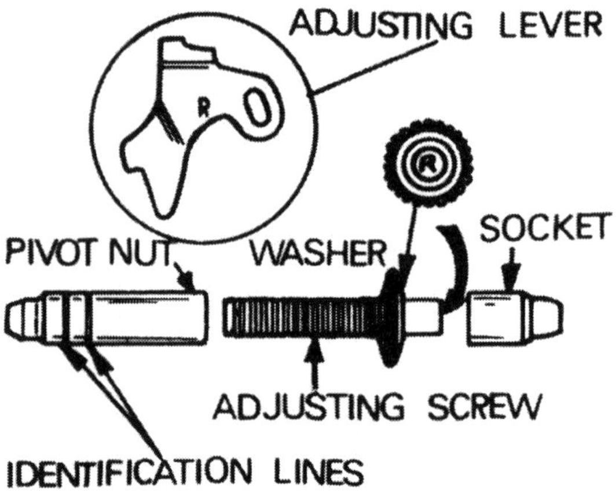

FIG 11:6 Adjusting screw and parts for a righthand side brake

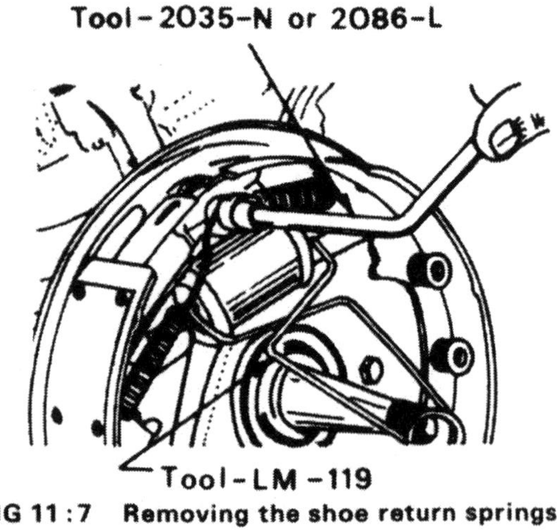

FIG 11:7 Removing the shoe return springs

Install the hose in the reverse order of removal. If seals are fitted at the end of the hose, they should be discarded and new seals fitted on installation. Check that the hose is clear of any obstructions, and clear of the muffler, when the suspension and steering moves through its range of movement.

11:4 Brake drums

Removal:

Slacken the road wheel attachment nuts slightly and jack up the appropriate side of the car. Remove the road wheel. On the front wheels it is best to remove the complete front hub assembly (see **Chapter 9, Section 9:3**) and then separate the drum from the hub. On the rear wheels, remove the three Timmerman nuts and draw the drum off.

If the drum is held in place by the brake shoes, slacken off the shoes to give clearance. Knock out the access slug at the back of the carrier plate and insert a thin screw-driver to disengage the adjusting lever from the adjusting screw, as shown in **FIG 11:2**. Use the brake adjusting tool to back off the adjusting screw as shown. **Take great care not to damage or burr the teeth on the adjusting screw otherwise the automatic adjustor may not operate correctly.**

Examination:

Wash down the carrier plate assembly using denatured alcohol and then brush and blow away any loose dirt. Wash the brake drum in any suitable degreasing solvent. On models where the drum is integral with the hub, avoid washing the grease retainer with solvent and wipe out the operating surface of the brake drum with a separate clean cloth and solvent to ensure that grease from the hub is not spread onto the operating surface. **Do not allow any solvent onto the brake linings themselves.** Take the opportunity to examine the wheel cylinder and hub assembly for leaks as well as checking the lining thickness.

Examine the operating surface of the drum for scores. Light damage may be machined off providing that not more than .060 inch of stock is removed. When machining a drum, be sure to mount it on an arbor and not in a chuck, as a chuck will slightly distort the drum. If the scoring is too deep to remove by machining, a new drum should be fitted. If the linings are only partially worn, leave the scored drum in place, provided that the scores are circumferential, and renew both drum and shoes when the linings are worn. Fitting a new drum to shoes that have mated into the scores in the old drum will reduce braking efficiency and probably cause scoring in the new drum as well.

Check the drum for cracks. Hang the drum on a wooden handle and tap it with a light metal tool. A flat note shows that the drum is cracked while if it rings true then the drum is satisfactory.

Do not use a cracked drum but discard it and fit a new one in its place. If the cracked drum is also found to be scored, new brake shoes must be fitted—irrespective of the wear left in the old linings.

Installation:

Measure the internal diameter of the brake drum with the Rotunda calipers, as shown in **FIG 11:3**. Clamp the calipers so that they do not alter and use them to set the brake shoes, as shown in **FIG 11:4**. The brake shoes are adjusted while holding the adjusting lever off the adjusting screw, as shown in **FIG 11:2**. Check that the adjusting screw rotates freely and if necessary lubricate it with a thin film of C1Az-19590-B grease on the threads.

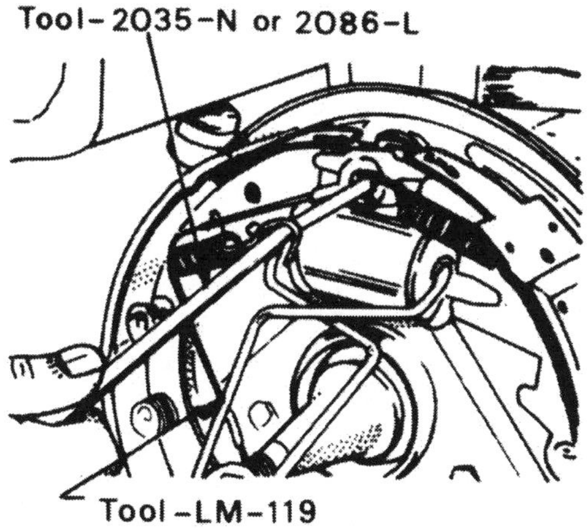

FIG 11:8 Installing the shoe return springs

Apply very small amounts of high temperature grease to points where the brake shoes contact the carrier plate and anchor pin. **Avoid getting any lubricant onto the linings themselves and avoid touching the shoes with dirty or greasy hands.**

When the shoes have been set, install the drum in the reverse order of removal. On the front wheels, set the correct hub clearance (see **Chapter 9, Section 9:3**). Install the road wheel.

When all the parts are installed and all drums checked, drive the vehicle forwards to ensure that the brakes are not binding. Accurately set the adjustment by backing the car and stopping it with a pedal pressure greater than 50 lbs. After every stop drive the vehicle forwards and apply the brakes before stopping the vehicle again while backing.

Do not apply the brake pedal while a brake drum is off, otherwise the pressure will force out the pistons from the wheel cylinder.

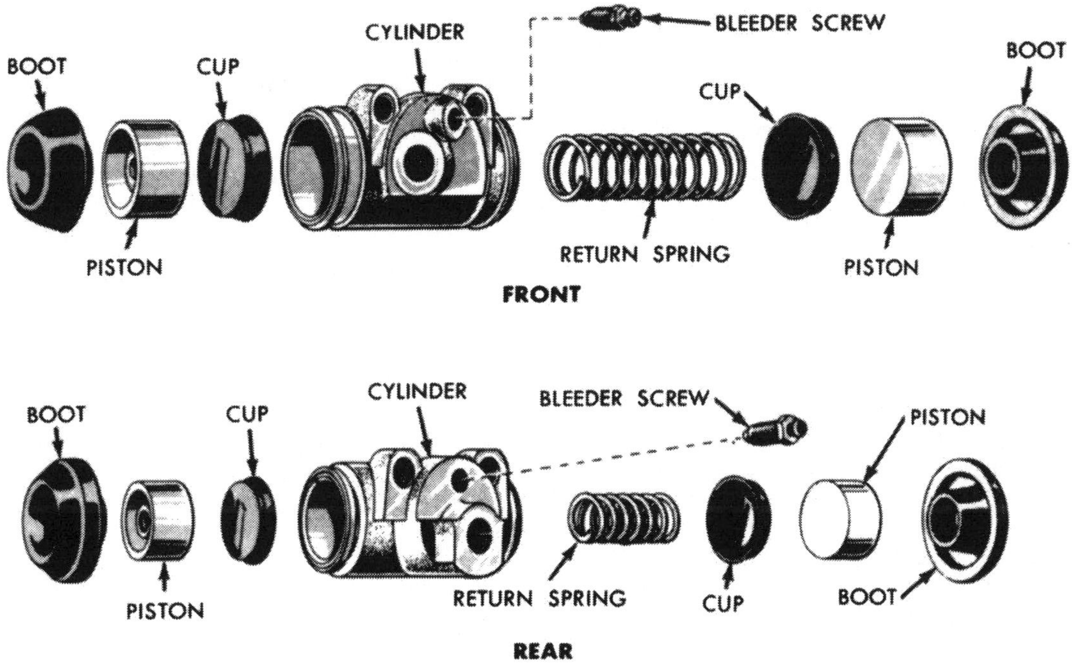

FIG 11:9 Front and rear wheel cylinders

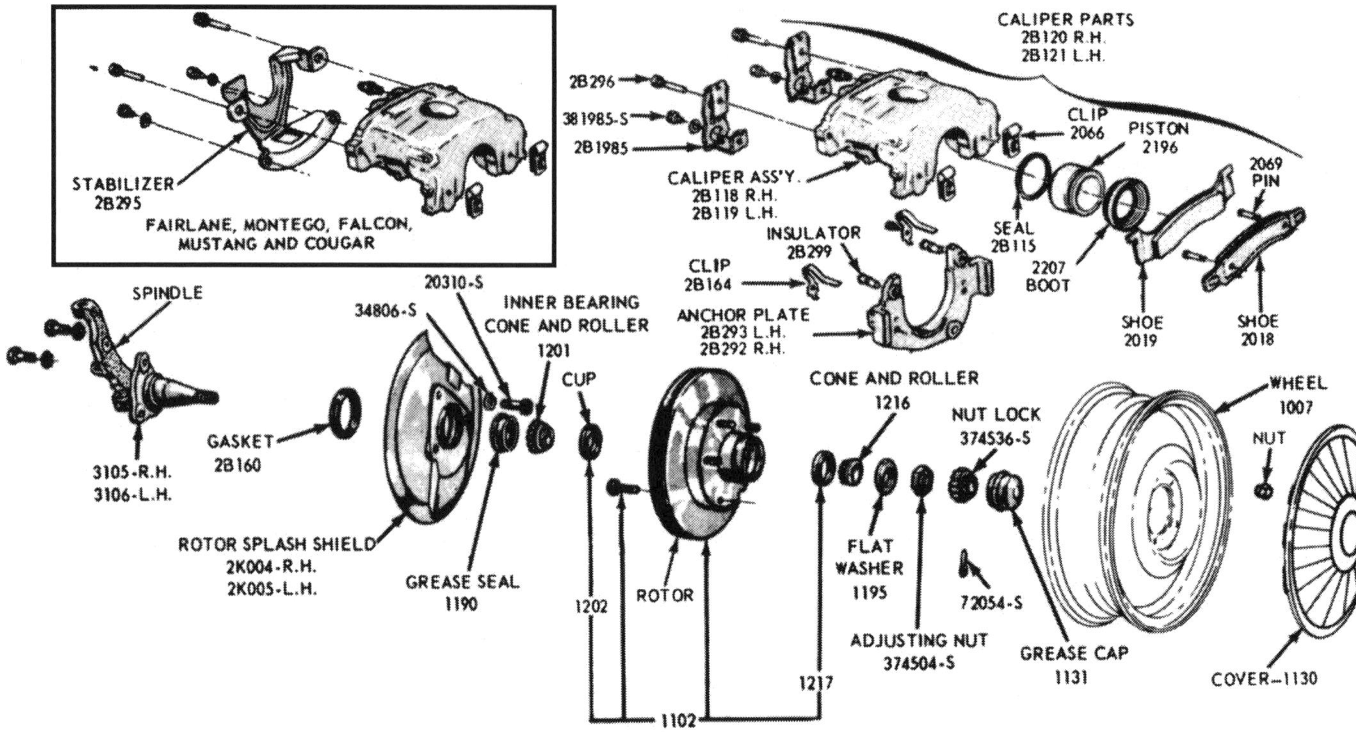

FIG 11:10 Typical disc brake parts, swinging caliper

11:5 Drum brake shoes and adjusting screws

The front and rear drum brakes are shown in **FIG 11:5.** The wheel hub assemblies are removed for illustrative purposes only. As the car backs and the brakes are applied, the movement of the shoes causes the anchor pin to pull the cable which acts on the adjusting lever. If there is sufficient play, the lever will move the adjusting screw by one tooth to take up the play. When the brakes are in adjustment there is insufficient movement for the adjusting lever to grip the tooth and turn it. To ensure that the parts actually adjust the brake, as fitting the parts to the incorrect side would cause them to slacken the shoes, the parts of the adjustor are marked, R for righthand side and L for lefthand. The righthand side adjustor parts are shown in **FIG 11:6** and the position of the identification letters can be seen in this figure. The adjusting nuts are also identified, two grooves for a righthand side nut and a single groove for a lefthand side one.

Removing parts:

1 Remove the brake drum as described in the previous section. Remove the shoe hold-down spring parts. Hold the pin in place with a finger behind the brake carrier plate and grip the washer with a pair of pliers. Press the washer in against the pressure of the spring and turn it through 90 deg. so that the washer can be slipped off the head of the retaining pin. Collect the spring and withdraw the retaining pin through the carrier plate.

2 As a safety precaution to keep the pistons in the wheel cylinder, clamp them in place. Free the return springs from the anchor pin, using the tool shown in **FIG 11:7.** If the special tools for removing and installing the springs are not available use a strong length of cord through the end of the spring to extend it and lift it off or onto the anchor pin.

3 Remove the shoes and adjusting screw as an assembly and then disassemble the parts on the bench. On the rear brakes, remove the parking brake link and spring and also disconnect the parking brake cable from the parking brake lever. After the shoes have been removed from the brake, remove the retaining clip and spring washers to free the parking brake lever from the secondary shoe.

Inspection:

Disassemble the components of the adjusting screw. Check the shoe linings for wear. **New linings must be fitted if the old ones are contaminated with oil or grease or if the linings are worn down to $\frac{1}{32}$ inch above the rivet heads.**

New linings must not be fitted to one wheel only as this will lead to inbalanced braking. **Always renew linings in complete axle sets.** When contaminated linings are found they must be renewed and no attempt made to restore the linings by baking or washing with solvents.

New linings are available already ground to shape but if the owner has not had experience of rivetting it is best to obtain exchange shoes and fit these. If new linings are fitted to the old shoes, use only the rivets supplied to secure them and make sure that the rivets are well clinched, leaving no gaps between shoe and lining. File the leading and trailing edges of the linings to a chamfer as this will help to prevent brake squeal.

Check the adjusting cable for damage or weakening. The cable must completely fill the crimped fittings or extend beyond them. If the cable appears weak or damaged, discard it and fit a new cable assembly, as a poor cable can be one cause of the automatic adjustor failing to operate correctly. Check the cable guide for

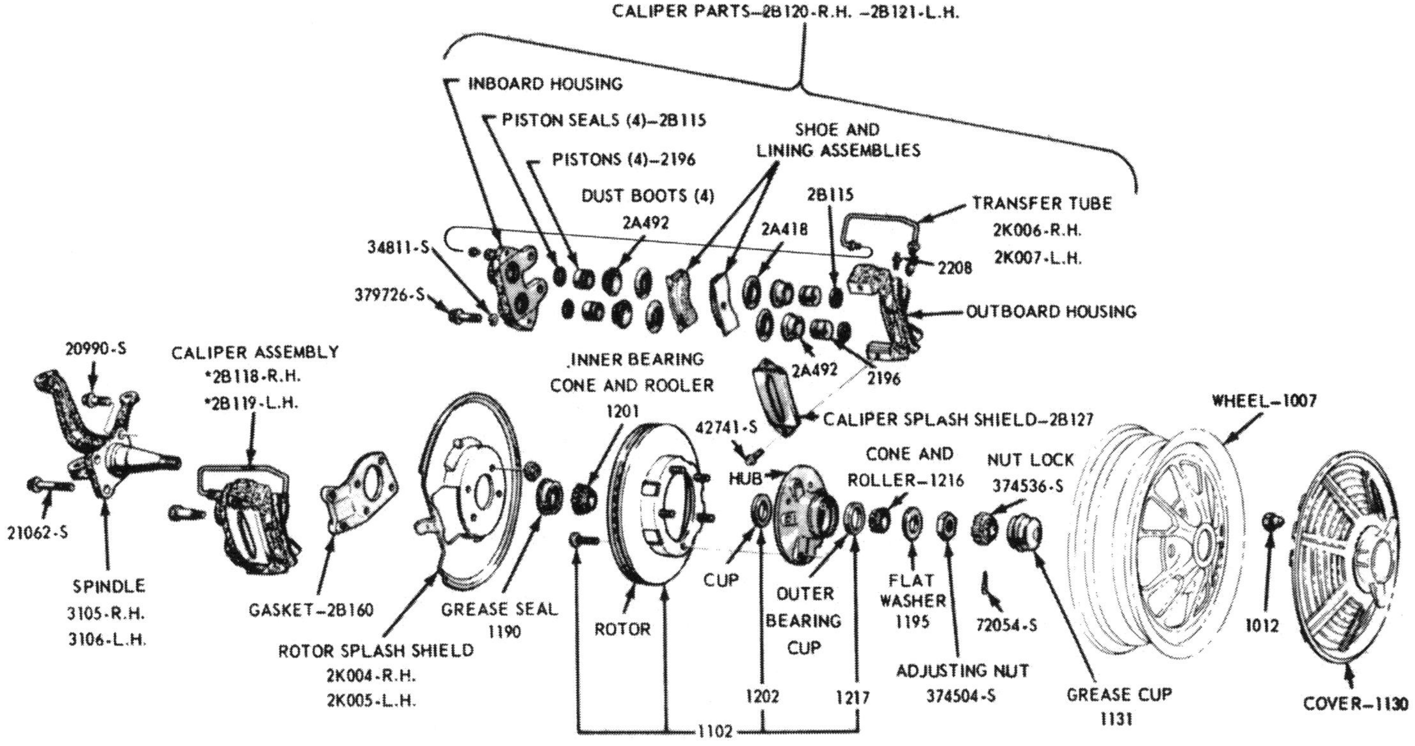

FIG 11 : 11 Fixed caliper disc brake fitted to some 1967 models

damage. The groove for the cable should be parallel to the shoe web and the body of the guide should lie flat against the web.

Check the pivot hook on the adjusting lever. The hook surfaces should be square with the body of the lever.

Renew any damaged or defective parts.

Installation:

The parts are installed in the reverse order of disassembly. Lubricate bearing points lightly with high temperature grease. The cable must be fitted with the crimped side of its eye towards the carrier plate and the cable must be routed in the guide and not between guide and carrier plate. Be sure that the parts all lie flat on the anchor pin and install the return springs as shown in **FIG 11 : 8**, noting that a length of cord can be used to extend the springs into position.

When the parts are in place, the adjusting lever should contact the adjusting screw teeth $\frac{3}{16} \pm \frac{1}{32}$ inch above the center line of the adjusting screw.

Check the operation of the adjustor. Pull on the adjustor cable, between the cable guide and anchor pin, toward the secondary shoe web far enough to lift the adjusting lever past a tooth on the wheel. The lever should spring into position behind the tooth and when the cable is released the return action of the lever should move the screw by one tooth.

Adjust the brakes and install the brake drums and road wheels. Set the final adjustment by backing the car and applying the service brake firmly.

11 : 6 Drum brake wheel cylinders

These are accessible after the brake drum has been removed (see **Section 11 : 4**) and the shoe assembly has been removed (see previous section). The components of the wheel cylinders are shown in **FIG 11 : 9**.

Disassembly and removal:

The parts can be disassembled without having to remove the wheel cylinder from the carrier plate. Only if the wheel cylinder bore is found to be scored or damaged need it be removed from the carrier plate.

Remove the links and boots from the wheel cylinder and then take out the cups, pistons and return spring from the cylinder.

If the cylinder has to be removed, take off the bleeder screw. On front wheels disconnect the flexible hose at the frame and then unscrew it at the wheel cylinder. On rear wheels, undo the union nut that secures the metal pipe to the wheel cylinder but **do not pull the pipe away from the cylinder**. Pulling the pipe will distort it, and the pipe will free anyway as the cylinder is removed from the backplate. Remove the wheel cylinder attaching bolts and take out the wheel cylinder.

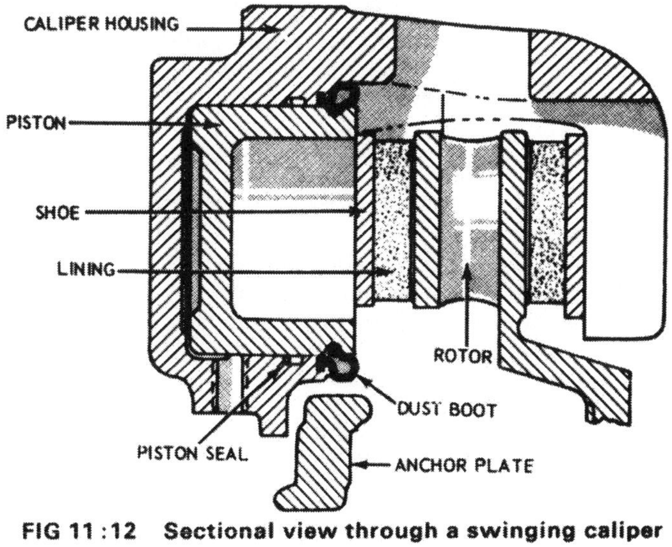

FIG 11 : 12 Sectional view through a swinging caliper

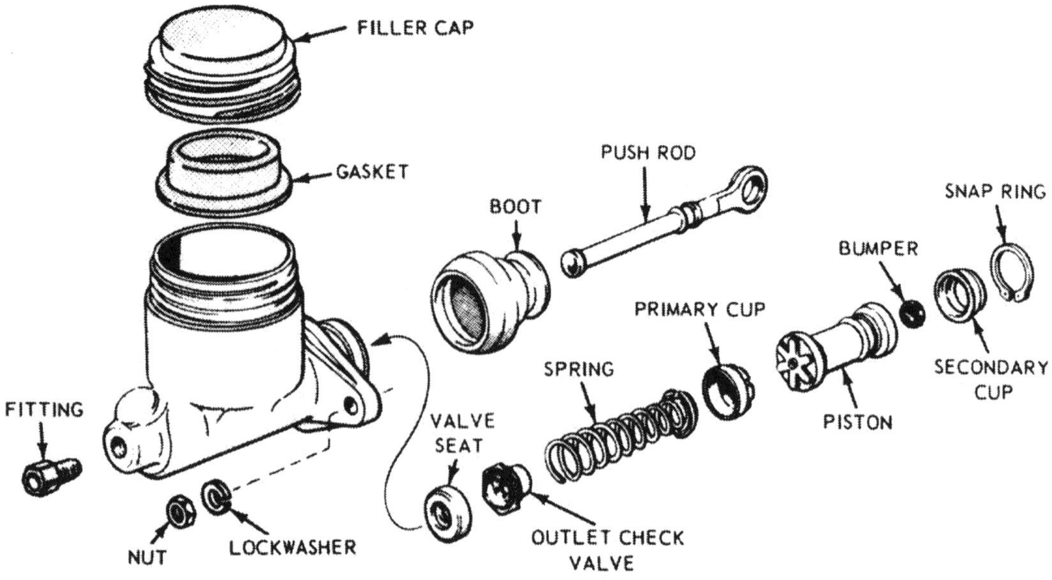

FIG 11:13 Single master cylinder fitted with a standard system

Inspection and cleaning:

Discard the old cup seals as well as rubber boots and fit new ones on assembly. Wash the parts with denatured alcohol. If the bore of the cylinder is scored, worn, or pitted a complete new assembly must be fitted.

Assembly and installation:

Wet the internal parts with hydraulic fluid. Fit one cup seal back into the bore, **taking great care not to bend back or damage the lips of the seal as it enters the bore.** From the open end, install the return spring and other cup seal, **taking care not to bend or damage the lips.** Be sure that the return spring is seated between the seals and insert the pistons, flat faces leading. Secure the parts in place with the rubber boots.

The components are installed in the reverse order of disassembly. Once the brake drums have been installed, fill the reservoir and bleed the system as described in **Section 11:10.**

Brake carrier plates:

Removal of the carrier plates entails removing the hub assembly. The complete brake assembly, apart from drum, can be removed as a unit after disconnecting the line to the wheel cylinder. The hydraulic system must be bled after the parts have been installed.

11:7 Disc brakes

A typical disc brake is shown disassembled in **FIG 11:10.** On some models a fixed caliper brake may be fitted and the parts are shown in **FIG 11:11.** The parts of the fixed caliper brake are shown fully disassembled but under no circumstances may the bolts that secure the two housings together be removed or slackened and all work must be carried out with the housing correctly bolted together.

A sectional view through a caliper is shown in **FIG 11:12.** The piston seal ensures that there is always a slight running clearance between the friction linings and rotor. When pressure is applied the piston moves in the bore to apply the friction linings to the rotor. As the piston moves it drags the outer edge of the seal with it so that the seal is deformed. If there is excessive clearance the piston slides through the seal, leaving the seal still distorted. As soon as the pressure is released, the seal returns to its normal shape and draws the piston slightly back with it.

A pressure control valve is fitted into the system to ensure that the hydraulic pressure to the rear brakes is limited when high pressures are required on the disc brakes. This valve ensures that the rear wheels do not lock under heavy braking. The proportioning valve cannot be dismantled or serviced and a new valve must be fitted if the old one is defective. The hydraulic system must be bled after the valve has been removed and refitted.

Caliper removal:

If the caliper is to be removed completely, disconnect the hydraulic line. If the caliper is being removed for the servicing of parts of the front suspension, leave the caliper attached to the line and, when it has been freed from the suspension, hang it safely out of the way so that the flexible hose is not strained or twisted.

Raise the front of the car on a jack and remove the road wheel. Take out the two bolts that secure the caliper to the suspension and slide the caliper off from the rotor.

Install the caliper in the reverse order of removal. The upper bolt should be tightened first. If the bolts are so equipped, lock them using a new length of safety wire after they have been torque loaded.

If the flexible hose has been disconnected, the hydraulic system must be filled and bled as described in **Section 11:10.**

Rotor:

Removal of the rotor entails removing the hub assembly and then separating the rotor from the hub (see **Chapter 9, Section 9:3**). Before separating rotor from hub, make alignment marks so that the rotor and hub will be reassembled into the original alignment.

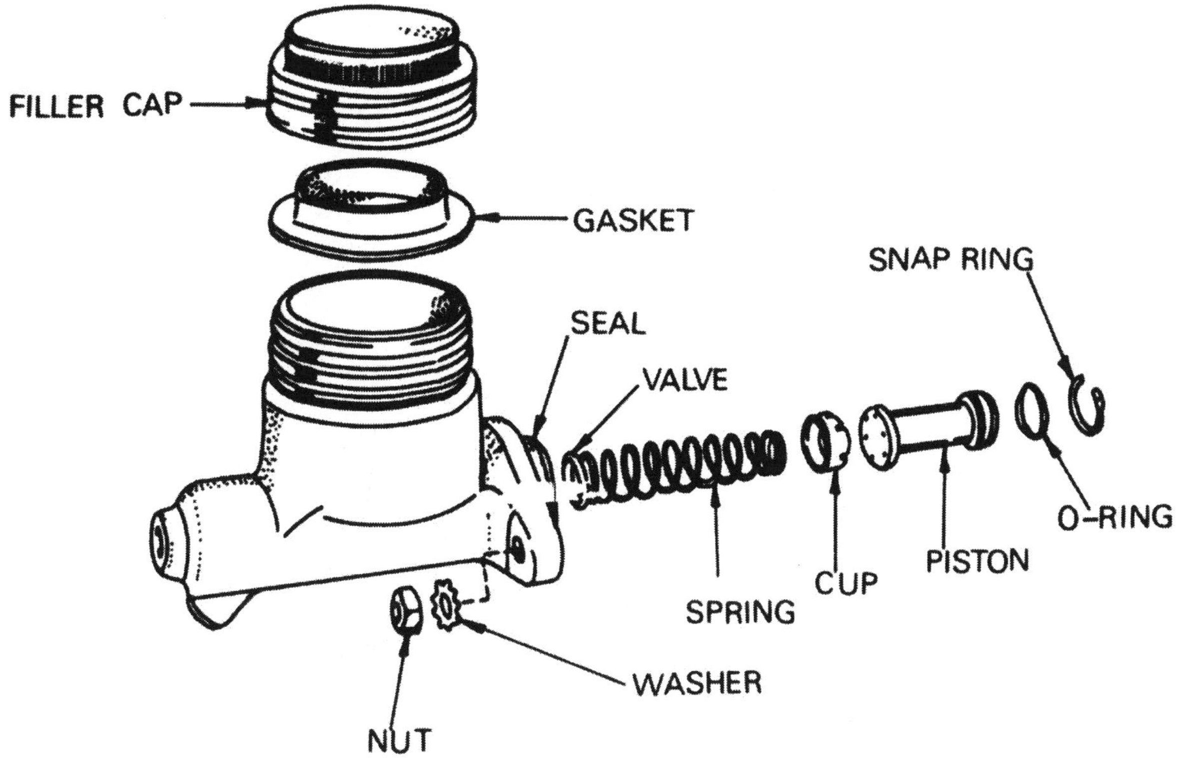

FIG 11:14 Single master cylinder fitted with a booster

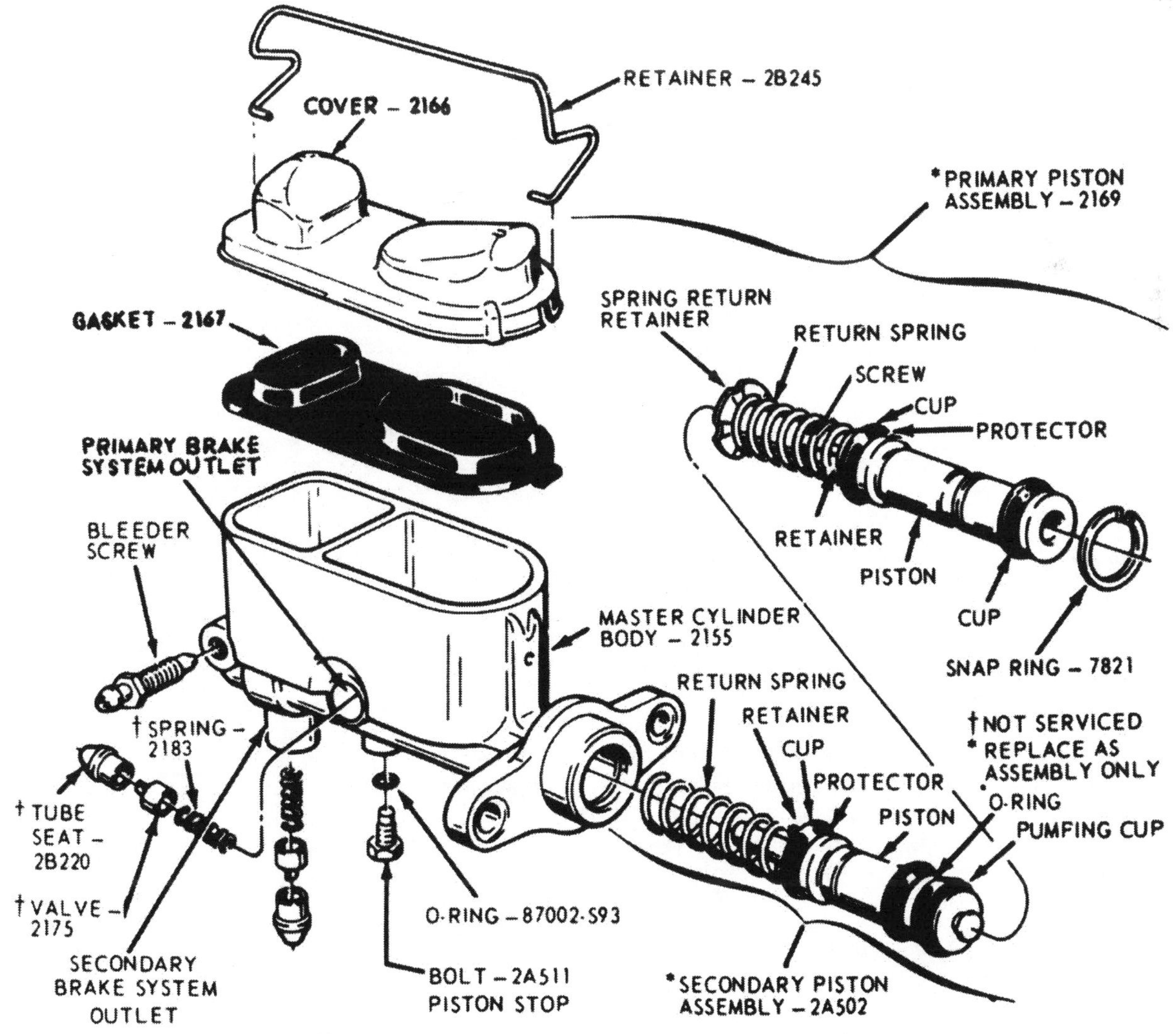

FIG 11:15 Dual master cylinder fitted with disc brakes

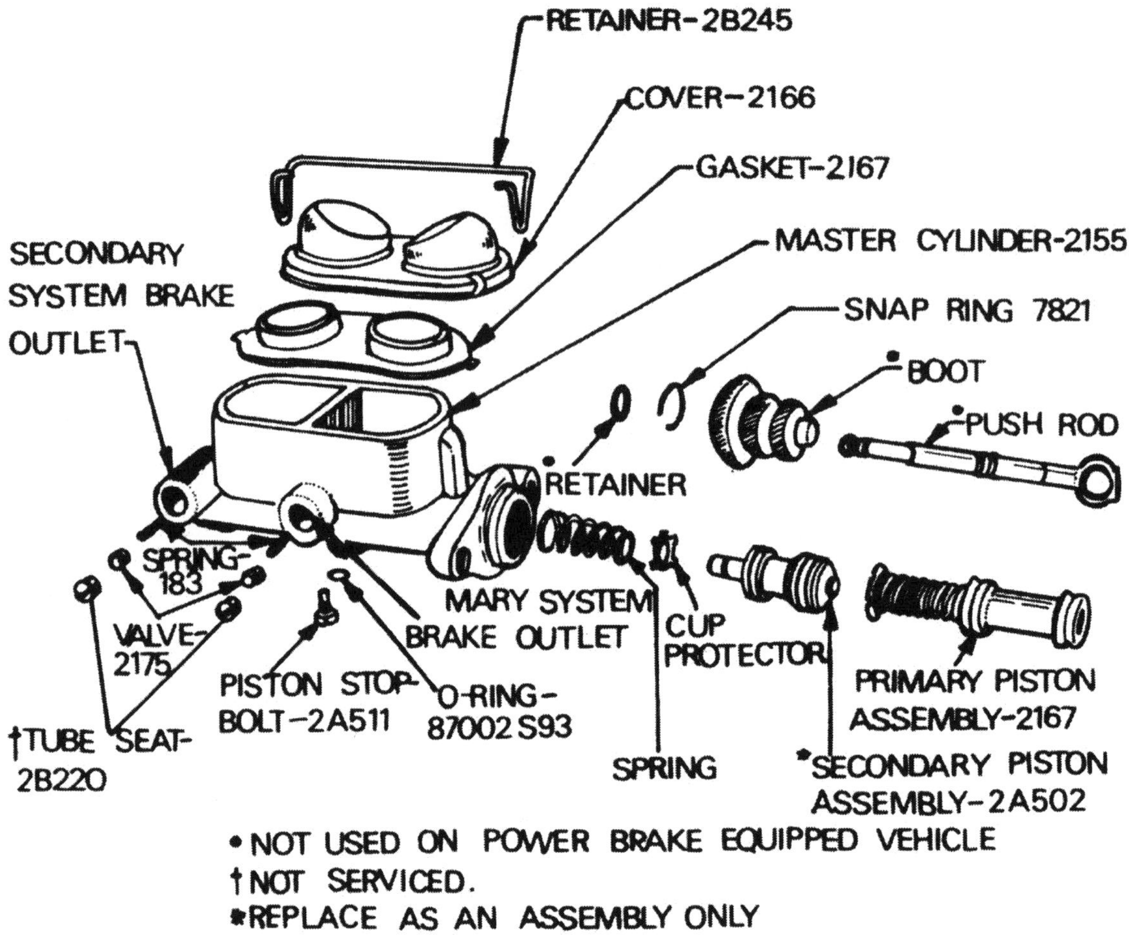

FIG 11:16 Dual master cylinder fitted with all drum brakes

If the rotor is scored or has excessive runout (measured with a dial indicator after the hub assembly has been correctly installed) the rotor may be machined down on a Rotunda rotor grinding machine. The rotor must be machined so that its operating faces are parallel and the thickness between them does not vary by more than .0007 inch. The rotor must not be machined below its minimum thickness, but this can be checked using a special ball and gauge. If the runout or scores cannot be cured by machining, a new rotor must be fitted.

Dirt and deposits, which affect braking efficiency, can be removed from the rotor faces by washing with trichlorethylene and gentle scrubbing with emerycloth.

Before installing the rotor to the hub, be sure that the mating faces are scrupulously clean, as specks of dirt will cause excessive runout on the rotor.

When the assembly has been installed on the spindle and the wheel bearings correctly set, use a dial indicator to measure the runout of the outer operating face near to the outer edge. If the runout is excessive, try rotating the rotor with respect to the hub to minimize the runout but is this fails the rotor must either be machined or renewed.

Renewing brake shoes and friction linings:

Exchange shoes are supplied with the linings bonded to them. **The owner should not attempt to bond new linings to the old shoes.**

New shoes must be fitted if the lining material has worn down to .030 inch at the front edge or if the lining material is contaminated with oil or grease.

1 Jack up the front of the vehicle and remove the road wheel.

2 Syphon out sufficient fluid from the reservoir of the master cylinder connected to the disc brakes so that the level is at least half-way down the reservoir. If this precaution is not taken, fluid may overflow from the reservoir as the pistons are pressed back into the caliper. **Discard the old fluid that was syphoned out**, and use fresh fluid for topping up.

2 Remove the bolts and clips that secure the shoes. Press the shoes carefully but firmly back into the caliper so that the pistons return into their bores. Pressure will need to be maintained for at least a minute and take great care not to damage or score either the caliper or the rotor.

3 Grip the lugs on each shoe with two pairs of pliers and pull each shoe in turn out of the caliper. Brush and blow out all loose dirt. If a solvent has to be used to wash out dirt use only denatured alcohol or hydraulic cleaning fluid as any other solvent will attach the material of the seals.

4 Install the new shoes in the reverse order of removal, taking care not to allow grease, oil or solvent onto the friction linings and not handling the parts with greasy or dirty hands. Secure the shoes in place.

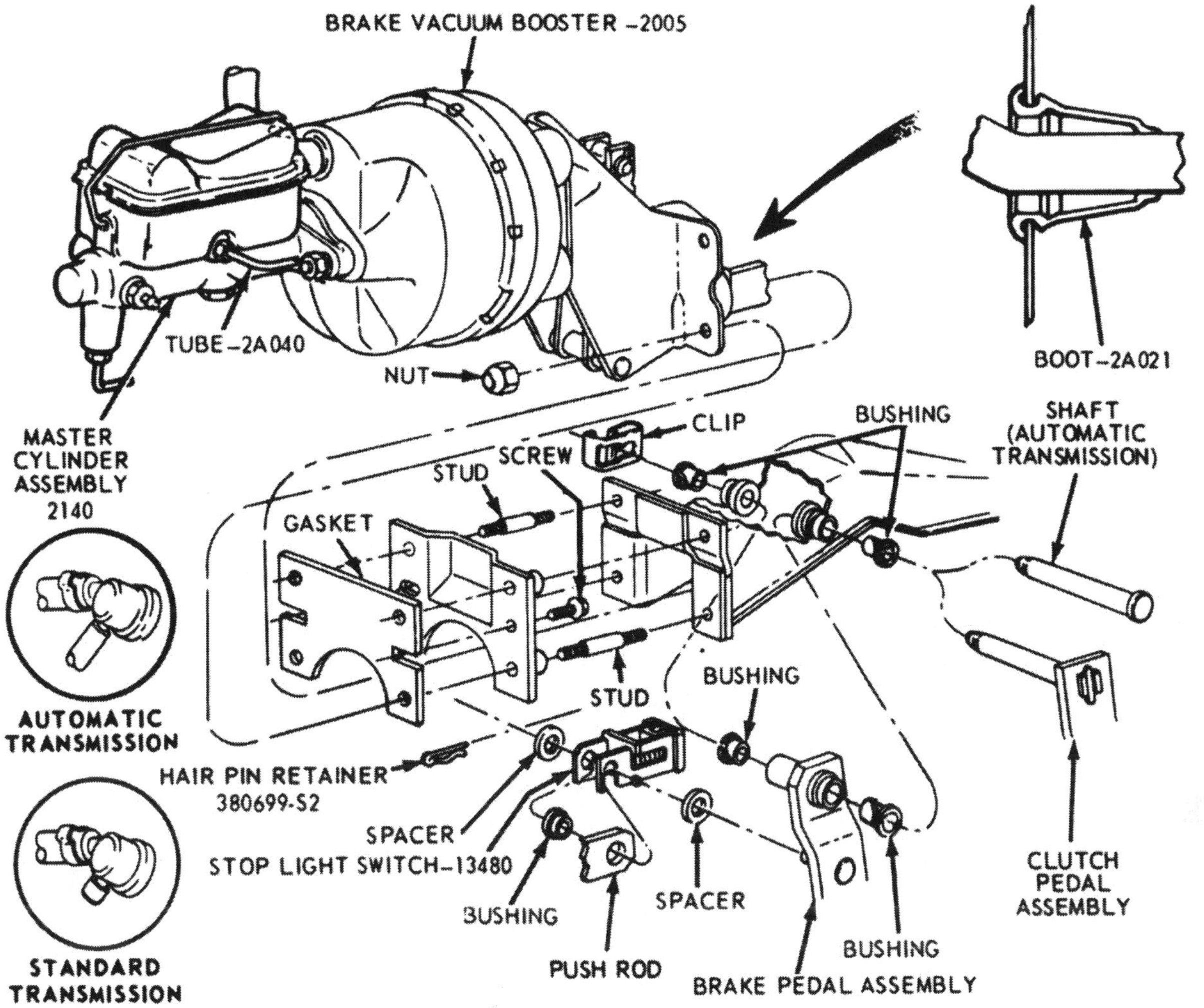

FIG 11 :17 Installation of dual master cylinder with a booster

5 Install the road wheel. Pump the brake hard several times to take up the adjustment before using the vehicle on the road. Top up the reservoir with fresh fluid until the level is $\frac{1}{4}$ inch below the top of the reservoir and fit the cap and diaphragm assembly.

Servicing brake caliper:

Remove the shoes from the caliper and the caliper from the suspension. Lay a piece of rag into the caliper and use air pressure through the inlet to gently blow out the pistons. Remove the dust covers from the pistons and carefully pry out the piston seals from the caliper bores. Discard the old dust covers and seals, fitting new ones on reassembly.

Wash all the parts in denatured alcohol or hydraulic cleaning fluid. Check the bores of the caliper for wear, scoring or pitting. If the bore is defective a complete new caliper assembly must be fitted.

Wet the piston seals with hydraulic fluid and insert them back into their recesses. **The seals should be refitted using the fingers only and they must seat fully and squarely into the recesses.**

Install the dust covers onto the pistons. Wet the pistons with hydraulic fluid. Carefully press the pistons back into their bores, taking great care to enter each piston squarely and make sure that it does not tilt. Secure the dust covers back in place on the caliper. The caliper is now ready for installation. After the hydraulic system has been bled, pump the brakes hard several times to take up the adjustment of the disc brakes.

11 :8 The master cylinder

Early models are fitted with a single master cylinder in which the one chamber operates all four brakes. On later models a dual master cylinder is fitted, one chamber operating the front brakes and the other chamber operating the rear brakes so that the hydraulic system is divided into two independent parts.

Care must be taken when ordering spare parts as the size and type of master cylinder varies with year and application.

The single master cylinder for models without power brakes is shown in **FIG 11 :13** and that for a system with power brakes in **FIG 11 :14**. A dual master cylinder for models with disc brakes is shown in **FIG 11 :15** while a dual master cylinder without disc brakes is shown in **FIG 11 :16**.

Removal:

The installation of a dual master cylinder with power brakes is shown in **FIG 11 :17** and that without power brakes in **FIG 11 :18**. The single master cylinders are installed in a very similar manner.

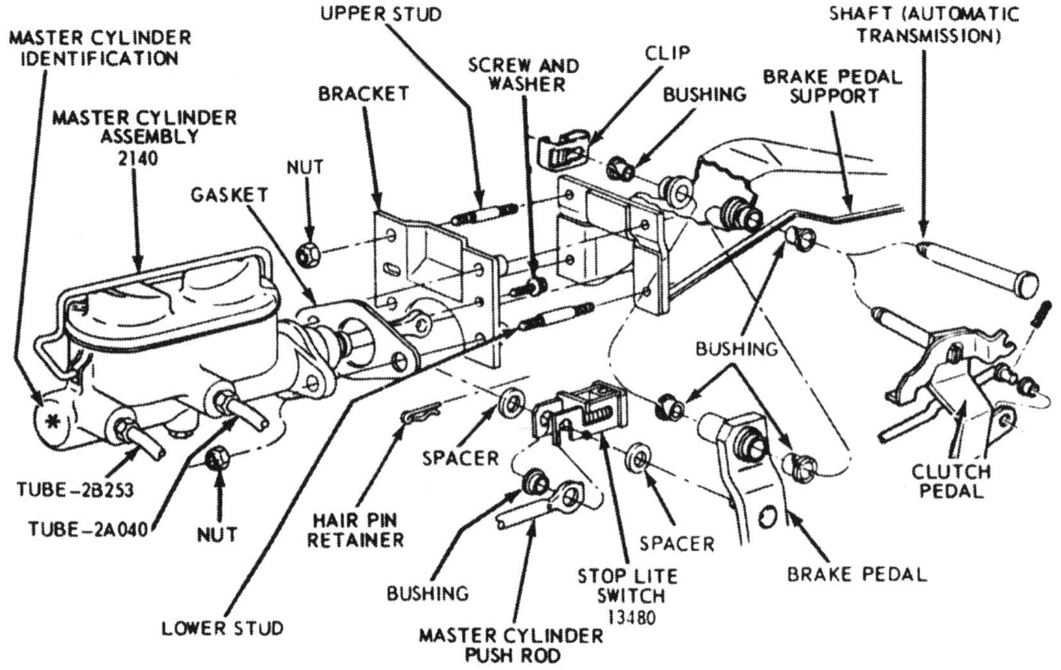

FIG 11:18 Installation of a dual master cylinder

Syphon out all the fluid from the reservoir(s). Disconnect the hydraulic lines from the master cylinder, using rags to catch any spillage. **Hydraulic fluid will rapidly damage paintwork if it is spilt.** On models not fitted with power brakes, the pushrod must be disconnected from the brake pedal. Disconnect the connectors at the stop light switch, remove the hairpin retainer, and slide the switch off the pivot and then up and out, taking care not to damage the switch. Remove the two attachment nuts and withdraw the master cylinder from the vehicle.

The master cylinder is installed in the reverse order of removal. Once the unit is in place the hydraulic system must be filled and bled as described in **Section 11:10.**

Servicing:

Wash the outside of the unit with denatured alcohol and pour out any remaining fluid. Pull back the rubber boot. Gently press the piston down the bore and remove the snap ring with snap ring pliers. Tap the open end of the master cylinder bore against the palm of the hand to free the internal parts. On dual master cylinders, remove the stop bolt that holds the secondary piston in place. Service kits are available and these contain all the parts necessary. On dual master cylinders, the primary piston parts must not be disassembled.

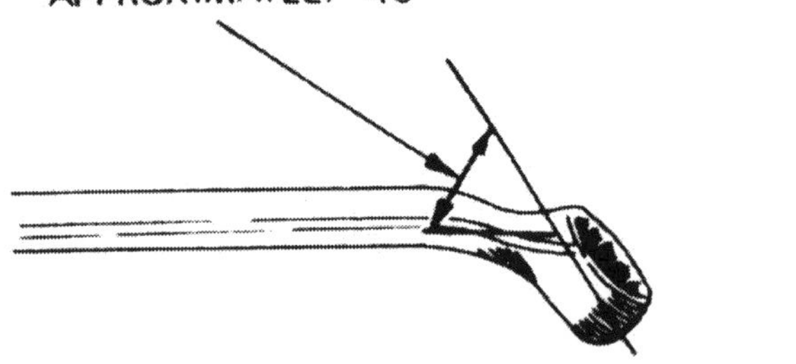

FIG 11:19 Modified wrench for bleeding the brakes

Wash the parts in denatured alcohol and blow through all passages to clean them. Lay the clean parts onto clean paper to prevent them picking up dirt. Inspect the bore. If there is only light etching, wear or pitting, the bore may be honed out by a specialist but if the damage is deep a new master cylinder must be fitted.

Wet the internal parts with hydraulic fluid and install them in the reverse order of disassembly. **Take great care when inserting the piston(s) not to damage or bend back the lips of the seals.** Secure the parts in place with the snap ring and install the rubber boot.

Pressure differential valve:

This is only fitted on models with dual master cylinders. If one half of the system fails, there will be a pressure differential which will move the plunger of the valve, causing it to operate the switch and illuminate the warning light. The unit cannot be serviced and if it is defective a new valve must be fitted in its place.

The method of centralizing the plunger if it has moved is dealt with in **Section 11:10.**

Removal of the valve is straightforward after disconnecting the electrical leads and hydraulic lines. When the valve has been installed, both halves of the brake system must be bled.

11:9 The power brake booster unit

Testing:

Apply the parking brake and pump the service brake several times to use up any vacuum. Apply the service brake, noting the pressure required. Start the engine and if the booster is functioning satisfactorily the pedal will gradually sink down and the pressure required to hold the brake will reduce.

If the unit does not operate, check the vacuum supply and the non-return valve on the unit. Disconnect the hose (two hoses on models fitted with automatic transmission)

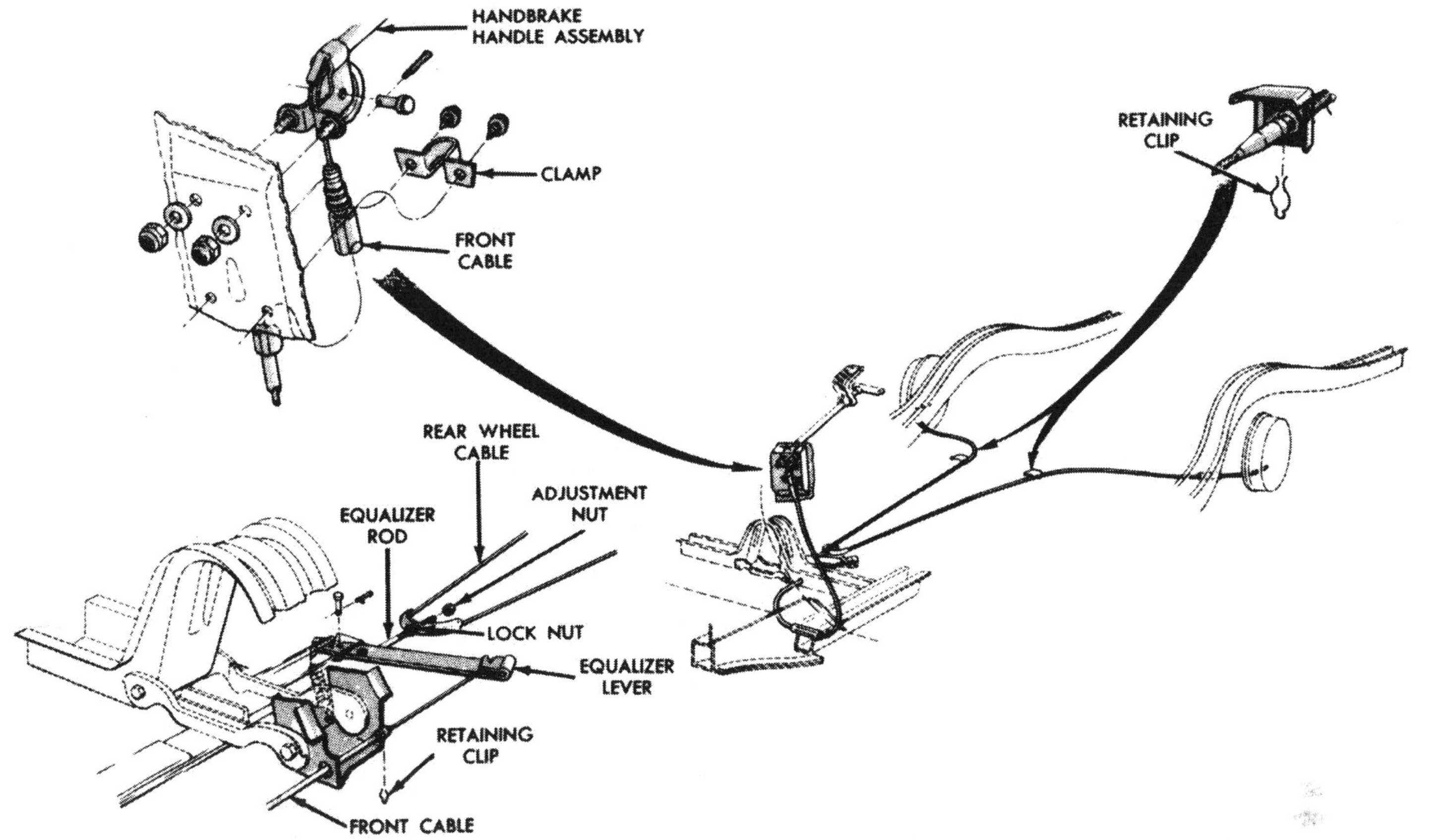

FIG 11 :20 The 1965 model parking brake linkage

and remove the check valve from the unit. Renew the valve if it is defective. Check the hoses for blockage or kinks.

If the unit still fails to operate, either fit a new unit or disassemble the old one and service it using a service kit.

Removal:

The installation with a dual master cylinder is shown in **FIG 11 :17.** Drain the master cylinder reservoir and dis-disconnect the hydraulic lines from the master cylinder. The master cylinder can then be removed from the booster unit. Disconnect the pushrod of the booster unit from the pedal, as described for removal of a master cylinder without booster. Take off the nut that secures the unit and remove it from the vehicle, after disconnecting the vacuum lines.

Install the unit in the reverse order of removal. Bleed the hydraulic system after installation of the master cylinder.

Servicing:

A service kit of seals and parts is available for servicing the unit.

Routine maintenance is confined to changing the air filter at the inlet, accessible after removing the unit from the car, at regular but long intervals.

Before installing the master cylinder or booster unit, check the protrusion of the booster unit pushrod at the front end of the unit. A special gauge is made for checking the protrusion, and the gauge bridges over the end of the unit. The pushrod is then adjusted so that it just touches the gauge.

11 :10 Bleeding the brakes

This is required when air has entered the system, either by allowing the fluid level to fall so low that air is drawn into the master cylinder or by dismantling the system.

On the models fitted with dual master cylinders, only the part of the system that has been dismantled requires bleeding as the other half of the system is hydraulically independent. On these systems, both systems must be bled if a component which is common to both systems has been removed and installed.

Fluid that has passed through the system should be discarded. Only if it is perfectly clean may it be used again but it must not be used until after it has been stored for at least 24 hours in a sealed clean container. The storage ensures that all air has risen out of the fluid. **Never return fluid directly from bleeding to the master cylinder reservoir. Never use fluid that has been stored in an open or dirty container.**

Before commencing the operation, remove the filler cap from the master cylinder and top up the reservoir as far as it will possibly go without spillage. Keep an eye on the level as the operation progresses and top up with more fluid when the level reaches the halfway mark. If the level is allowed to drop too far, air will be drawn into the master cylinder.

Bleed the brakes in the order of decreasing pipe runs, starting at the brake with the longest run. Disc brakes will require more bleeding than conventional drum brakes to remove all trapped air. The master cylinder itself should be bled first, if it is fitted with a bleeder screw. On the models fitted with disc brakes, it will be necessary to

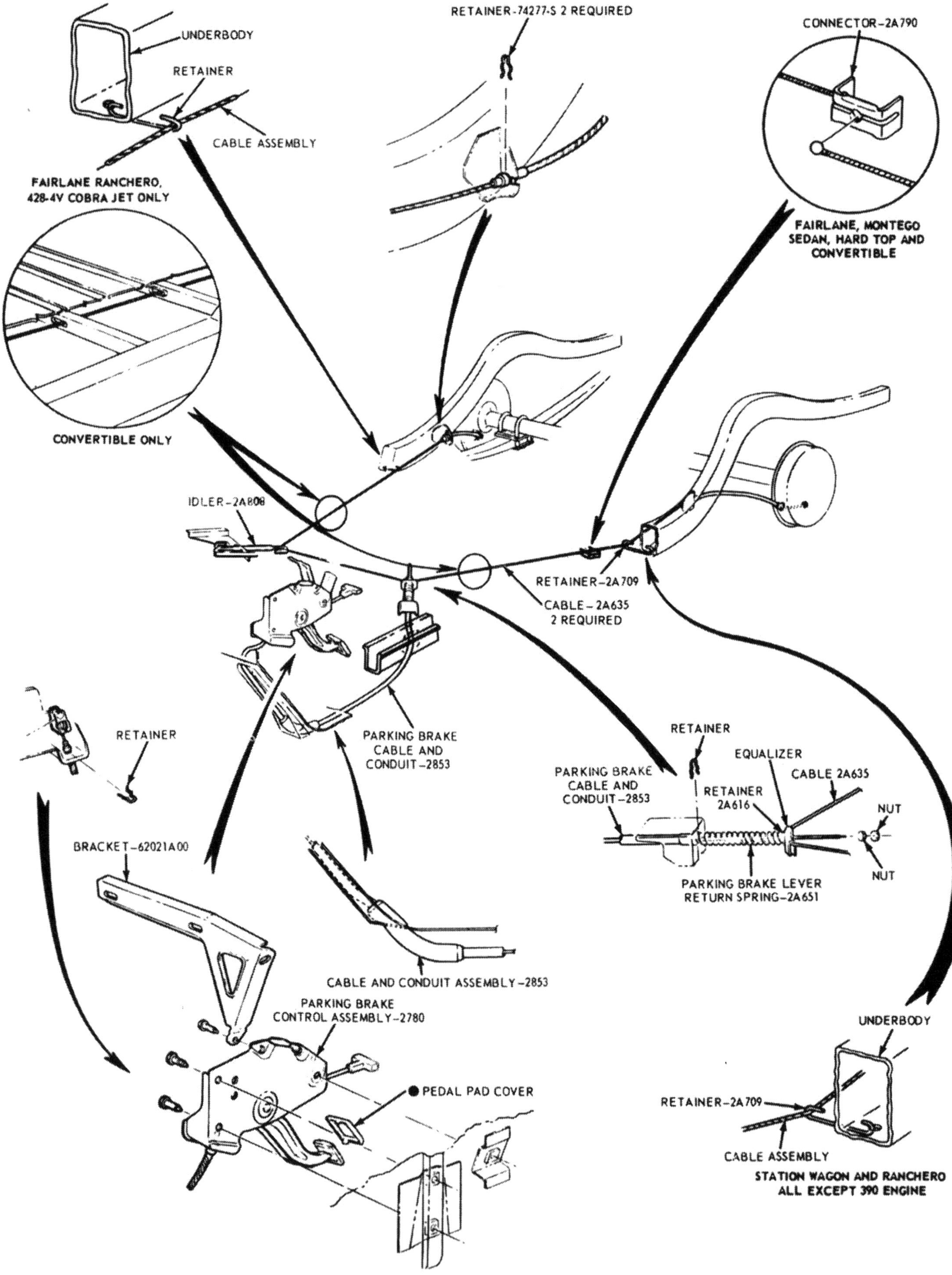

FIG 11:21 Typical later parking brake linkage, 1969 model

jack up the front of the vehicle and remove the road wheels in order to gain access to the disc brake bleeder screws.

Attach a length of plastic or rubber tube to the bleeder screw of the brake to be bled. Be sure that the tube is a snug fit on the bleeder screw. Dip the free end of the tube into a little clean hydraulic fluid in a clean glass container.

Open the bleeder screw by $\frac{3}{4}$ turn. The task will be easier if a $\frac{3}{8}$ inch box wrench is bent to the angle shown in **FIG 11 :19** and slipped onto the bleeder screw before installing the bleeder tube. Have an assistant slowly press the brake pedal right down through its full range of travel. If a power brake is fitted, exhaust the unit by pumping the service brake several times and carry out the bleeding without the engine running. When the pedal has reached the end of its travel, close the bleeder screw and have the pedal released. Repeat this cycle of opening, pressing down pedal, closing and releasing pedal until the fluid comes out of the bleeder tube in an unbroken stream with no air bubbles in it. Once the fluid is air free, transfer the bleeder tube to the next bleeder screw to be bled and repeat the process.

If the system has been fully dismantled, it will be difficult to remove all the air. Bleed the system until a firm pedal pressure is obtained and the brakes are safe to use. Use the vehicle normally for a few days and then rebleed the complete system. All the minute air bubbles will then have had a chance to coalesce into larger bubbles that are easier to bleed out.

In extreme cases, take the vehicle to a service garage for bleeding using pressure equipment.

Centralizing pressure differential valve:

If the valve is accidentally moved over during the bleeding operation, or the valve operated because of a leak in one system, it can be centralized to turn off the warning light. Make sure that the fault that caused the valve to move has been repaired.

A reverse pressure will be required to move the valve spool back into the central position. Switch on the ignition to ACC or On position but do not start the engine. Open a bleed screw on the half of the system other than the half which leaked and caused the spool to move. Apply steadily increasing pressure to the brake pedal and stop further movement of the pedal the instant that the warning light goes out. Close the bleed screw and release the brake pedal. If excessive pressure is applied to the pedal, the warning lamp will go out and then come on again as the spool moves past the center position. The original system will then have to have a bleeder screw opened and the pressure applied to move the spool back again.

Before using the vehicle, be sure that a firm pedal pressure is obtained.

11 :11 The parking brake

On the 1965 models a hand operated parking brake is fitted. The brake is applied by pulling on the handle in the vehicle and released by turning the handle clockwise and allowing it to return. The linkage is shown in **FIG 11 :20.** To adjust the cables, pull the handle out by three notches and raise the rear of the vehicle of the ground so that the rear wheels are free. Slacken the nuts on the equalizer rod and screw them forwards along the rod, so

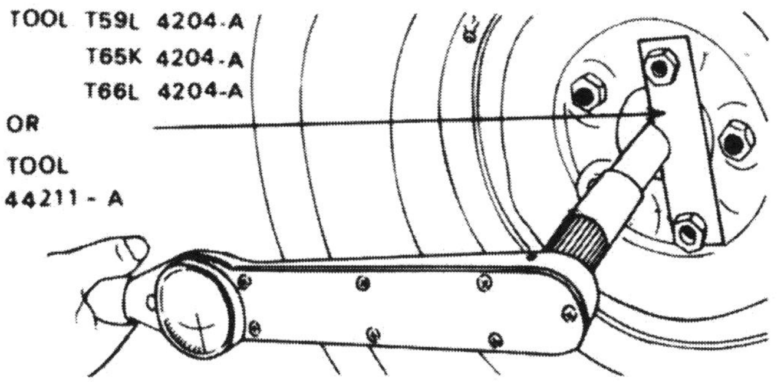

FIG 11 :22 Adaptor fitted to the rear wheel

that the rear nut pushes the equalizer forwards, until moderate drag is felt on the rear wheels when they are turned in a forward direction. Tighten back the front nut to the equalizer. Release the handle and check that both rear wheels rotate freely, allowing for the drag of the rear axle.

On later models, a foot-operated parking brake is fitted. The brake is applied by pressing down on the parking brake pedal and it is released by pulling on the release handle. A typical linkage is shown in **FIG 11 :21.**

To adjust the later linkage, jack up the rear of the vehicle and fit an adaptor to each rear wheel, as shown in **FIG 11 :22.**

Check that the wheels rotate freely, allowing for the drag of the rear axle, when the parking control is fully released. Apply the parking brake pedal to the first notch. Slacken the nuts on the equalizer and wind the nut forward to tension the rear cables. Use a torque spanner to check the force required to rotate the rear wheels. The adjustment at the equalizer is correct when there is a torque of 100 lb ft at break-away. Tighten the jam nut at the equalizer to hold the adjustment.

Release the parking brake and check that the rear wheels rotate freely. Depress the parking brake pedal to the third notch and check that this holds the vehicle. Release the brake and check that the brake shoes return fully. If the shoes do not fully release, check the cables for binds or kinks.

11 :12 Trouble shooting guide

(a) Spongy pedal

1 Air in the hydraulic system
2 Fluid leak in the hydraulic system
3 Gap between lining and shoes on drum brakes
4 Excessive hub end play on disc brakes

(b) Excessive pedal movement

1 Check 1, 2 and 4 in (a)
2 Excessively worn friction linings
3 Defective automatic adjustors
4 Very low fluid level in reservoir

(c) Brakes grab or pull to one side

1 Wet or oily friction linings
2 Distorted or damaged brake rotor
3 Excessive rotor runout
4 Seized piston in wheel cylinder or caliper
5 Seized parking brake cable
6 Mixed linings of different grades
7 Uneven tire pressures or wear
8 Defective steering or suspension

Inches	Decimals	Millimetres	Inches to Millimetres: Inches	Inches to Millimetres: mm	Millimetres to Inches: mm	Millimetres to Inches: Inches
1/64	.015625	.3969	.001	.0254	.01	.00039
1/32	.03125	.7937	.002	.0508	.02	.00079
3/64	.046875	1.1906	.003	.0762	.03	.00118
1/16	.0625	1.5875	.004	.1016	.04	.00157
5/64	.078125	1.9844	.005	.1270	.05	.00197
3/32	.09375	2.3812	.006	.1524	.06	.00236
7/64	.109375	2.7781	.007	.1778	.07	.00276
1/8	.125	3.1750	.008	.2032	.08	.00315
9/64	.140625	3.5719	.009	.2286	.09	.00354
5/32	.15625	3.9687	.01	.254	.1	.00394
11/64	.171875	4.3656	.02	.508	.2	.00787
3/16	.1875	4.7625	.03	.762	.3	.01181
13/64	.203125	5·1594	.04	1.016	.4	.01575
7/32	.21875	5.5562	.05	1.270	.5	.01969
15/64	.234375	5.9531	.06	1.524	.6	.02362
1/4	.25	6.3500	.07	1.778	.7	.02756
17/64	.265625	6.7469	.08	2.032	.8	.03150
9/32	.28125	7.1437	.09	2.286	.9	.03543
19/64	.296875	7.5406	.1	2.54	1	.03937
5/16	.3125	7.9375	.2	5.08	2	.07874
21/64	.328125	8.3344	.3	7.62	3	.11811
11/32	.34375	8.7312	.4	10.16	4	.15748
23/64	.359375	9.1281	.5	12.70	5	.19685
3/8	.375	9.5250	.6	15.24	6	.23622
25/64	.390625	9.9219	.7	17.78	7	.27559
13/32	.40625	10.3187	.8	20.32	8	.31496
27/64	.421875	10.7156	.9	22.86	9	.35433
7/16	.4375	11.1125	1	25.4	10	.39370
29/64	.453125	11.5094	2	50.8	11	.43307
15/32	.46875	11.9062	3	76.2	12	.47244
31/64	.484375	12.3031	4	101.6	13	.51181
1/2	.5	12.7000	5	127.0	14	.55118
33/64	.515625	13.0969	6	152.4	15	.59055
17/32	.53125	13.4937	7	177.8	16	.62992
35/64	.546875	13.8906	8	203.2	17	.66929
9/16	.5625	14.2875	9	228.6	18	.70866
37/64	.578125	14.6844	10	254.0	19	.74803
19/32	.59375	15.0812	11	279.4	20	.78740
39/64	.609375	15.4781	12	304.8	21	.82677
5/8	.625	15.8750	13	330.2	22	.86614
41/64	.640625	16.2719	14	355.6	23	.90551
21/32	.65625	16.6687	15	381.0	24	.94488
43/64	.671875	17.0656	16	406.4	25	.98425
11/16	.6875	17.4625	17	431.8	26	1.02362
45/64	.703125	17.8594	18	457.2	27	1.06299
23/32	.71875	18.2562	19	482.6	28	1.10236
47/64	.734375	18.6531	20	508.0	29	1.14173
3/4	.75	19.0500	21	533.4	30	1.18110
49/64	.765625	19.4469	22	558.8	31	1.22047
25/32	.78125	19.8437	23	584.2	32	1.25984
51/64	.796875	20.2406	24	609.6	33	1.29921
13/16	.8125	20.6375	25	635.0	34	1.33858
53/64	.828125	21.0344	26	660.4	35	1.37795
27/32	.84375	21.4312	27	685.8	36	1.41732
55/64	.859375	21.8281	28	711.2	37	1.4567
7/8	.875	22.2250	29	736.6	38	1.4961
57/64	.890625	22.6219	30	762.0	39	1.5354
29/32	.90625	23.0187	31	787.4	40	1.5748
59/64	.921875	23.4156	32	812.8	41	1.6142
15/16	.9375	23.8125	33	838.2	42	1.6535
61/64	.953125	24.2094	34	863.6	43	1.6929
31/32	.96875	24.6062	35	889.0	44	1.7323
63/64	.984375	25.0031	36	914.4	45	1.7717

UNITS	Pints to Litres	Gallons to Litres	Litres to Pints	Litres to Gallons	Miles to Kilometres	Kilometres to Miles	Lbs. per sq. In. to Kg. per sq. Cm.	Kg. per sq. Cm. to Lbs. per sq. In.
1	.57	4.55	1.76	.22	1.61	.62	.07	14.22
2	1.14	9.09	3.52	.44	3.22	1.24	.14	28.50
3	1.70	13.64	5.28	.66	4.83	1.86	.21	42.67
4	2.27	18.18	7.04	.88	6.44	2.49	.28	56.89
5	2.84	22.73	8.80	1.10	8.05	3.11	.35	71.12
6	3.41	27.28	10.56	1.32	9.66	3.73	.42	85.34
7	3.98	31.82	12.32	1.54	11.27	4.35	.49	99.56
8	4.55	36.37	14.08	1.76	12.88	4.97	.56	113.79
9		40.91	15.84	1.98	14.48	5.59	.63	128.00
10		45.46	17.60	2.20	16.09	6.21	.70	142.23
20				4.40	32.19	12.43	1.41	284.47
30				6.60	48.28	18.64	2.11	426.70
40				8.80	64.37	24.85		
50					80.47	31.07		
60					96.56	37.28		
70					112.65	43.50		
80					128.75	49.71		
90					144.84	55.92		
100					160.93	62.14		

UNITS	Lb ft to kgm	Kgm to lb ft	UNITS	Lb ft to kgm	Kgm to lb ft
1	.138	7.233	7	.967	50.631
2	.276	14.466	8	1.106	57.864
3	.414	21.699	9	1.244	65.097
4	.553	28.932	10	1.382	72.330
5	.691	36.165	20	2.765	144.660
6	.829	43.398	30	4.147	216.990

CHAPTER 12

THE ELECTRICAL SYSTEM

12:1 Description

A 12-volt electrical system is fitted to all models covered by this manual. **The correct polarity must be observed, as otherwise some of the components such as alternator and radio, may be irreparably damaged.**

Power for the system is produced by an alternator driven by a belt from the engine and the battery acts as storage. The battery supplies the very heavy current required by the starter motor for starting the engine and it also supplies current to the accessories when the engine is not running. A lead/acid battery is used. **It is essential that the charging circuit is disconnected when boost-charging the battery or carrying out arc welding repairs to the frame or bodywork.**

A test lamp and leads, or a 0-20 voltmeter, may be used to test the continuity of a circuit, but cheap instruments must not be used for checking the performance of, or adjusting components in, the electrical system. Cheap instruments will not measure to the accuracy required and for such tests high-grade, preferably moving-coil, instruments should be used.

Typical schematic wiring diagrams are given in Technical Data at the end of the manual to enable those with electrical experience to trace and rectify wiring faults.

Instructions for servicing electrical components are given in this chapter but it must be pointed out that it is a waste of time and money to try repairing items which are seriously defective, either mechanically or electrically. Such items should either be exchanged for new or reconditioned units, or discarded and new units fitted in their place.

12:2 The battery

Regular maintenance is essential for long battery life and to ensure satisfactory engine starting. If the battery is neglected, it will rapidly fail and, in addition to producing starting difficulties, will affect the performance of the whole electrical system.

In cold weather, the battery loses efficiency and at the same time more demands are made on it as the engine is stiffer to turn and the accessories are used more often. A battery that has limped through the summer will not be able to cope with the winter. If the battery is satisfactory, efficiency can be restored in extremely cold weather by removing the battery and storing it in a warm room overnight.

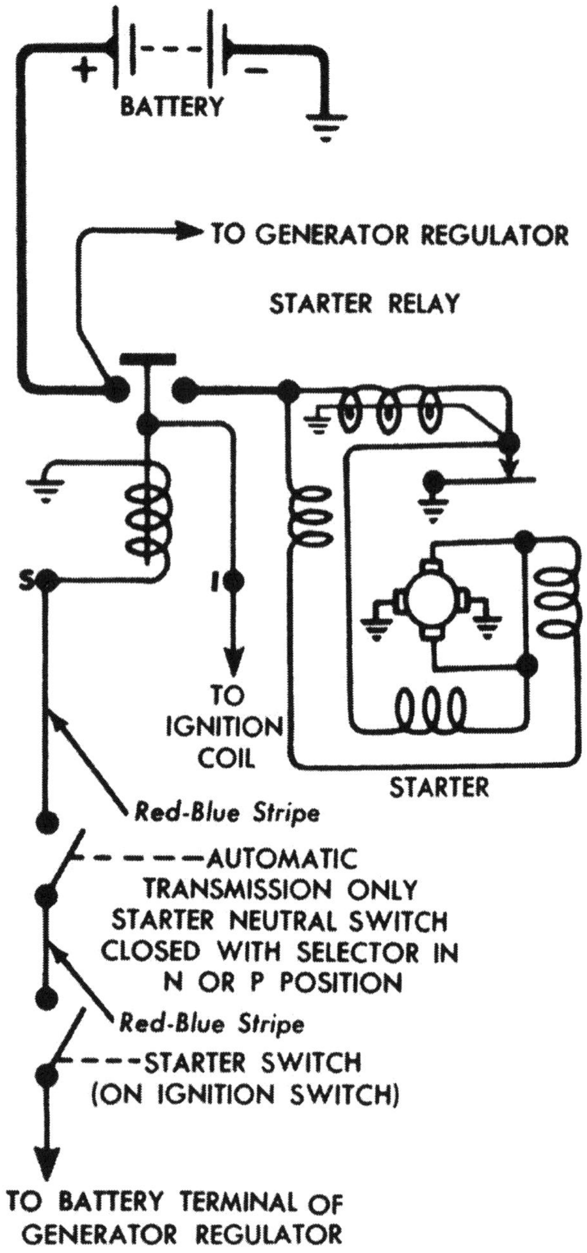

FIG 12:1 The starter circuit

The alternator produces a higher charging current at all speeds than a conventional generator, but if the vehicle is used for short-mileage journeys only in winter, a boost charge from a trickle charger at monthly intervals will keep the battery at full charge and make starting easier.

Maintenance:

Always keep the top of the battery clean and dry. Wipe away any spillage when refilling the cells with distilled water and at regular intervals wipe away any dirt or moisture from the top.

If spillage has caused corrosion to form on the battery surrounds, remove the battery and clean the area. Wash with a dilute solution of ammonia or baking powder and then flush liberally with clean water. When the parts are dry, paint them with anti-sulphuric paint.

Keep the battery terminal connectors clean and tight. High resistance at this point is often a cause of difficult starting. Disconnect the connectors. Use special wire brushes to clean the connectors and battery terminal posts. Corrosion may be washed off by using dilute ammonia and then flushing with clean water. Smear the connectors and posts with a non-metallic grease (such as petroleum jelly) to prevent further corrosion and make sure that the connectors are tight on the posts.

At regular intervals, approximately monthly, check the electrolyte level in the cells. Remove the cell vent caps and top up with pure distilled water until the level is just over the ring in the filler well. **Do not inspect the level with a naked light and do not add anything but pure distilled water.** Ready-mixed electrolyte should only be added to replace spillage or leakage.

Testing:

A hydrometer provides the easiest method available to the owner of testing the battery. If it is suspected that the battery or starter circuit is defective, take the vehicle to a service garage where they will check the battery under heavy discharge and also measure the voltage drop in the heavy duty starter motor leads.

Draw sufficient electrolyte up from each cell in turn to ensure that the float of the instrument is clear of the sides and bottom. Take the reading at eye level, check the color of the electrolyte and return it to its original cell before testing the next cell. If the electrolyte appears cloudy or filled with specks. the cell is probably defective and this will be confirmed by the reading of that cell varying by more than .050 from the readings of the other cells.

If the level in the cell is so low that electrolyte cannot be drawn up into the instrument, top up with distilled water and charge the battery for ½ hour at 35 amps to mix the electrolyte thoroughly.

The readings of the float give the following indications:

For climates below 32°C (90°F):

1.270 to 1.290	Cell fully charged
1.190 to 1.210	Cell half-charged
1.110 to 1.130	Cell discharged

Replace spillage with electrolyte of 1.270 specific gravity.

For climates above 32°C (90°F):

1.120 ro 1.230	Cell fully charged
1.130 to 1.150	Cell half-charged
1.050 to 1.070	Cell discharged

Replace spillage with electrolyte of 1.120 specific gravity.

The figures are given assuming a standard electrolyte temperature of 16°C (60°F). To convert the actual reading to standard, add .002 for every 3°C (5°F) increment above standard temperature and subtract for every increment below.

Charging:

The maximum charging current and period for the various battery capacities and state of charge are given in the following table:

Specific gravity reading	*Charge rate amperes*	*BATTERY CAPACITY—AMPERE HOURS* 45	55	70	80
1.125 to 1.150*	35	65 min.	80 min.	100 min.	115 min.
1.150 to 1.175	35	50 min.	65 min.	80 min.	95 min.
1.175 to 1.200	35	40 min.	50 min.	60 min.	70 min.
1.200 to 1.1225	35	30 min.	35 min.	45 min.	50 min.
Above 1.225	5	**	**	**	**

*If the specific gravity is below 1.125, use the indicated high rate of charge for the 1.125 specific gravity, then charge at 5 amperes until the specific gravity reaches 1.250 at 80°F.

**Charge at 5 ampere rate only until the specific gravity reaches 1.250 at 80°F.

The rates given are the maximum and no harm will be done if the battery is charged at a low rate for a correspondingly longer time.

A battery is fully charged when it is gassing freely and the specific gravity does not rise in three consecutive readings taken at hourly intervals. **The gases given off by the battery while charging are explosive, so do not bring naked lights near the battery and avoid sparks when installing a newly charged battery.**

A poor battery may sometimes be partially reclaimed by giving it three or four cycles of discharge and charge. Discharge the battery down to a cell voltage of 1.8 volts by connecting it to a lamp bank taking approximately 5 amps (approximately 60 watts). Recharge the battery at a 5 amp rate and then repeat the cycle. At the end of a few cycles charge the battery up ready for use.

Storage:

Short term storage of a well maintained battery presents no problems.

If the battery is to be stored for long periods, be sure that it is fully charged, with the top dry and all exposed metal well covered with non-metallic grease. Store the battery in a cool dry room well away from any temperature extremes.

At monthly intervals give the battery a freshening-up charge and at intervals of three months discharge it on a lamp bank and fully recharge it. If this precaution is not taken, the battery will slowly self-discharge and the plates sulphate up.

12:3 Servicing electrical motors

All electrical motors for automobiles follow the same general design principles and therefore there are many similar constructional features. To save repetition, general instructions for servicing motors are gathered into this section. It should be noted that some motors may be sealed on manufacture and therefore cannot be serviced but must be renewed if defective.

Brushgear:

Dismantle the motor sufficiently to expose the brushes. If the armature has to be removed, or the brush holder assembly pulled off the commutator, take great care not to allow the brushes to be damaged on the edge of the commutator.

Check the brushes for wear and renew them if they are excessively worn. Check that the brushes move freely in their holders and that the springs press them firmly down

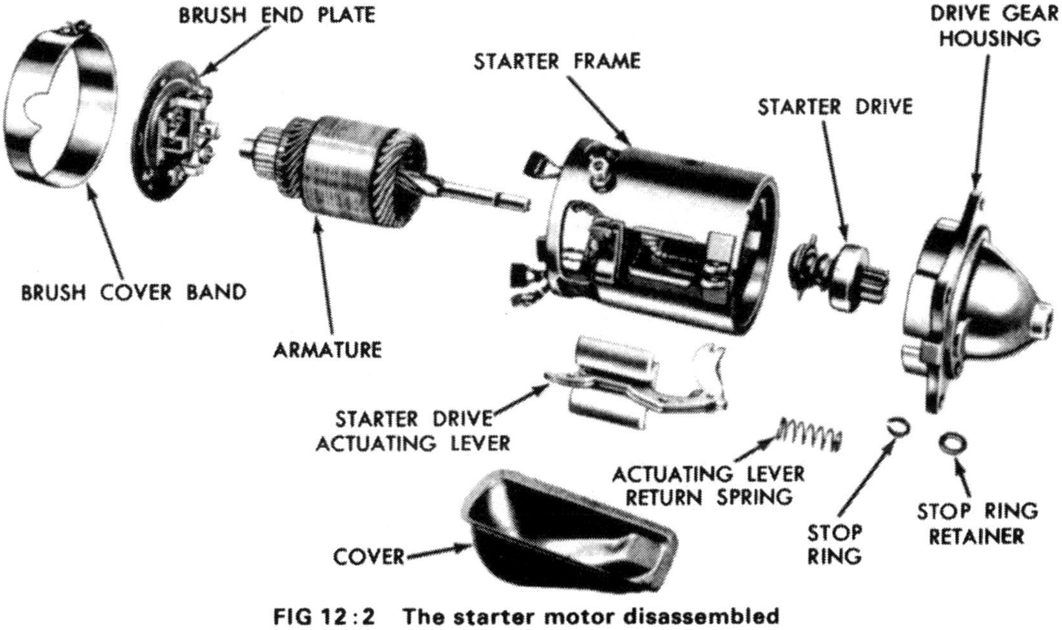

FIG 12:2 The starter motor disassembled

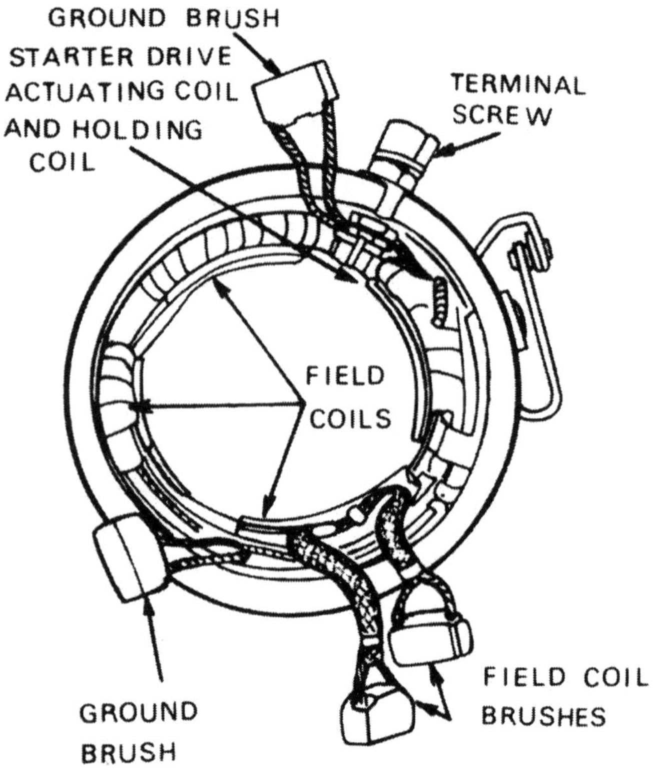

FIG 12:3 **The starter motor field coil assembly**

onto the commutator. Renew the springs if they are weak or damaged. If the brushes stick, remove them from their holders and lightly polish the sides of the brushes with a fine file. Clean the brush holders with a piece of cloth dipped into fuel. Install the brushes into their original positions so that the bedding-in is not altered.

Brushes are usually supplied with their ends ground to shape but if further bedding-in is required, make up a wooden mandrel the same diameter as the commutator. The commutator itself may be used. Wrap the mandrel with glasspaper (never emerycloth) as this leaves particles embedded in the copper) and rotate the mandrel with the brushes lightly pressed against the glasspaper.

When new brushes are to be attached by soldering, grip the connector with a pair of needle-nosed pliers to act as a heat sink and to prevent the solder from creeping up the connector and making it stiff.

Commutator:

Normally the commutator should have a darkened surface that is polished and free from scores or burn marks. Cleaning can be carried out by wiping the commutator with a piece of lint-free cloth dipped in denatured alcohol or gasoline.

Light burn marks or scores can be polished out using a strip of fine-grade glasspaper. **Never use emerycloth as this leaves hard particles embedded in the copper.**

If the damage is deeper, it may be machined off in a lathe. Use the highest speed possible and a very sharp tool and remove only sufficient stock to take out the damage. On some motors the insulation between the segments is undercut. If undercutting is required, either use a special tool or grind a hacksaw blade to the exact width of the insulation and use it to undercut the insulation squarely to a depth of $\frac{1}{32}$inch. Take a light final cut using a diamond- or carbide-tipped tool and clean off metal dust before installing the armature. If a diamond-tipped tool is not available, the commutator may be polished instead, using a very fine grade of glasspaper.

Armature:

Apart from skimming the commutator, very little can be done to the armature apart from visual examination. A service garage can check the windings for shortcircuits and these can be suspected if individual commutator segments are badly burnt.

Check the laminations for scoring or similar damage. Possible causes of scoring are loose polepieces, excessively worn bearings or a bent armature shaft. **Do not attempt to machine or straighten the armature.**

Field coils:

On some smaller motors a permanent magnet may be fitted in place of field coils. Care must be taken in dismantling such motors as the armature will be drawn out with the cover, with possible damage to the brushes.

The field coils may be tested for continuity by connecting a test lamp and battery across them. The resistance can be measured using a suitable resistance meter. It should be noted that the resistance can also be measured using a 12-volt battery and ammeter. Measure the current flow and use Ohm's law to calculate the resistance (Resistance=Voltage divided by Current).

The field coils on the starter motor are held in place by the polepieces and the polepieces in turn are secured by special screws. These screws must only be removed and installed using a special machine screwdriver and therefore if the field coils are defective renewal should be left to a service garage.

Insulation:

Wipe away all carbon and metal dust. If necessary use a cloth moistened with denatured alcohol or gasoline. The insulation may be checked using a 12-volt battery and test lamp. The lamp should not light when connected across the insulation. A better method is to use a 110 AC supply and neon test bulb.

12:4 The starter motor

The starting circuit is shown in **FIG 12:1** and the components of the motor in **FIG 12:2.** The starter relay is only energized when the ignition switch is turned to the START position and on models fitted with automatic transmission, the selector is in the N or P position. When the relay is energized, the coils draw in the plunger and close the heavy duty contacts allowing current to flow to the starter motor. One of the field coils is connected directly to ground and therefore a heavy current will flow through this coil. The coil pulls across the movable pole shoe and this presses the drive pinion into engagement with the teeth on the flywheel ring gear. When the movable polepiece is fully seated, the linkage opens the contacts which ground the field coil. A holding coil keeps the movable polepiece in position, against the pressure of its return spring, while current is flowing through the motor.

Starter fails to operate or operates sluggishly:

1. Check the condition and charge of the battery, paying special attention to the connectors and battery posts. Check the grounding points for corrosion or dirt that will cause a high resistance. If need be, take the vehicle to a service garage for a voltage-drop check on the starter circuit.
2. The starter may fail to operate because the pinion is jammed in mesh. On a vehicle fitted with manual transmission, select a gear and rock the vehicle backwards and forwards. If this fails, remove the cover over the movable pole and try to tap the pinion out of engagement. If all fails, remove the starter motor for further checks and repair. Worn teeth on pinion and ring gear can cause the pinion to jam in mesh.
3. If the motor fails to operate at all, switch on some lights that can be seen from the driver's seat. Select Start on the ignition switch and if the starter motor is taking current the lights will go dim. If the lights do not go dim then there is a fault in the wiring, starter relay, switch or starter motor itself. Trace through using a test lamp or voltmeter to find the point at which current is no longer available. If current is available up to the starter relay, short across the heavy duty terminals of the relay with a thick piece of metal and if the starter motor operates correctly then the contacts are burnt and a new relay should be fitted. If current is reaching the starter motor but it fails to operate, the motor must be removed for further examination.

Removal:

Disconnect the battery and raise the front of the vehicle. Disconnect the starter cable at the starter motor terminal. Remove the starter motor attachment bolts, withdraw the motor and lower it down and out of the vehicle.

Install the motor in the reverse order of removal. Hold the motor firmly in place while snugging up the bolts, to ensure that the motor is mounted squarely against its mounting surface.

Disassembly:

The parts of the starter motor are shown in **FIG 12:2**. Slacken off the brush cover band retaining screw and remove the brush cover and starter drive cover. Note the positions of the brushes and remove them from the brush holders. Remove the through bolts (not shown in the figure) and take off the drive end housing and plunger return spring. If needle roller bearings are fitted, insert a dummy shaft into the bearing to keep the needle rollers in place. Drive out the pivot pin for the starter drive plunger lever and remove the lever as well as the armature. Remove the brush end plate from the starter frame.

If the starter drive requires removal, remove the stop ring retainer so that the stop ring can be taken off and discarded.

The motor is assembled in the reverse order of disassembly. Lubricate the armature shaft splines with a thin coating of Lubriplate 777 and apply a little grease to needle roller bearings. Note that a fiber thrust washer is fitted to the armature shaft between the commutator and

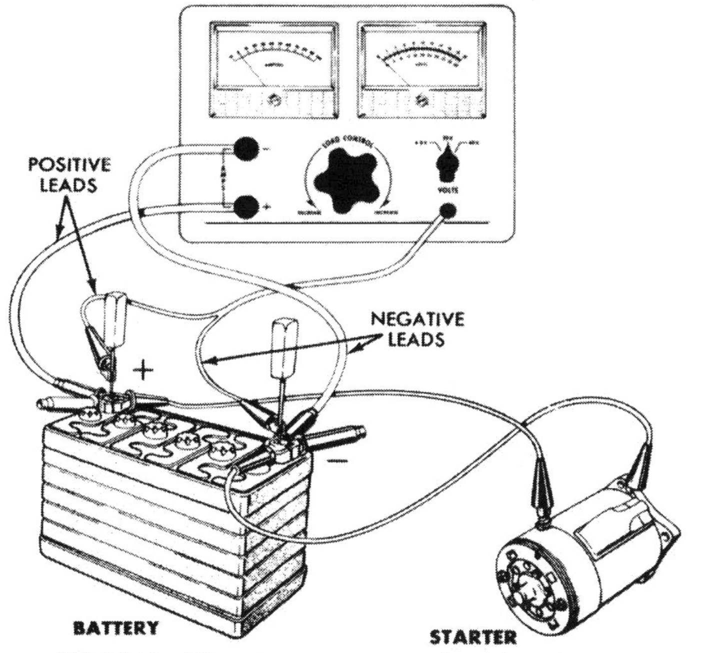

FIG 12:4 The starter motor no-load bench test

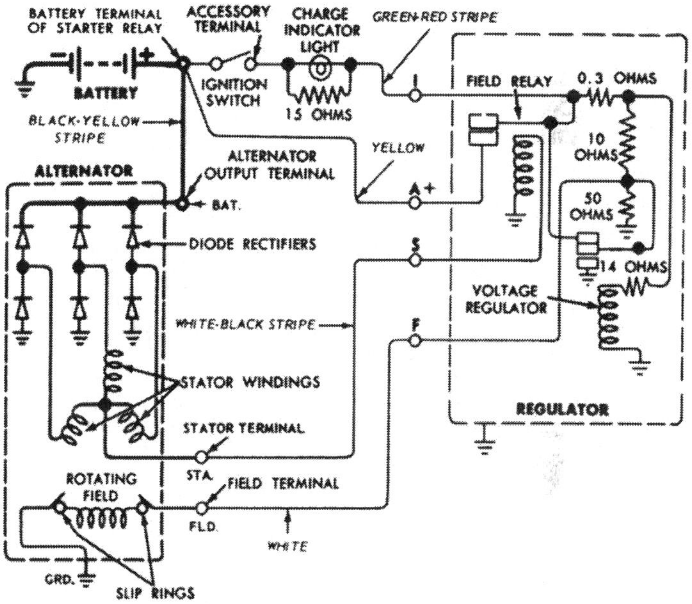

FIG 12:5 Alternator system with indicator light

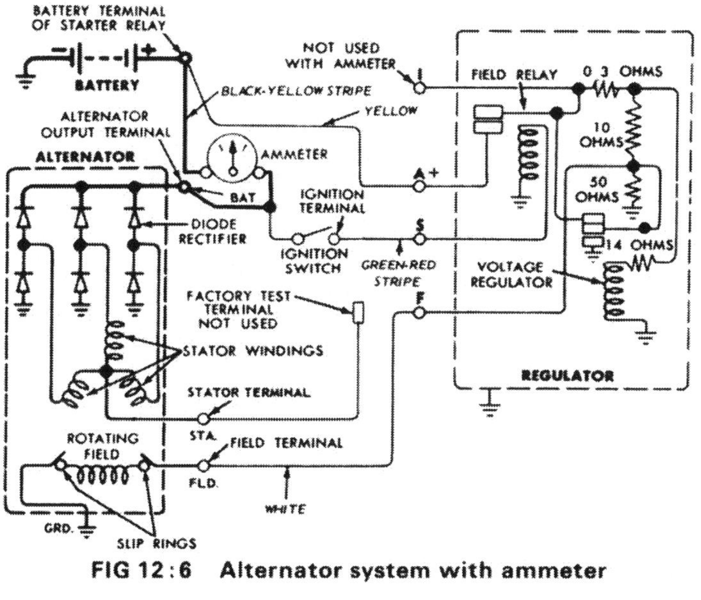

FIG 12:6 Alternator system with ammeter

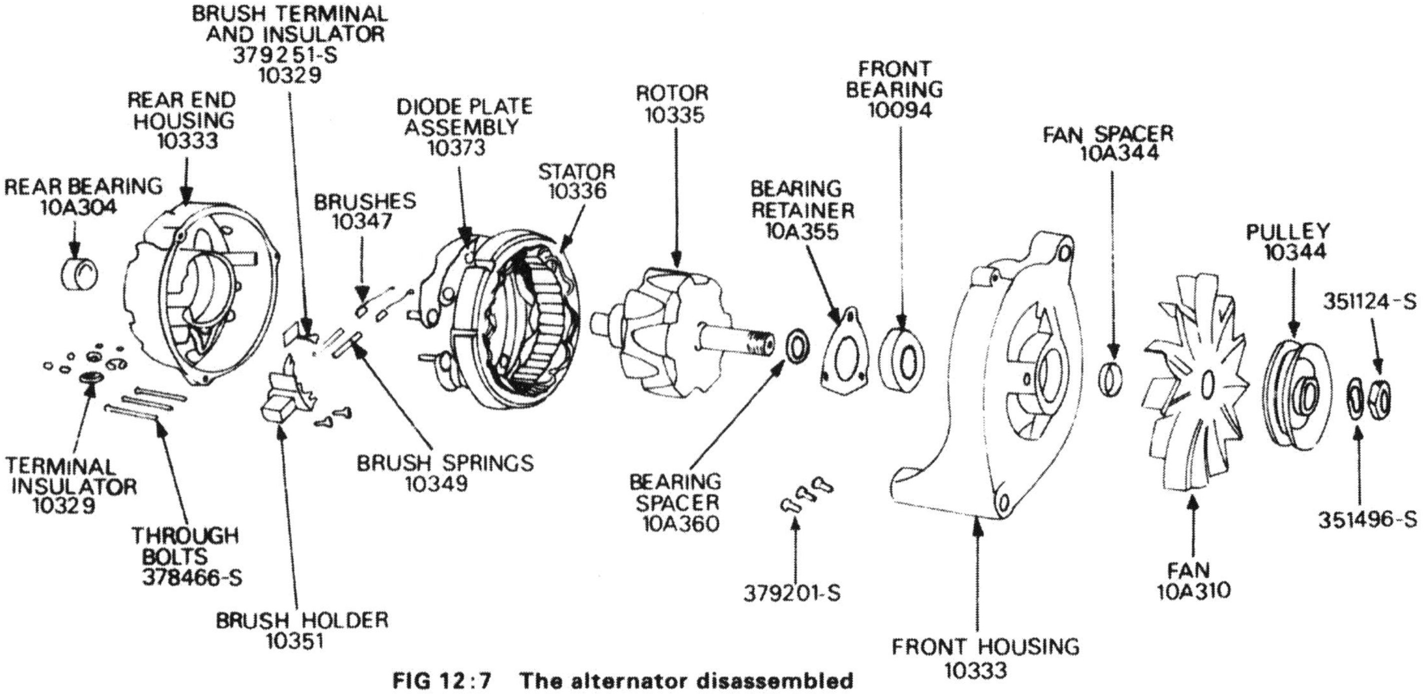

FIG 12:7 The alternator disassembled

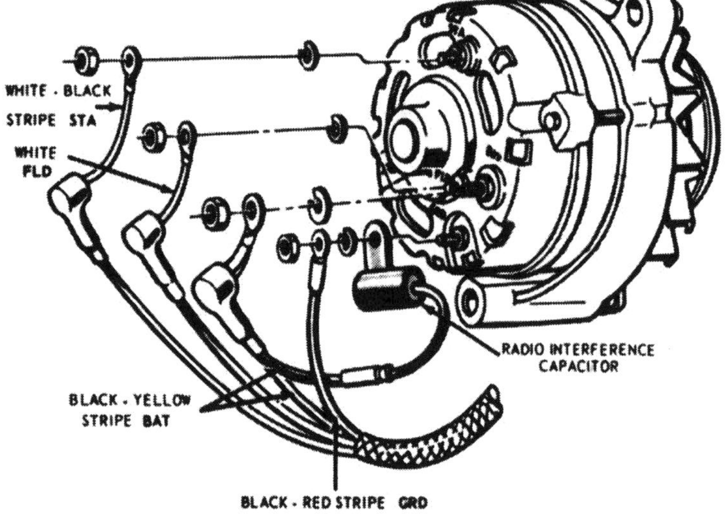

FIG 12:8 Typical early alternator wiring connections

the starter frame. The brushes fitted are shown in **FIG 12:3**, and note they must all be removed from their brush holders before removing the brush end plate.

Starter drive:

Check that the starter drive moves freely along the splines of the armature shaft. Dirt will cause the drive to jam. The teeth must mesh to a depth of at least $\frac{1}{2}$ inch to ensure trouble-free operation. Light damage to the teeth may be smoothed off with an oil stone or smooth file but if the teeth are milled, worn or broken a new drive must be fitted. Examine the teeth on the ring gear through the starter motor mounting hole while slowly turning the engine.

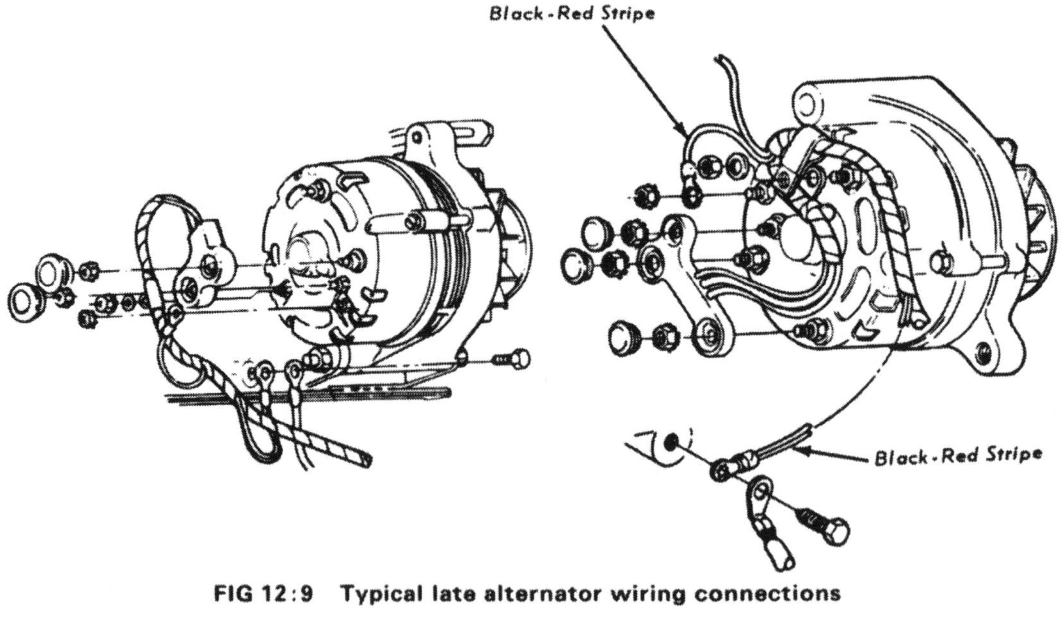

FIG 12:9 Typical late alternator wiring connections

To check the drive with the motor fitted, flood the engine by pumping the accelerator pedal 8 to 10 times. Set the ignition switch to start and hold it there. The engine should fire but not start while the starter motor should continue to crank the engine. If possible, remove the cover and observe the operation of the drive plunger. If the engine stops turning and the rotor spins at high speed then there is a fault in the drive.

Starter motor tests:

The circuit for testing the motor on the bench is shown in **FIG 12:4. The starter motor should be firmly held down as it will kick vigorously on start up.** With the battery connected as shown the starter motor should spin freely and at high speed. Heavy duty leads must be used because even on the no-load condition the starter motor will take 70 amps at least.

12:5 The alternator

Typical alternator charging circuits are shown in **FIGS 12:5** and **12:6**. Typical alternator components are shown in **FIG 12:7**.

The alternator produces an AC current. Heavy stator windings carry the main current and as these are stationary no allowance need be made for centrifugal forces so heavy wiring can be safely used. The rotor carries the light excitation current. The current does not require to change direction and the varying magnetic fields are set up by the polepieces, which give alternate North and South poles. The rotor windings are therefore light and the rotor can be constructed for high rotational speeds. Slip rings and light brushes carry the excitation current to and from the rotor windings and these are far more reliable than the commutator and brushes on a conventional generator, which not only carry a heavy current but also have to change the direction of flow.

Diode rectifiers change the AC current produced by the alternator to a DC current acceptable to the battery. The diodes also prevent a reverse current from flowing so no cut-out is required. A separate regulator limits the maximum voltage to the demands of the system. It should be noted that a defective diode will lower the maximum current output by approximately 10 amps and cause the alternator to operate with a whining noise.

Alternator fails to charge:

1 Check the drive belt tension (see **Chapter 4, Section 4:4**) and be sure that the belt is not excessively glazed or slipping.

2 Be sure that all the connections are clean and tight. **The charging circuit connections must not be made or broken while the engine is running.** Typical earlier alternator connections are shown in **FIG 12:8** and those on later models in **FIG 12:9**. It should be noted that if the alternator is removed from the engine, the battery must be disconnected.

3 A typical output test circuit is shown in **FIG 12:10**. If an ammeter and voltmeter are used instead of the proper test equipment, apply load to the alternator by switching on the headlamps. The engine should be run for at least 20 minutes to allow the alternator

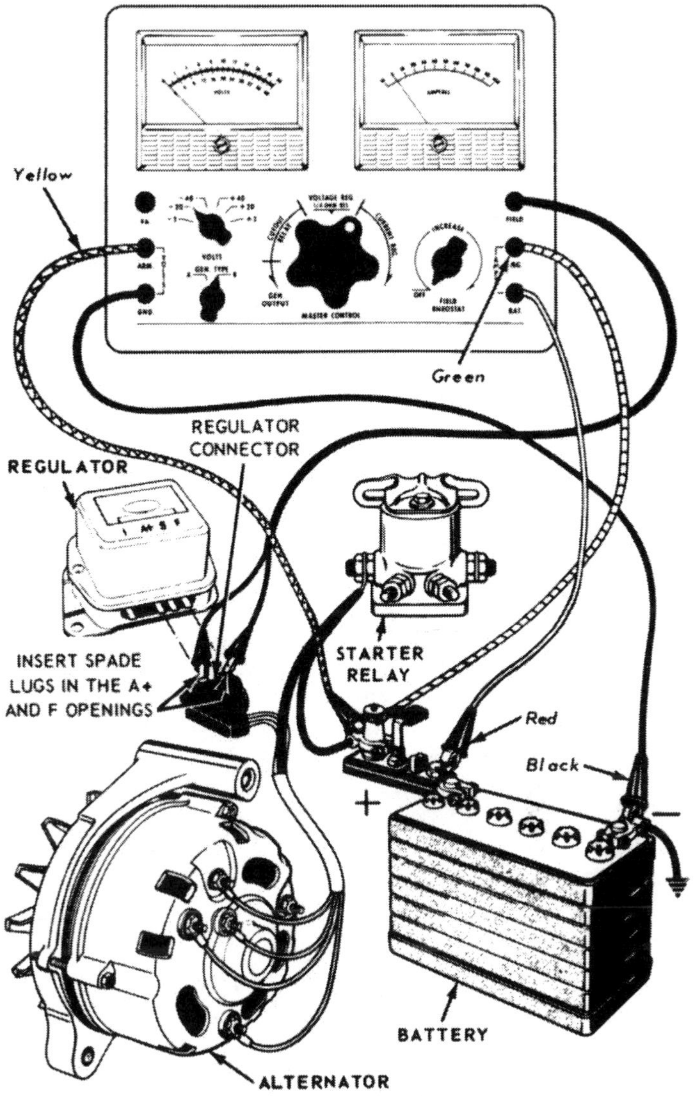

FIG 12:10 Alternator output test

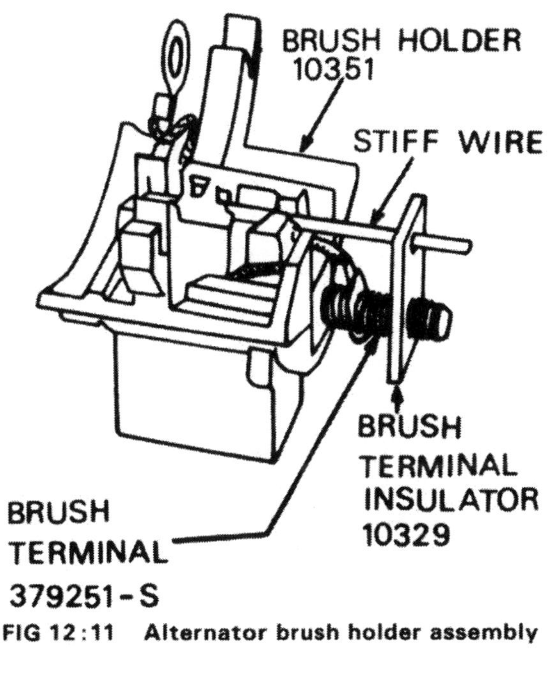

FIG 12:11 Alternator brush holder assembly

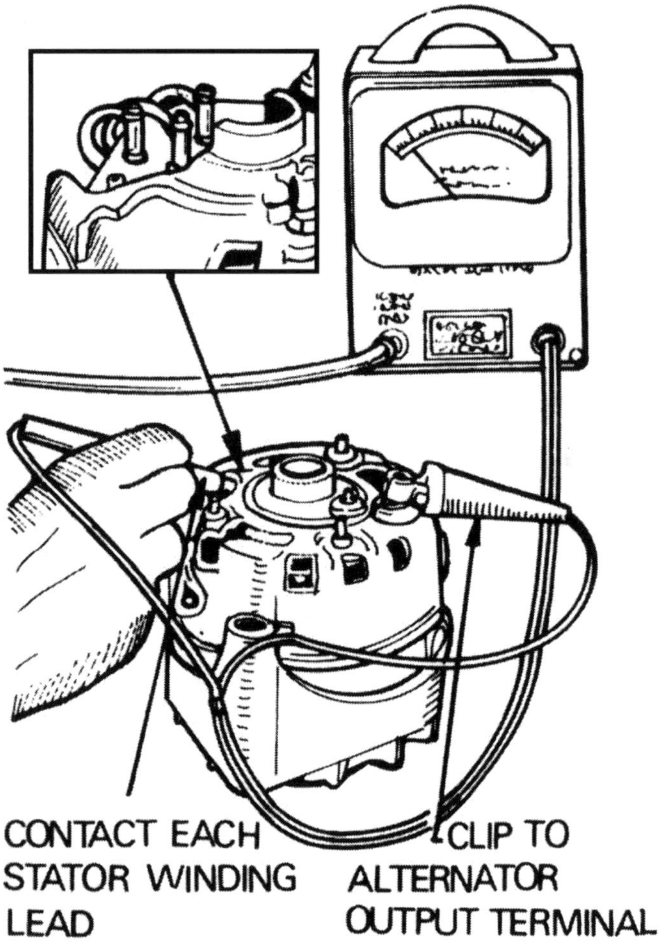

FIG 12:12 Positive diode test

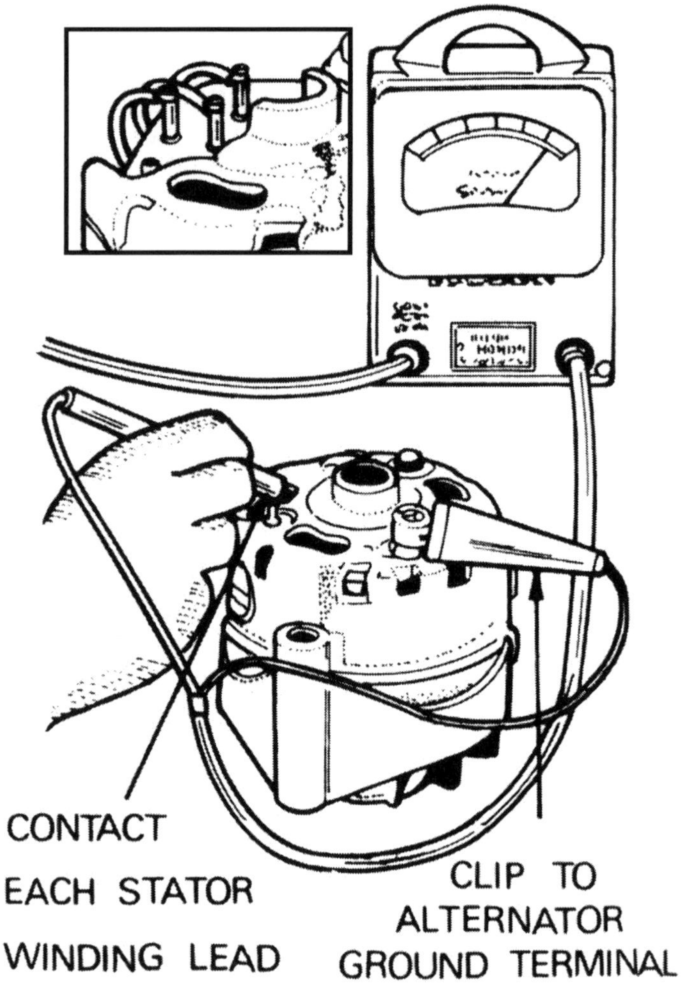

FIG 12:13 Negative diode test

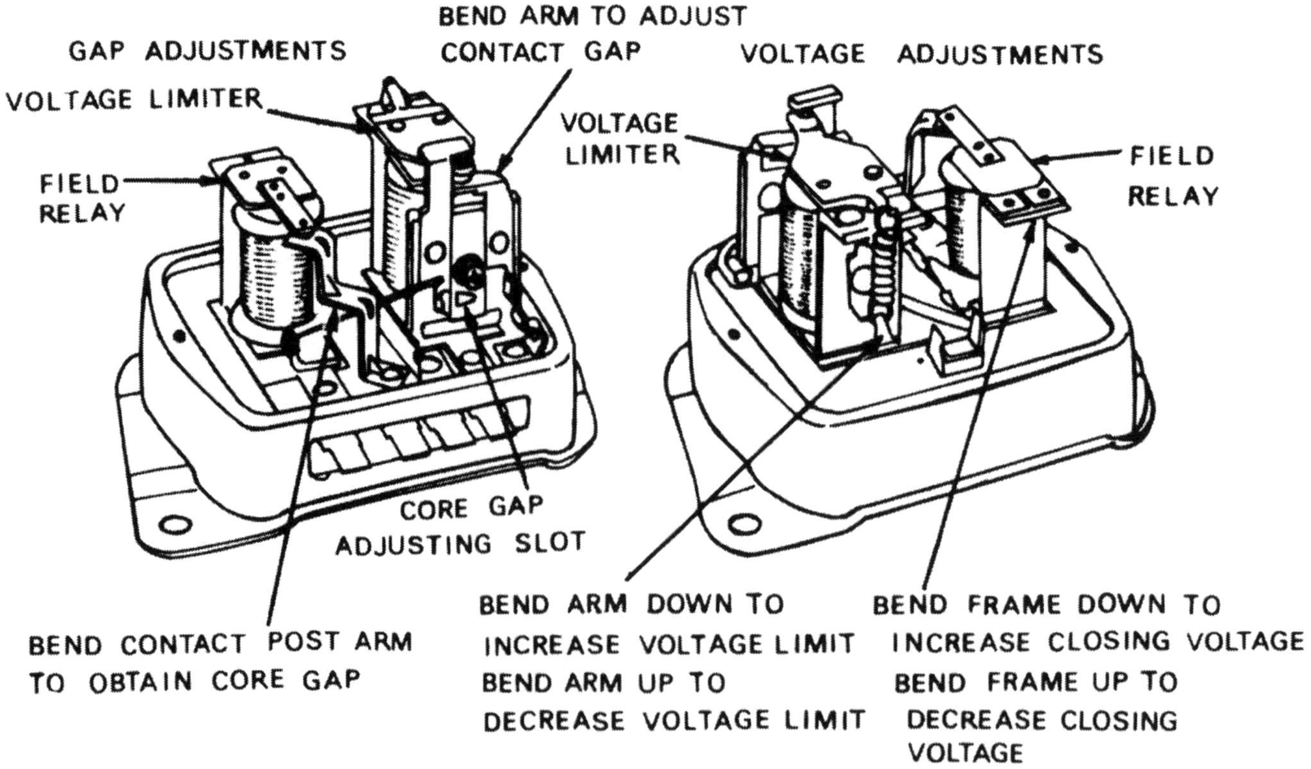

FIG 12:14 Typical regulator adjustments

to reach its normal operating temperature. Adjust the load until the voltmeter reads 15 volts with the engine running at 2500 rev/min. The alternator should then be delivering its rated output.

Disassembly:

Remove the alternator from the engine, after slackening the drive belt and disconnecting the battery.

Mark both end housings and stator with light aligning scribe marks. Take out the three through-bolts and remove the front end housing and rotor from the stator and rear end housing. Take off the nuts and washers so that the rear end housing can be removed. Take out the brush holder mounting screws and remove the brush holder, shown in **FIG 12:11.**

Brush gear and slip rings:

Check the brushes for wear and renew them if they are short. Check the springs and renew the assembly if the springs are weak.

The slip rings should only require cleaning with a fine grade of glasspaper or cloth dipped in denatured alcohol. They may be machined down if badly worn or burnt but do not go beyond the minimum diameter of 1.22 inch.

Rotor and stator:

Check the continuity of the stator, using a test lamp and 12-volt battery between any pair of stator wires, noting that the wires must be unsoldered from the diode assembly. Again check for continuity between one of the wires tested and the third wire.

Measure the resistance of the field coils between the slip rings. Wipe off dirt from both stator and rotor using rag moistened with denatured alcohol. Check the insulation using a 110 AC supply and neon test lamp.

Diode tests:

The test circuit for the diodes is shown in **FIGS 12:12** and **12:13.** A test lamp may be used in place of the meter. Press the probe firmly through the varnish to make good contact. With the polarity in one direction the meter should read or the test lamp illuminate, but with the polarity reversed there should be no reading. No reading in either direction or a reading in both directions indicates a faulty diode.

Voltage regulator:

A typical regulator is shown in **FIG 12:14.** Dirt on the contacts can be a cause of trouble. If the contacts require cleaning, take a sheet of 400 grade silicone carbide paper and partially wear it by folding and rubbing together. Fold a strip so that the abrasive is outwards and clean the contacts by pulling the strip between them. Blow away all loose dust and dirt. If the contact gaps are being set, use hospital-clean feeler gauges. The circuit for testing the regulator is shown in **FIG 12:15.**

Too high a voltage setting will cause bulbs to blow, fuses to fail and the battery will use excessive amounts of distilled water. Too low a setting will prevent the alternator from charging the battery correctly.

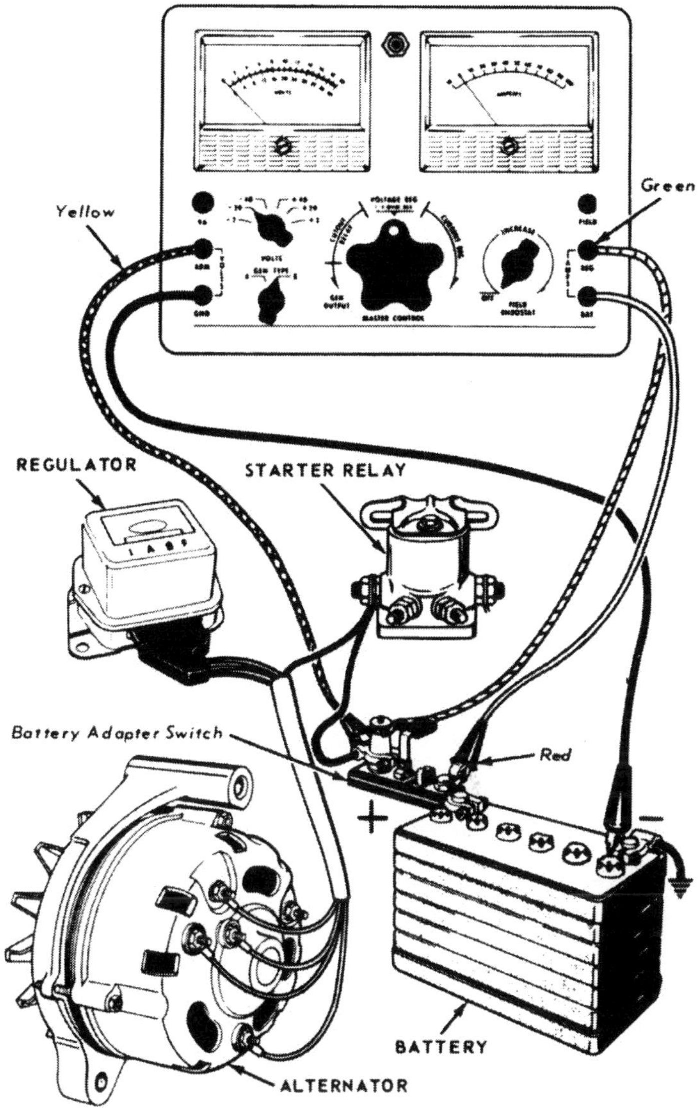

FIG 12:15 **Regulator test**

12:6 The windshield wiper

A typical single-speed wiper motor is shown disassembled in **FIG 12:16** and a two-speed wiper motor in **FIG 12:17.**

Parking:

The motor is self-parking and the wiper arms should be lying flat along the bottom edge of the windshield. If the arms are not flush along the lower edge when the motor stops, remove the arms from the wheelbox spindles and install them so that they lie in the correct position. With the arms removed, operate the wiper switch briefly so that the motor runs and returns to its stop position.

Typical connections for testing the motor are shown in **FIG 12:18.** The motor should run to the park position and stop. If the motor operates satisfactorily there is a fault in the switch or wiring. If the motor is defective, remove the through-bolts and take off the motor cup and armature. Rotate the output shaft until the contacts, shown in **FIG 12:19,** are firmly closed. Rotate the adjusting screw clockwise until the contacts just open and then rotate the adjusting screw counterclockwise by

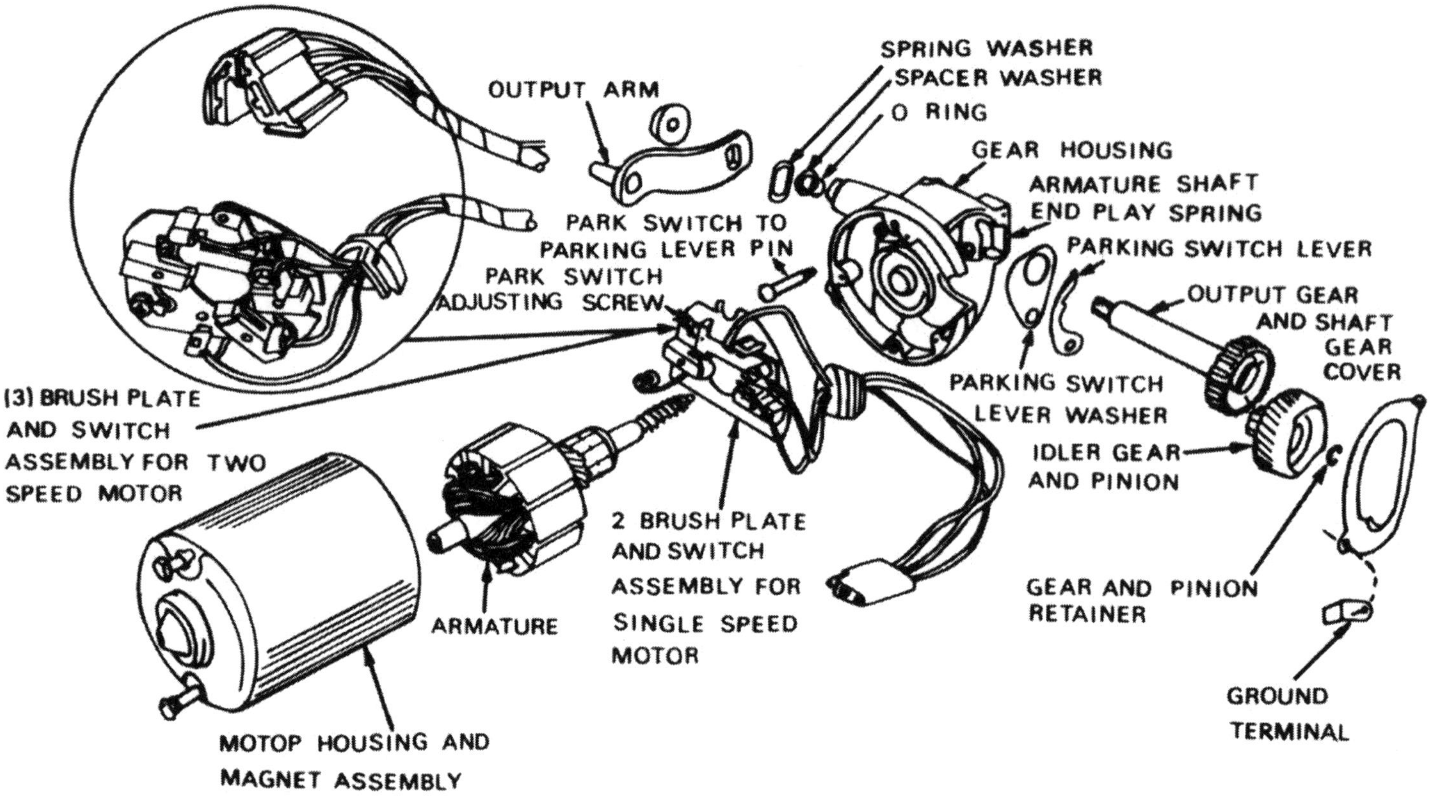

FIG 12:16 Permanent magnet type windshield wiper motor

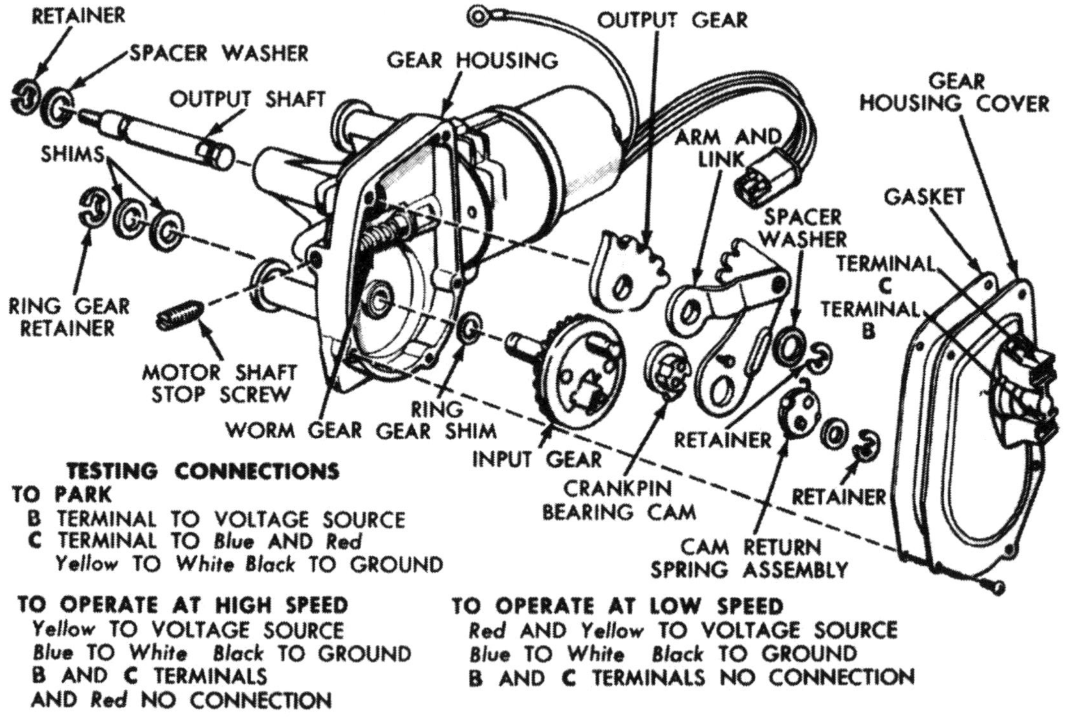

FIG 12:17 Two-speed cranked wiper motor

one complete turn. Check the bridge to ensure that the legs are contacting the brush plate when the contacts are closed.

Install the armature and motor cup in the reverse order of removal and check that it now operates satisfactorily.

Poor wiper operation:

Note that the wiper blades will drag on a dry windshield so wet the windshield before carrying out tests.

Disconnect the linkage from the motor and run the motor so that it can be checked separately. Move the linkage by hand and check for binding or defects.

Check that full battery voltage is reaching the terminals of the motor. The test circuits for the motor are shown in **FIG 12:20** and the test circuit for the switch circuit breaker is shown in **FIG 12:21**. If the motor operates satisfactorily on the test circuit then the wiring or switch is at fault. **If the motor is operated with the gearbox cover removed, take great care as the gears have sufficient torque to cause injuries to the fingers.**

Likely faults of poor motor operation are: defective brushgear, insufficient armature end float (.003 inch approximately), insufficient lubricant and worn or damaged gears.

12:7 The temperature and fuel gauges

Both these gauges operate by the bending of a bi-metallic strip when it is heated by current passing through it. To ensure that voltage fluctuations in the electrical system do not cause the gauges to read incorrectly they take their supply from a voltage stabilizer unit. This unit makes and breaks the circuit by the action of another bi-metallic strip to give over a period of time the same current flow as would a constant voltage over the same period of time. For accurate testing a special test unit is required and the sender units must not be connected directly to full battery voltage. If a voltmeter is connected to the output of the voltage stabilizer it should oscillate between 10 and zero volts. If the supply is steady the unit is defective.

Intermittent or incorrect readings can be caused by wiring faults so trace through the wires and check for breaks, faulty insulation or poor connections. No readings indicate a defective voltage stabilizer unit (mounted on the back of the instrument cluster).

If the fuel gauge operates incorrectly, borrow a new tank sender unit. Disconnect the lead from the sender unit fitted to the tank and connect it to the new sender unit. Connect a jumper lead between a good ground and the new sender unit. Switch on the ignition and leave the float arm at its lowest position. The needle on the gauge should settle on the Empty mark. Raise the float arm and wait a minute for the needle to settle and it should then read Full. If the reading is incorrect and the wiring is satisfactory then the fault lies with the gauge itself. If the reading is satisfactory the sender unit in the tank is defective and should be exchanged for a new unit.

If temperature warning lights are fitted, they operate from full battery voltage. The sender unit contains a bi-metallic strip which closes one contact when the temperature is low and the other contact when the temperature is high.

It should be noted that on some models a low fuel level warning light is fitted. The system operates when

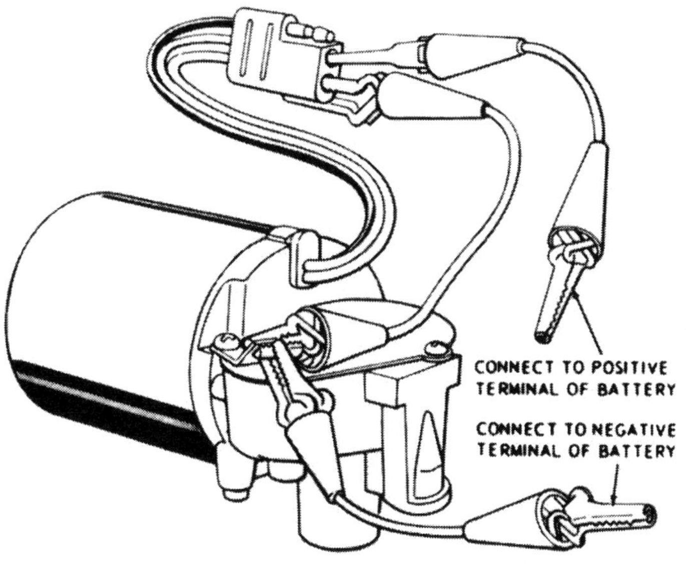

FIG 12:18 Wiper motor park switch test

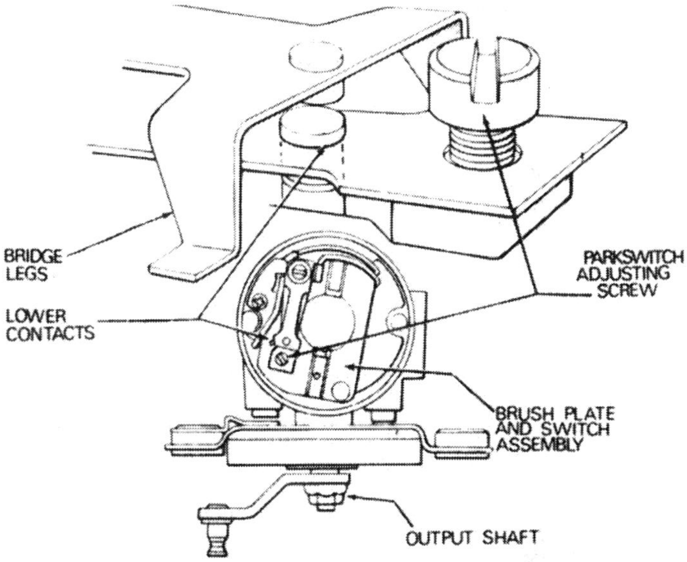

FIG 12:19 Wiper motor park switch adjustment

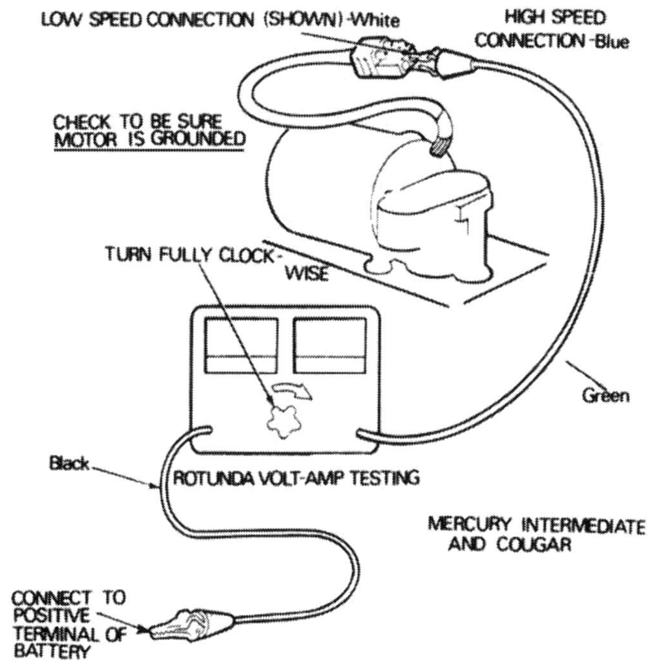

FIG 12:20 Wiper motor test running circuit

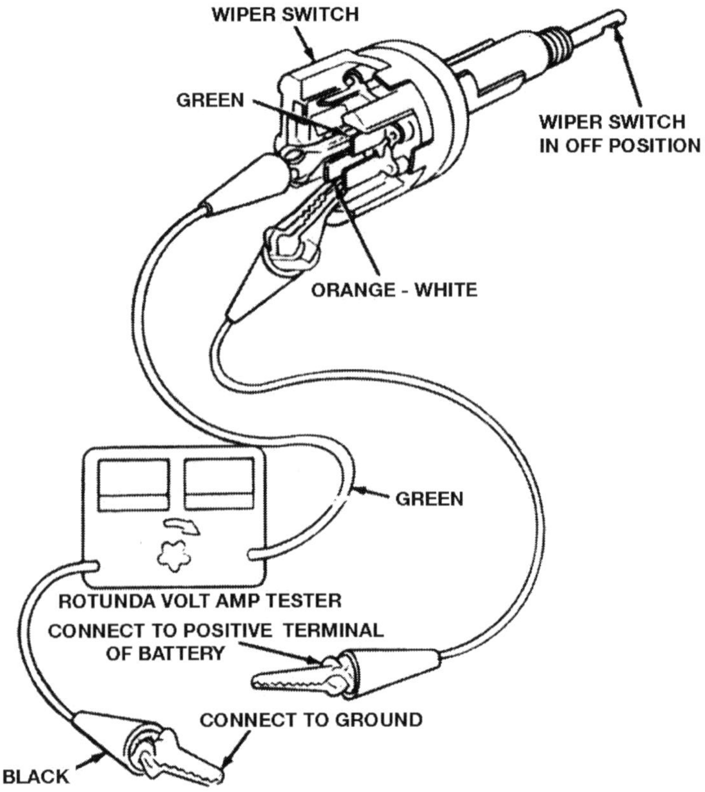

FIG 12:21 Wiper switch circuit breaker test

there is approximately $4\frac{1}{2}$ gallons of fuel left in the tank. The tank unit is shown in **FIG 12:22**. The thermistor is normally covered by fuel and kept cool. When the level is low the thermistor is uncovered, so that it warms up and operates a relay to illuminate the warning light.

If a temperature gauge is fitted, the system can be checked using two resistors. Disconnect the lead from the sender unit and connect it to ground in series with a 10 ohm resistor. The gauge should give a high reading.

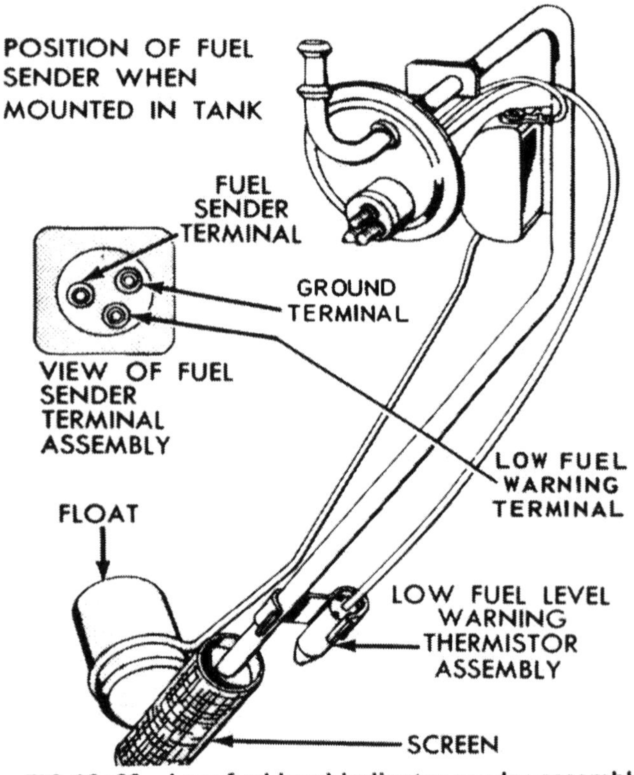

FIG 12:22 Low fuel level indicator sender assembly

Substitute a 73 ohm resistor and the gauge should give a low reading. If the gauge reads incorrectly still, try substituting another voltage stabilizer unit. **The sender unit must be sealed in place with a water-proof sealant that is also electrically conducting.**

12:8 The turn indicators

The turn indicators use the dual-filament parking lights to indicate the direction of turn. A flasher unit located in the back of the instrument panel makes and breaks the circuit to provide the flashing at the indicator lamps. A separate emergency flasher unit comes into action when the emergency button is pressed and this makes all four turn indicator lamps flash.

If the indicators do not operate correctly, check the bulbs for blown filaments as if a filament is blown the flash rate will alter.

Reach under the instrument panel and push in and then twist the flasher unit to free it from its mounting. Disconnect the wires. Switch on the ignition and use a test lamp or voltmeter to check that battery voltage is reaching the flasher unit. Switch off the ignition and connect the wires together. Switch on the ignition and operate the turn indicator switch. The appropriate lamps should come on and stay on. If the lamps do not light, check through the wiring until the fault is found. If current is passing through the wires and switch, check the earth points on the lamps. If the lamps operate correctly, the flasher unit is defective and must be renewed.

Handle the flasher unit with care as it is delicate and will be damaged by mishandling.

12:9 The headlights

Two sealed-beam headlights are fitted. A typical attachment is shown in **FIG 12:23** and access is made after removing the headlamp door.

To remove the headlights, take out the screws which secure the door. Hold the headlight assembly by hand and unhook the spring from the retaining ring. Remove the retaining ring by unhooking it from the clip on the opposite side to the spring. Pull the headlight bulb out and disconnect it from the wiring assembly plug.

Install the headlight in the reverse order of removal.

Beam setting:

Two adjusting screws are fitted to each headlight and they are used to set the beams correctly. The vehicle should be taken to a service garage for accurate setting using special equipment, as the owner cannot set the beams with accuracy.

12:10 Fuses

A fuse panel containing the fuses is fitted. The panel is marked with the circuits protected by the individual fuses as well as the correct rating of the fuse to be used.

A fuse blowing is not necessarily an indication of a fault in the system, as fuses do weaken with age. If a fuse does blow, briefly check the circuits for damage and fit a new fuse of the correct rating. If the new fuse blows immediately a circuit is switched on, check the circuit thoroughly. Shake the wires in case insulation has frayed

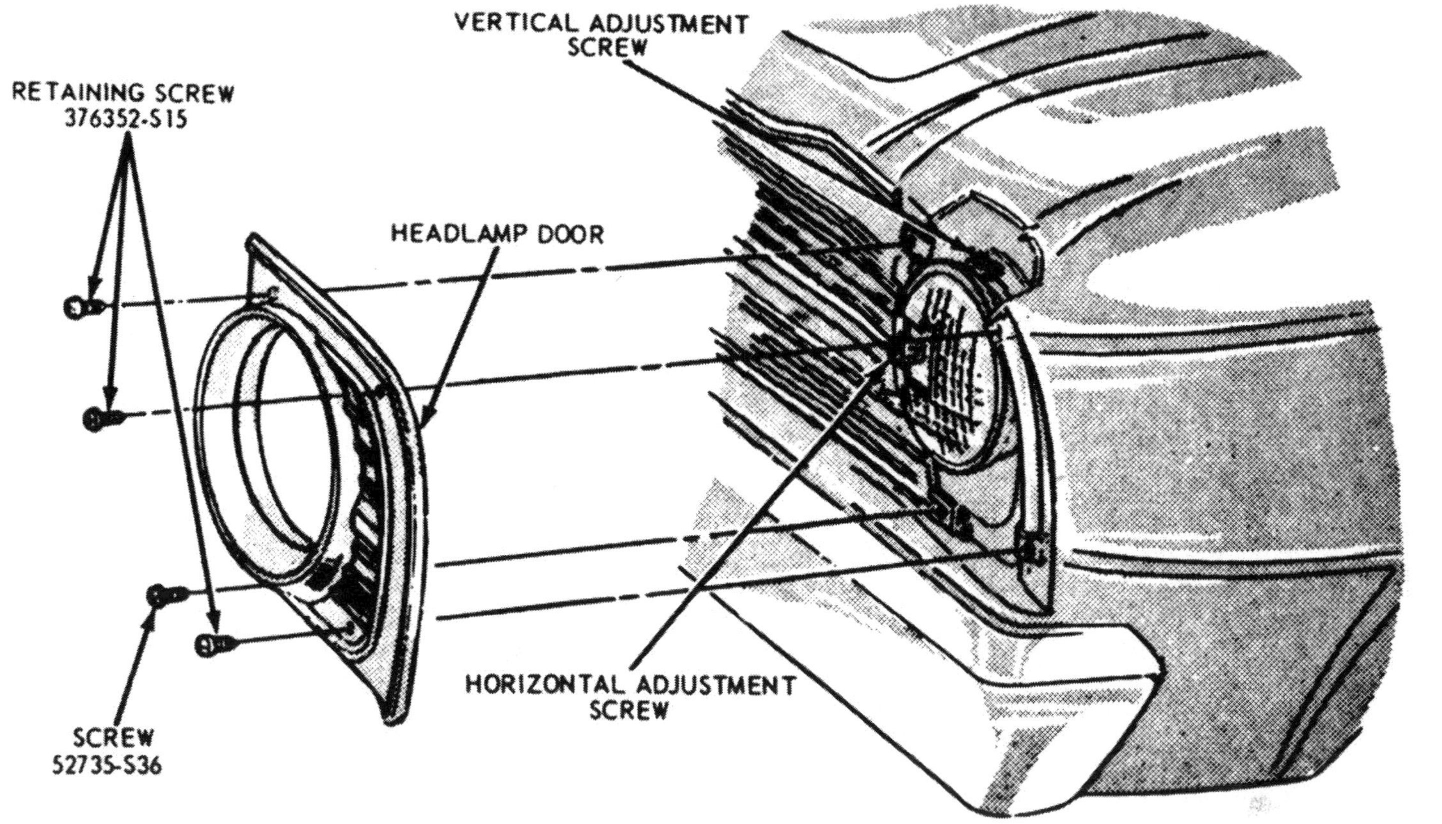

FIG 12:23 Typical headlight attachments

and is causing a ground. **Do not install a fuse of higher rating than called for, as this could prevent the fuse from blowing and the wiring may burn out instead.**

Circuit breakers:

These are fitted to some circuits, such as headlights and windshield wipers. In the event of an excessively high current being taken the circuit breaker will operate. The circuit breakers can be tested using a 12-volt battery, variable resistance, and ammeter. The current is adjusted to the normal value and the circuit breaker should not operate. When the current is adjusted to a high value, the circuit breaker should operate within 30 seconds. Renew defective circuit breakers as they cannot be repaired.

12:11 Trouble shooting guide

(a) Battery discharged

1 Loose or dirty terminal connectors
2 Insufficient charging current
3 Shortcircuit in part of circuit not protected by fuse
4 Defective alternator diode
5 Battery internally defective
6 Accessories left on

(b) Battery will not hold charge

1 Low electrolyte level
2 Battery plates sulphated or separators ineffective
3 Electrolyte leakage or spillage

(c) Alternator output low or nil

1 Broken or slipping drive belt
2 Defective or incorrectly set regulator
3 Loose connectors
4 Defective diodes
5 No battery supply
6 Defective rotor or stator windings

(d) Starter motor lacks power or will not operate

1 Battery discharged or loose and dirty terminals
2 Starter pinion jammed in mesh
3 Defective ignition switch
4 Defective starter relay
5 Defective brushgear or commutator
6 Defective armature or field coil windings
7 Starter motor mechanically defective
8 Engine abnormally stiff

(e) Starter runs but does not turn engine

1 Badly worn or broken teeth on drive

Inches	Decimals	Millimetres	Inches to Millimetres: Inches	Inches to Millimetres: mm	Millimetres to Inches: mm	Millimetres to Inches: Inches
1/64	.015625	.3969	.001	.0254	.01	.00039
1/32	.03125	.7937	.002	.0508	.02	.00079
3/64	.046875	1.1906	.003	.0762	.03	.00118
1/16	.0625	1.5875	.004	.1016	.04	.00157
5/64	.078125	1.9844	.005	.1270	.05	.00197
3/32	.09375	2.3812	.006	.1524	.06	.00236
7/64	.109375	2.7781	.007	.1778	.07	.00276
1/8	.125	3.1750	.008	.2032	.08	.00315
9/64	.140625	3.5719	.009	.2286	.09	.00354
5/32	.15625	3.9687	.01	.254	.1	.00394
11/64	.171875	4.3656	.02	.508	.2	.00787
3/16	.1875	4.7625	.03	.762	.3	.01181
13/64	.203125	5·1594	.04	1.016	.4	.01575
7/32	.21875	5.5562	.05	1.270	.5	.01969
15/64	.234375	5.9531	.06	1.524	.6	.02362
1/4	.25	6.3500	.07	1.778	.7	.02756
17/64	.265625	6.7469	.08	2.032	.8	.03150
9/32	.28125	7.1437	.09	2.286	.9	.03543
19/64	.296875	7.5406	.1	2.54	1	.03937
5/16	.3125	7.9375	.2	5.08	2	.07874
21/64	.328125	8.3344	.3	7.62	3	.11811
11/32	.34375	8.7312	.4	10.16	4	.15748
23/64	.359375	9.1281	.5	12.70	5	.19685
3/8	.375	9.5250	.6	15.24	6	.23622
25/64	.390625	9.9219	.7	17.78	7	.27559
13/32	.40625	10.3187	.8	20.32	8	.31496
27/64	.421875	10.7156	.9	22.86	9	.35433
7/16	.4375	11.1125	1	25.4	10	.39370
29/64	.453125	11.5094	2	50.8	11	.43307
15/32	.46875	11.9062	3	76.2	12	.47244
31/64	.484375	12.3031	4	101.6	13	.51181
1/2	.5	12.7000	5	127.0	14	.55118
33/64	.515625	13.0969	6	152.4	15	.59055
17/32	.53125	13.4937	7	177.8	16	.62992
35/64	.546875	13.8906	8	203.2	17	.66929
9/16	.5625	14.2875	9	228.6	18	.70866
37/64	.578125	14.6844	10	254.0	19	.74803
19/32	.59375	15.0812	11	279.4	20	.78740
39/64	.609375	15.4781	12	304.8	21	.82677
5/8	.625	15.8750	13	330.2	22	.86614
41/64	.640625	16.2719	14	355.6	23	.90551
21/32	.65625	16.6687	15	381.0	24	.94488
43/64	.671875	17.0656	16	406.4	25	.98425
11/16	.6875	17.4625	17	431.8	26	1.02362
45/64	.703125	17.8594	18	457.2	27	1.06299
23/32	.71875	18.2562	19	482.6	28	1.10236
47/64	.734375	18.6531	20	508.0	29	1.14173
3/4	.75	19.0500	21	533.4	30	1.18110
49/64	.765625	19.4469	22	558.8	31	1.22047
25/32	.78125	19.8437	23	584.2	32	1.25984
51/64	.796875	20.2406	24	609.6	33	1.29921
13/16	.8125	20.6375	25	635.0	34	1.33858
53/64	.828125	21.0344	26	660.4	35	1.37795
27/32	.84375	21.4312	27	685.8	36	1.41732
55/64	.859375	21.8281	28	711.2	37	1.4567
7/8	.875	22.2250	29	736.6	38	1.4961
57/64	.890625	22.6219	30	762.0	39	1.5354
29/32	.90625	23.0187	31	787.4	40	1.5748
59/64	.921875	23.4156	32	812.8	41	1.6142
15/16	.9375	23.8125	33	838.2	42	1.6535
61/64	.953125	24.2094	34	863.6	43	1.6929
31/32	.96875	24.6062	35	889.0	44	1.7323
63/64	.984375	25.0031	36	914.4	45	1.7717

UNITS	Pints to Litres	Gallons to Litres	Litres to Pints	Litres to Gallons	Miles to Kilometres	Kilometres to Miles	Lbs. per sq. In. to Kg. per sq. Cm.	Kg. per sq. Cm. to Lbs. per sq. In.
1	.57	4.55	1.76	.22	1.61	.62	.07	14.22
2	1.14	9.09	3.52	.44	3.22	1.24	.14	28.50
3	1.70	13.64	5.28	.66	4.83	1.86	.21	42.67
4	2.27	18.18	7.04	.88	6.44	2.49	.28	56.89
5	2.84	22.73	8.80	1.10	8.05	3.11	.35	71.12
6	3.41	27.28	10.56	1.32	9.66	3.73	.42	85.34
7	3.98	31.82	12.32	1.54	11.27	4.35	.49	99.56
8	4.55	36.37	14.08	1.76	12.88	4.97	.56	113.79
9		40.91	15.84	1.98	14.48	5.59	.63	128.00
10		45.46	17.60	2.20	16.09	6.21	.70	142.23
20				4.40	32.19	12.43	1.41	284.47
30				6.60	48.28	18.64	2.11	426.70
40				8.80	64.37	24.85		
50					80.47	31.07		
60					96.56	37.28		
70					112.65	43.50		
80					128.75	49.71		
90					144.84	55.92		
100					160.93	62.14		

UNITS	Lb ft to kgm	Kgm to lb ft	UNITS	Lb ft to kgm	Kgm to lb ft
1	.138	7.233	7	.967	50.631
2	.276	14.466	8	1.106	57.864
3	.414	21.699	9	1.244	65.097
4	.553	28.932	10	1.382	72.330
5	.691	36.165	20	2.765	144.660
6	.829	43.398	30	4.147	216.990

CHAPTER 13

THE BODYWORK

13:1 Bodywork repairs

Large-scale repairs to bodywork should be left to experts. Even small dents can be tricky as injudicious hammering will cause the metal to stretch and make the dent worse instead of better.

Acrylic finishes are used as standard on the exterior of the vehicle and Nitrocellulose lacquers must not be used.

Spotting or minor blemishes on the paintwork can be removed by polishing. This method can also be used to restore the finish if it has suffered by weathering and lost its gloss. Machine polishing will make the task easier. Remove the defects using 600 grit paper with water or white spirits as a lubricant. **Remove only sufficient paint to take out the defect.**

Paint on a white or light colored, medium grit, machine polishing compound (such as Sno-Flake No. 16) with a brush. Polish the entire surface at 1850 rev/min and using a carpet pad ($\frac{5}{8}$ inch nap) or lambswool pad. Remove the compound and buff the surface with a clean lambswool pad.

Self-spraying cans of matching paint and touch-up kits are available. The touch-up kit should be used for small stone chips and minor scratches. After hardening, polishing can be carried out using a mild metal polish. Regular use of the touch-up kit will prevent corrosion forming on the metal under chips and scratches.

For deeper dents, filling and then spraying is the best method of repair available to the owner. It is better to spray a complete panel rather than just a patch as any mismatch in the paint will then not be so obvious. Remember that the finish will alter color slightly and fade with age. Remove all trim and handles from the area, as a better finish will be obtained than after masking them over. Wash off all wax polish using white spirits. **If silicone-based polishes have been used it is essential to use a special cleansing fluid to remove them.** Traces of polish will leave the finish with pits, craters and pop-ups. Lightly scuff the area to give a good key for new finish. If there are any traces of corrosion, rub down to bright metal and apply a coat of primer.

Apply paste stopper or filler as required to bring the damaged area just slightly proud of the original finish. Leave the filler to harden and then rub it down using 400 grit paper with water as a lubricant. Add further stopper or filler as required to achieve the best possible surface finish. Wash off all slurry and leave the surface to dry.

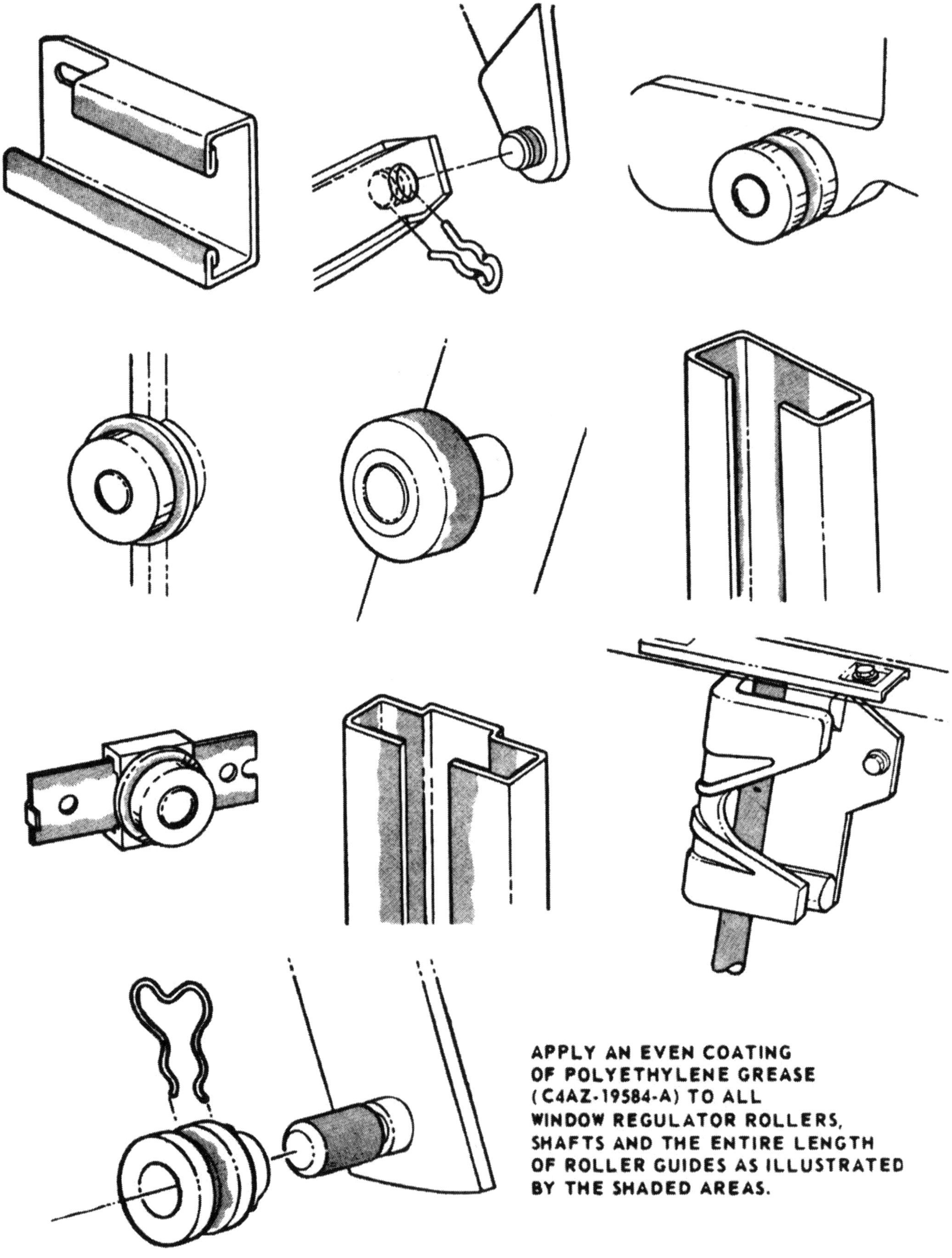

FIG 13:1 Window mechanism lubrication points

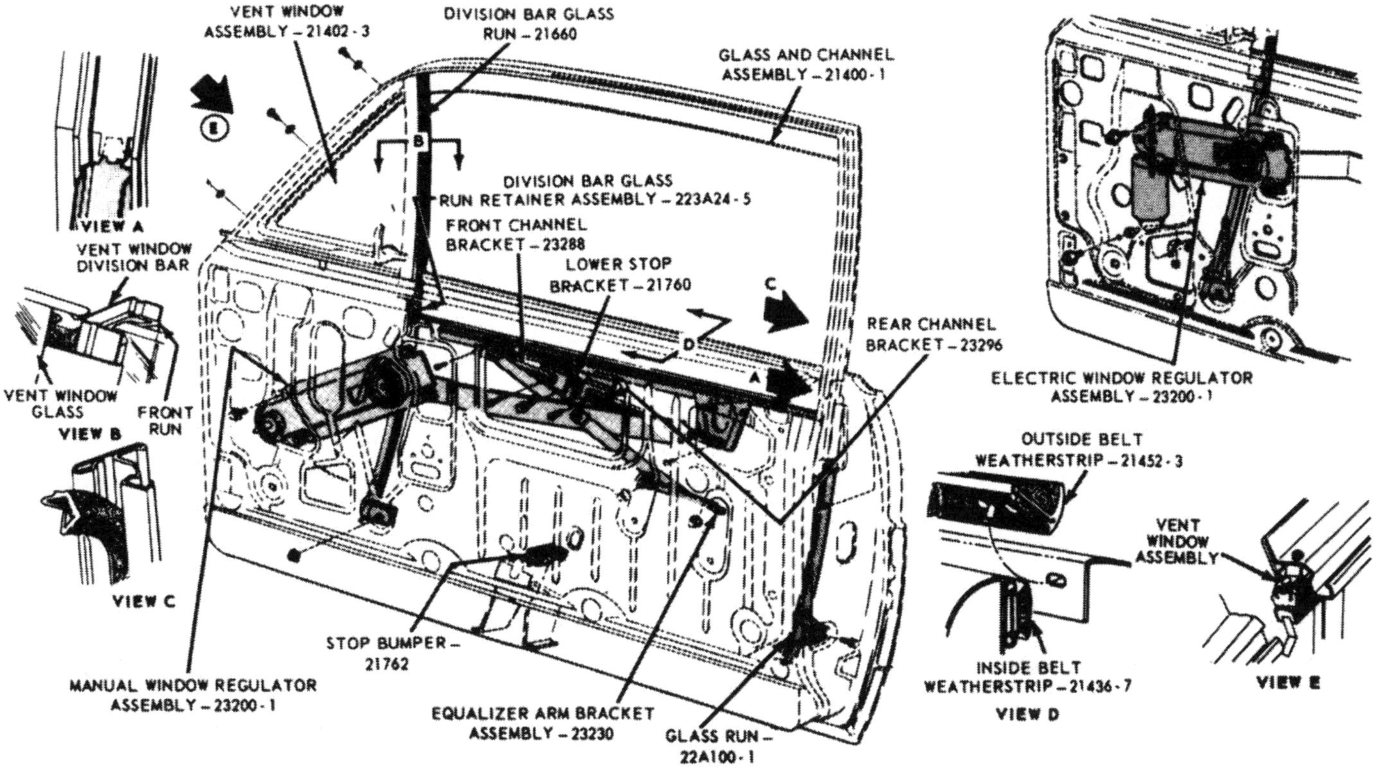

FIG 13:2 Front door window mechanism

A second wash will probably be required to remove slurry missed in the first wash. When the surface is dry, wipe it over with a proprietary 'tacky' cloth to remove any remaining dust.

Mask off surrounding areas with paper and masking tape. Spray the surface evenly and spray over onto the masking to ensure that the edges of the panel are also evenly sprayed. Apply the lacquer in thin layers, gently rubbing down between each coat, rather than in one thick coat which will probably run. Remove the masking and leave the finish to dry for at least 24 hours. Polish the surface as described earlier to remove any runs or sags and then **leave the lacquer to harden for several weeks before applying wax polish.**

13:2 Door components

The door components vary slightly from model to model and year to year, so only typical assemblies will be shown.

Whenever the door mechanism parts are removed, all bearing surfaces and pivots should be lubricated with Polyethylene grease. This grease is available in aerosol containers with a long nozzle for reaching awkward spots. The shaded areas in **FIG 13:1** show the points of application.

Electrically operated windows:

Before carrying out checks, make sure that the battery is fully charged and turn the ignition key to the accessory position. Connect a test lamp or voltmeter between ground and the 20 amp circuit breaker on the starter relay. Full battery voltage should be available. On models fitted with power operated seats and hood, these accessories use the same circuit breaker and if they do not operate as well then the fault most likely lies with the circuit breaker. Renew a defective circuit breaker. Similarly test the supply from the relay and renew the relay if it is defective. Check the wires for breaks or poor connections if the components are satisfactory.

If only one window fails to operate, activate the switch and listen for the motor running free. If the motor is running but the window does not operate the most likely fault is a sheared rubber drive. If the motor does not run at all, disconnect the wires and use a test lamp or voltmeter to check the feed, tracing back through the wiring until the fault is found. If feed is available to the motor and it does not run then it is likely that the motor itself is defective. It should be noted that if the window will operate in one direction only the motor is satisfactory and the fault lies in wiring or switch.

If a window operates sluggishly, first check the wiring for poor connections or partially grounded wires. Disconnect the supply to the motor and reconnect it through an ammeter. Under normal running the motor should take approximately 12 amps and 20 amps at stall. A low stall current indicates poor connections in the circuit. The motor should take approximately 5 amps when disconnected from the winder mechanism and running free.

Regulator mechanism:

The front door regulator mechanism on a 1969 Model 62 is shown in **FIG 13:2**. The rear door adjustment points are shown in **FIG 13:3** while the quarter window mechanism is shown in **FIG 13:4**. Before access can be made to the parts, the interior trim panel and handles must first be removed and the plastic sealing sheet taken off.

Lock mechanism:

Typical lock mechanism components and their attachments are shown in **FIG 13:5**.

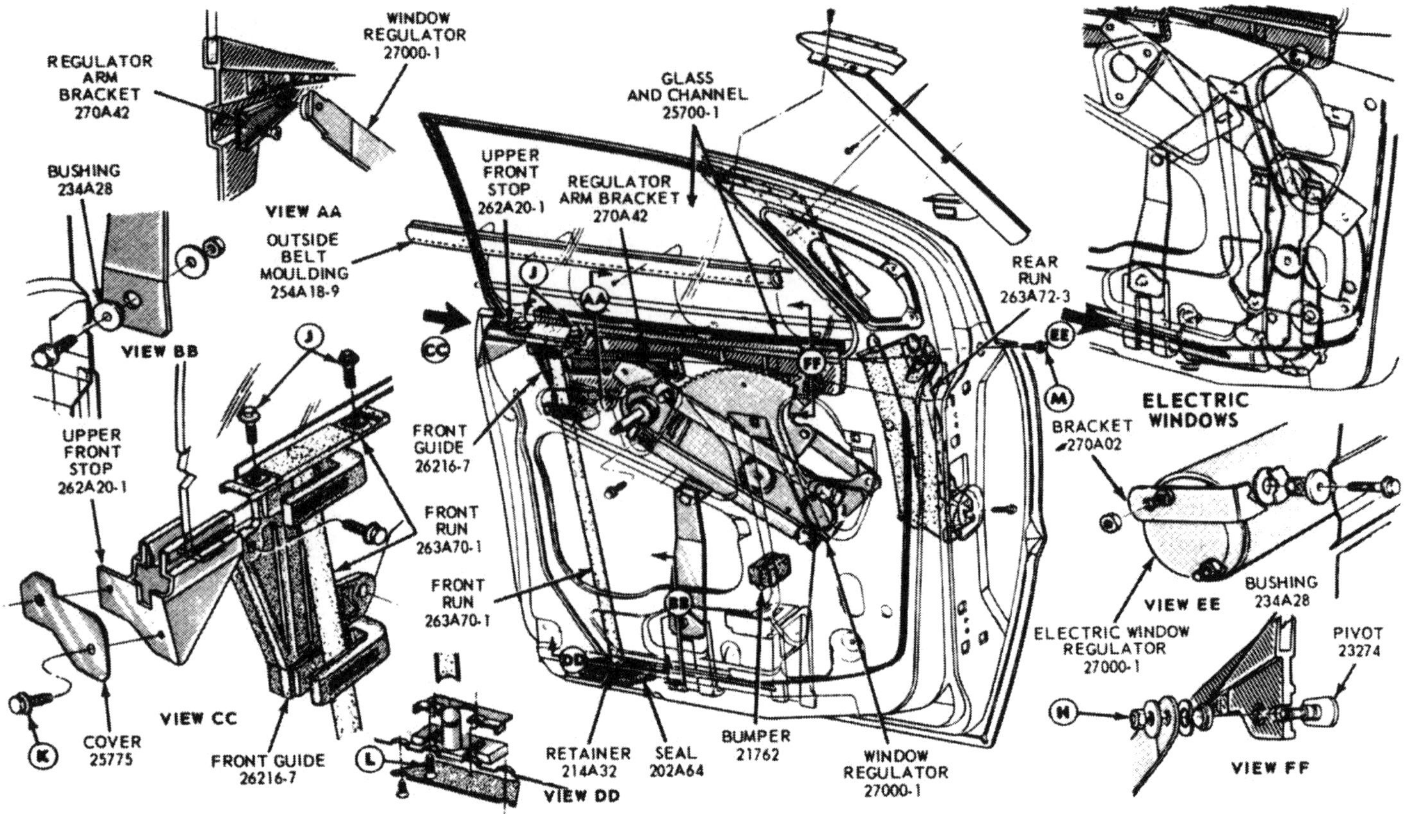

FIG 13:3 Rear door window adjustment points

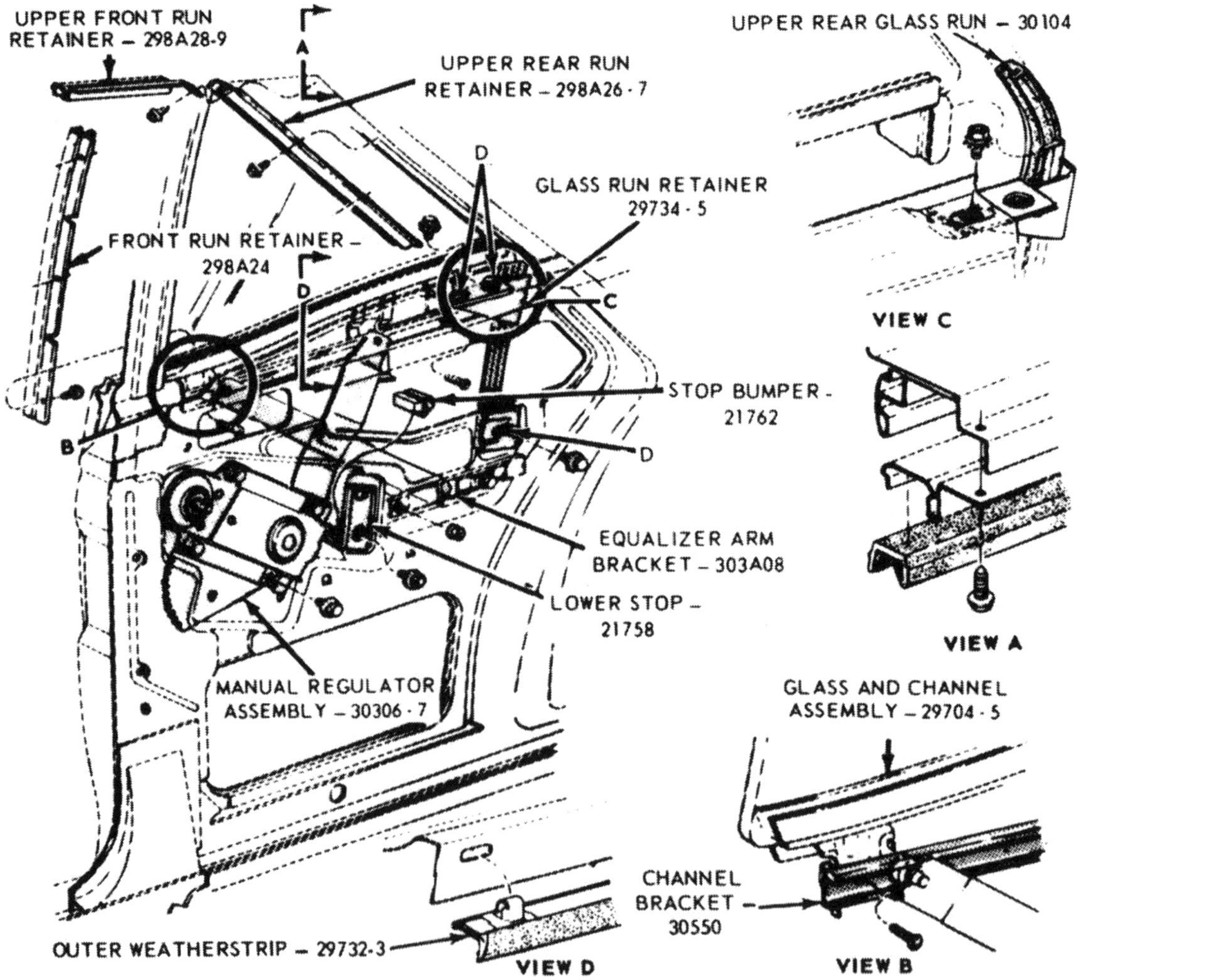

FIG 13:4 Quarter window mechanism

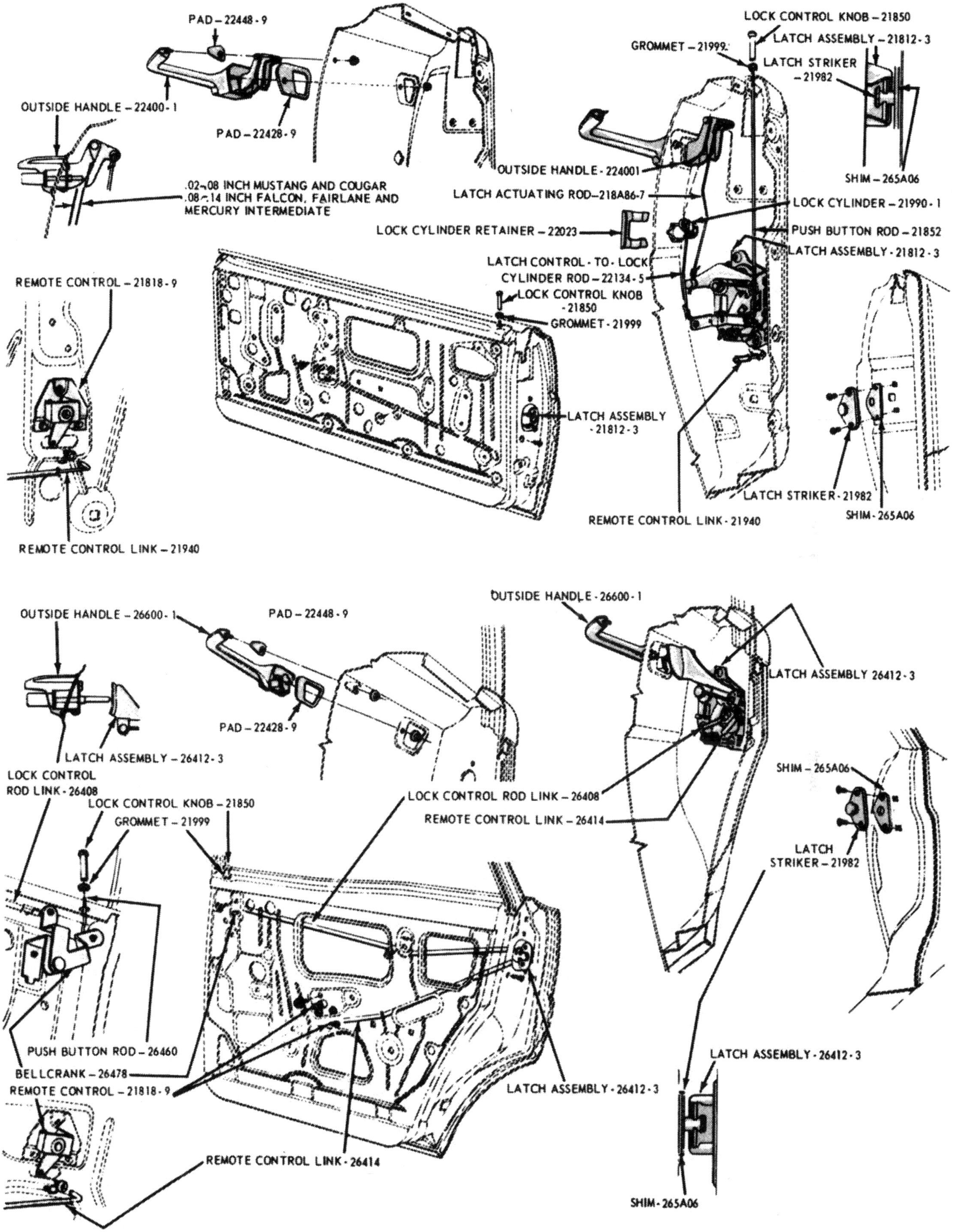

FIG 13 : 5 Typical door handle and latch installation

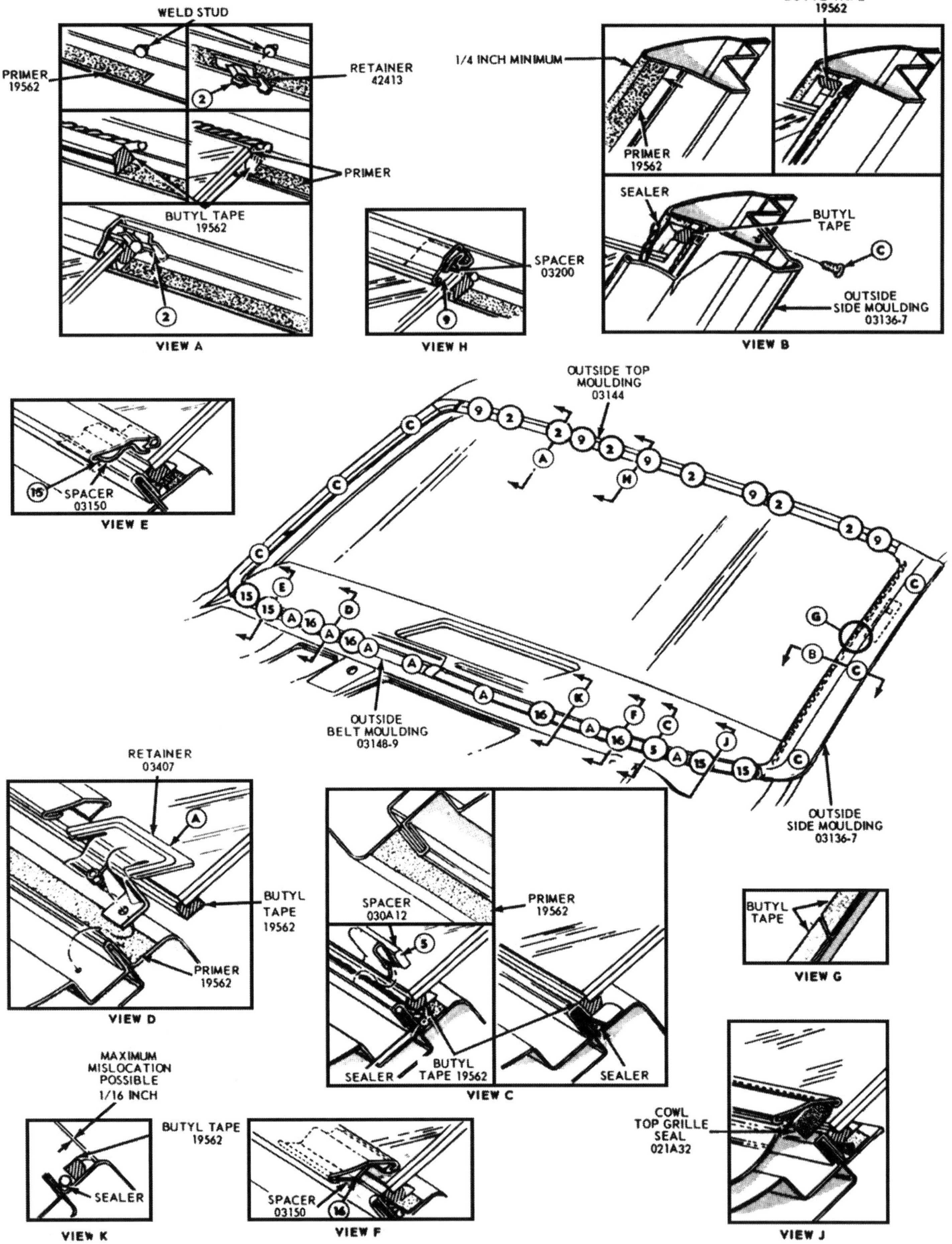

FIG 13:6 Late model windshield installation

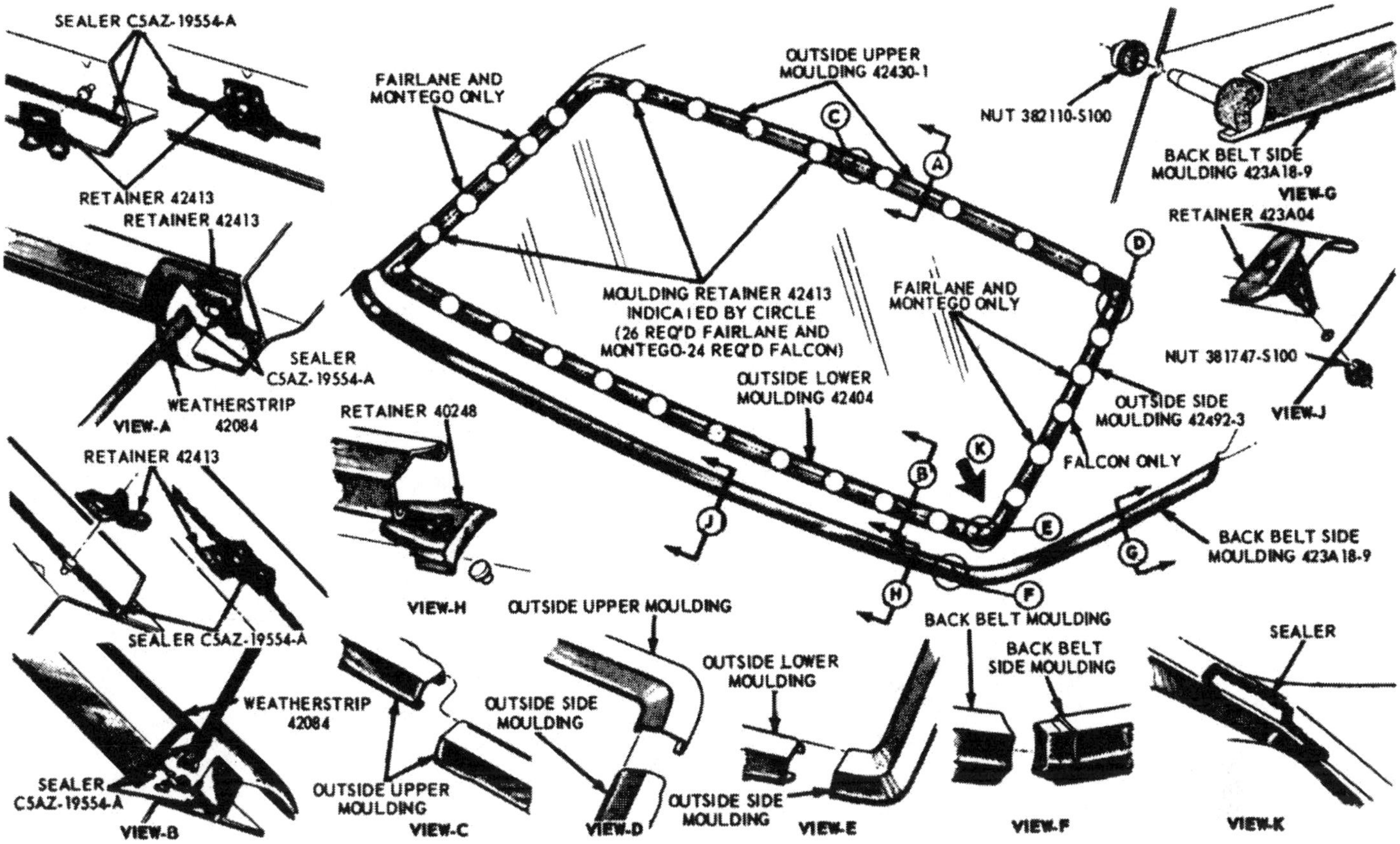

FIG 13:7 Late model back window installation

13:3 Windshield and back window

Two differing methods of securing and sealing the glass in place are used. Earlier models use a weatherstrip type seal where the glass fits into a slit in the rubber weatherstrip. On some later models Butyl tape is used.

If the glass has broken but it is still in place, it will be found easiest to remove it after sticking paper onto both sides and laying cloth or paper over the hood or luggage locker lid to protect the paint. If the windshield has broken, **make sure that all particles are removed from the windshield demist system,** otherwise they may be blown into the faces of the front seat occupants by the heater.

Once the glass, mouldings and sealer have been removed, check the aperture flange for dents or protrusions. Dents should be dressed out using a block of metal on one side and a hammer to tap them flat while protrusions should be smoothed down with a file. **If the flange is left in a damaged state it may set up stress points in the glass which may cause it to fail in service.**

Removal when Butyl strip is used:

Remove the moulding pieces. If the windshield is to be removed, first take off the wipers and their arms. A typical late model windshield installation is shown in **FIG 13:6,** and a back window installation in **FIG 13:7.** Insert a 2 foot length of thin piano wire between the flange and glass. Wrap each end of the wire around a piece of wooden dowl as a handle and with the aid of an assistant cut through the seal all the way around the glass, using a sawing motion. Remove the glass. Pry the remains of the Butyl strip away from one corner of the aperture, grasp the strip near to the flange and steadily pull it away from the flange. Collect the two spacers from the lower edge of the aperture.

Removal when weatherstrip is fitted:

Remove the moulding pieces. If the windshield is to be removed, take off the wipers and their arms. A typical early model windshield installation is shown in **FIG 13:8** and a back window installation in **FIG 13:9.** Remove any interior garnish mouldings. If a heated back window is fitted, disconnect the wires. If necessary use a thin wooden wedge to free the inner lip of the weatherstrip from the aperture. Apply pressure on a corner of the glass, using a well-padded foot or flat of hand, and press the glass and weatherstrip outwards from the aperture. Have an assistant outside the vehicle to take the glass as it comes free. Lay the glass onto a padded bench and strip off the weatherstrip. It is unlikely that the weatherstrip will be removed without causing damage to it, so it is always best to fit a new weatherstrip on installation and discard the old one.

Installation using Butyl strip:

Install the spacers in the lower corners of the aperture. Check the moulding retainers and repair or renew as required if they are defective. Temporarily position the glass in the aperture and move it into the best position for even overlap. Handling of the glass will be easier if rubber suckers are attached to the outside and used as hand holds. When the glass is satisfactorily positioned, mark the location with wax crayon.

Remove the glass and apply primer (furnished with the kit) around the entire perimeter of the aperture flange and

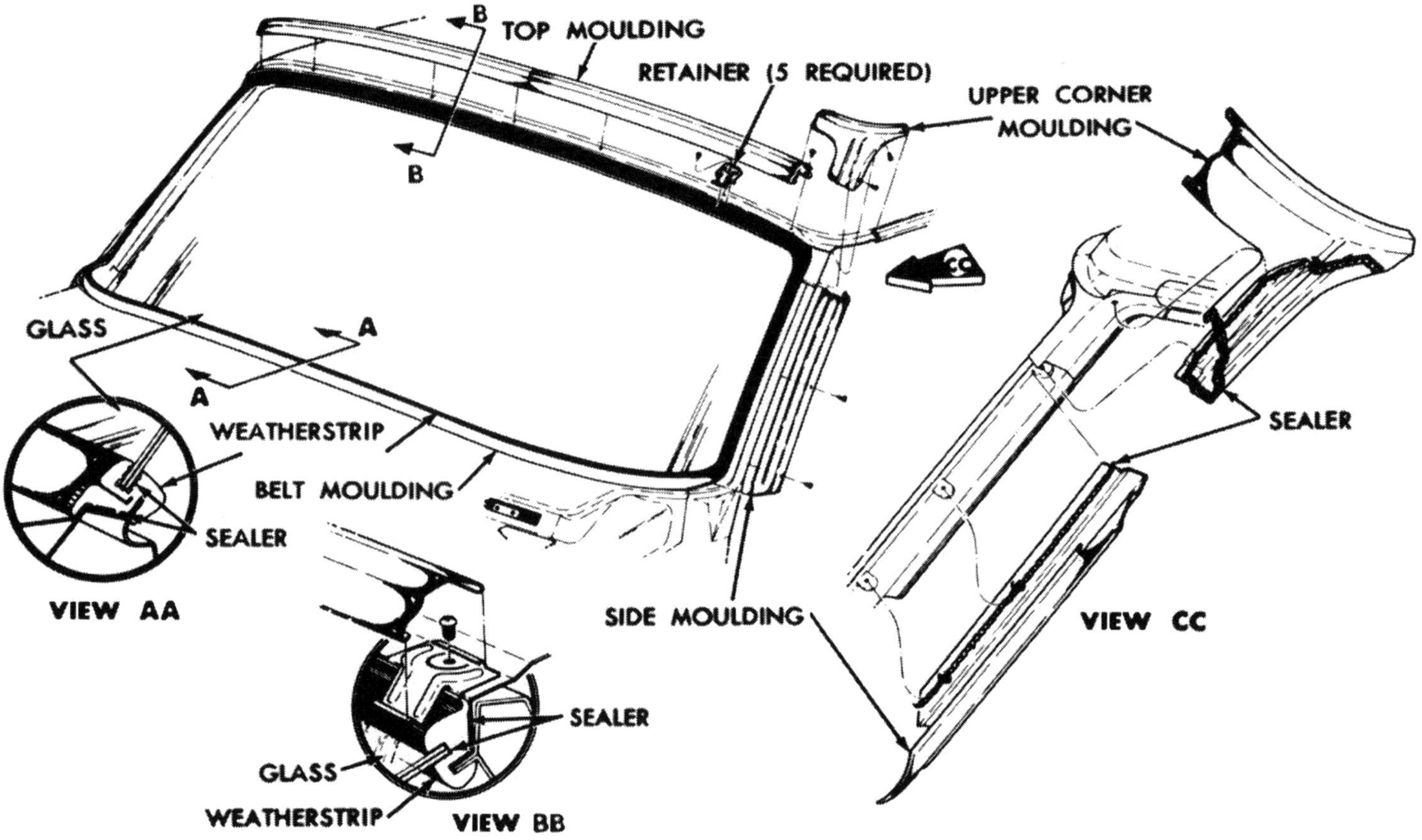

FIG 13:8 Early model windshield installation

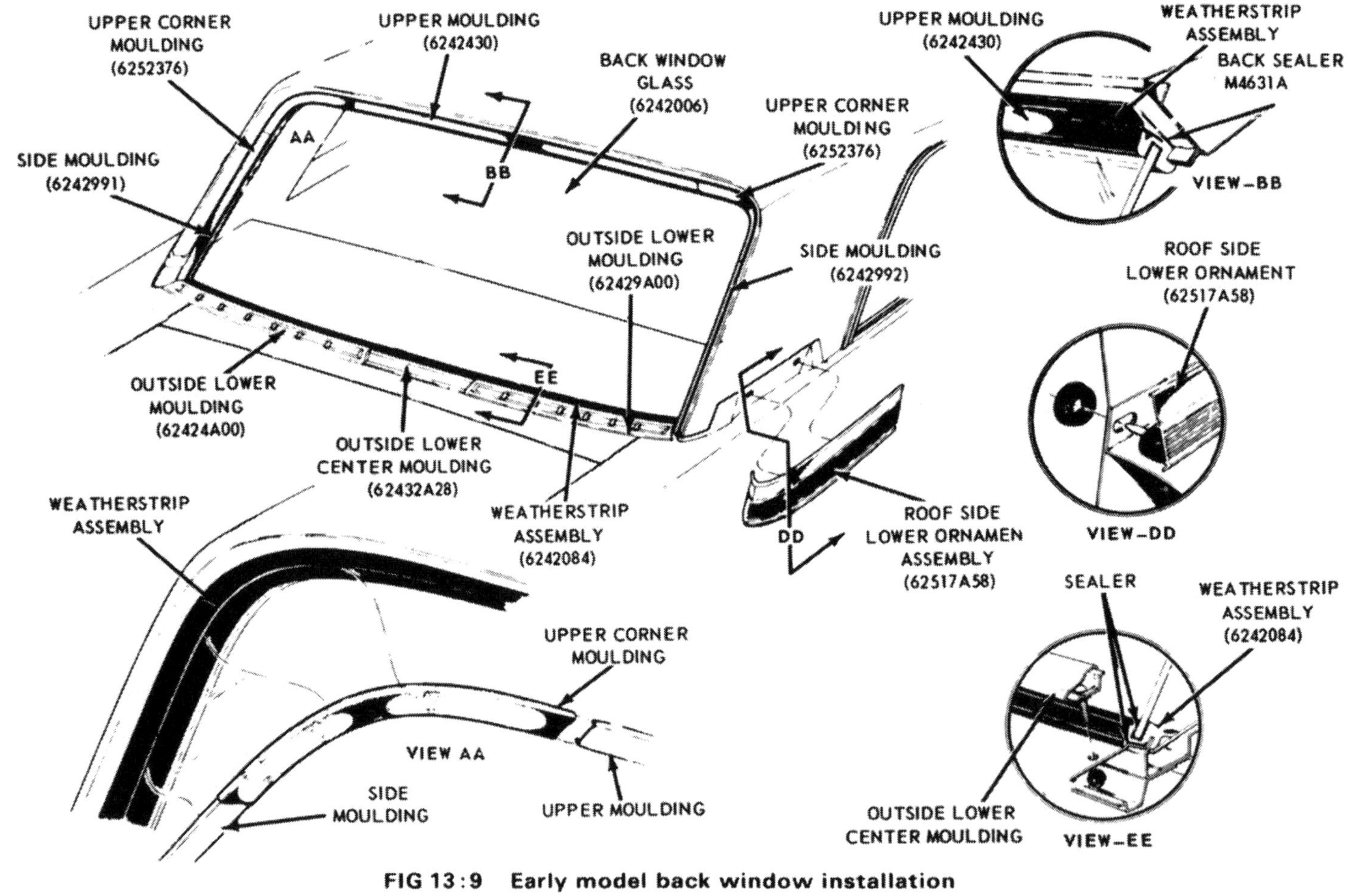

FIG 13:9 Early model back window installation

the edge of the glass where it will contact the Butyl strip. Allow the primer to dry for 10 minutes.

Position the Butyl tape around the aperture, so that its edge is within $\frac{1}{16}$ inch of the edge of the flange, but do not allow the tape to overhang the edge of the flange. **Do not stretch the tape at the corners but guide it gently around and into place.** Carefully splice the two ends of the tape where they meet.

Place the glass against the aperture, using the previously made crayon marks to align it accurately. When the glass is in position, press it firmly against the tape using hand pressure. If the tape appears dull through the glass, the glass and tape are not in contact so apply hand pressure at that point until the tape appears shiny. Remove surplus primer from the glass with a razor blade and wipe clean with a cloth moistened with naptha. Pour water over the outside, or use a hose-pipe, to check for leaks. Leaks can be repaired by applying liquid Butyl around the edge of the glass.

Install the mouldings and any other parts removed.

Installation using weatherstrip:

Inject sealant around the weatherstrip glass opening groove. Position the weatherstrip around the glass. Insert a length of cord into the body aperture flange groove in the weatherstrip making sure that there is sufficient cord to leave the ends overlapped and hanging free. Tape the ends of the cord to the inside of the glass, as shown in **FIG 13:10.** Apply a bead of sealer around the body aperture flange.

With the aid of an assistant, position the glass (from the outside of the vehicle) against the aperture. While the assistant applies firm pressure to hold the glass in place, pull the cord up and out of the groove, parallel to and towards the center of the glass, so that as the cord comes free it lifts the lip of the weatherstrip up and over the flange in the aperture.

On models fitted with heated back windows, position a piece of tape over the flange where the wires for the glass pass. Install the mouldings and any parts removed. Surplus sealer should be removed with a cloth moistened with solvent. Do not use excessive solvent as it may creep into the weatherstrip and remove the sealer that should be there.

Rear view mirror:

On some models the rear view mirror is bonded directly to the glass, using a vinyl pad. Heat from a hot air gun is used to soften the vinyl pad for removal or installation. Insulate any chips or scratches with wet rags or puttty. Blow hot air over the windshield and after approximately 8 to 10 minutes the rear view mirror may be removed by levering carefully on the mirror. Gently reduce the heat and then allow the glass to cool naturally.

Preheat the glass in the area before installing the mirror and then heat the button. Press the button firmly into place and keep heating until vinyl can just be seen extruding between glass and mirror support. Pressure must be applied to the mirror mount while bonding it in place.

Maintain pressure for at least 2 minutes while reducing the heat and then allow the glass to cool naturally. Do not remove insulation while the glass is hot.

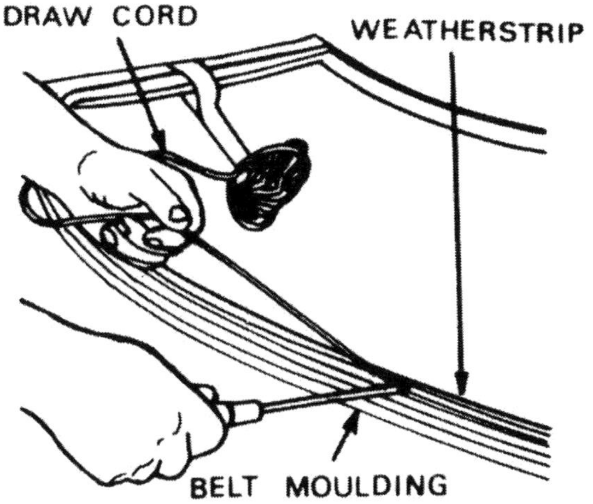

FIG 13:10 Fitting the draw cord to the weatherstrip

13:4 The hood

The hood hinge on an early model is shown in **FIG 13:11** and on a late model in **FIG 13:12.** The arrows indicate the directions in which the parts can be moved to align the hood accurately with the body.

Before removing the hood, lightly mark around the hinge with pencil to ensure that the hood will be installed back in its original position. Lay pads of cloth onto the paintwork to protect it from damage and have enough assistants to support the hood as it is freed. Open the hood and remove the bolts that secure the hinge to the hood. Lift off the hood.

Install the hood in the reverse order of removal. If the hood is not in accurate alignment with the body, slacken the attaching bolts sufficiently to hold the weight of the hood but allow it to be moved under pressure. Move the hood into alignment and tighten the attaching bolts fully.

An early type hood lock mechanism is shown in **FIG 13:13.** Adjust the lock until the lock dowel enters it centrally and the safety catch engages on the striker. The lock dowel can be screwed up or down to adjust the tension on the hood when it is closed. If the hood is loose and rattles when closed screw the dowel upwards, while if it is stiff and difficult to close screw the dowel slightly downwards.

FIG 13:11 Early hood hinge

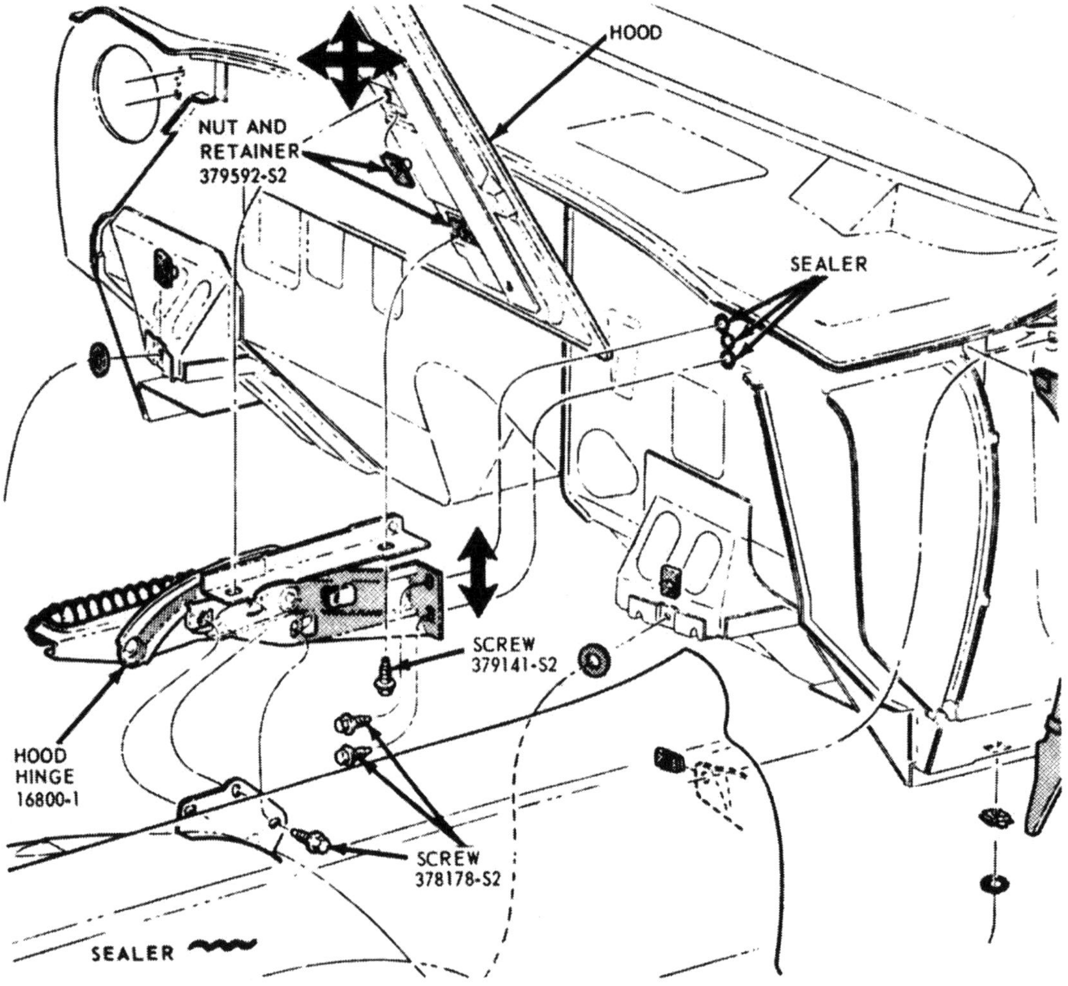

FIG 13:12 Late hood hinge adjustment

13:5 The luggage locker door

A typical late model hinge attachment is shown in **FIG 13:14.** Removal and installation of the door is very similar to the method used for the hood (see previous section) though any shims fitted should be carefully collected and installed back in their original positions.

Torsion bars are used to counterbalance the weight of the door. The method of altering the torsion bar tension is shown in **FIG 13:15.**

13:6 Air conditioning

A typical air conditioning system is shown in **FIG 13:16.** The system is charged with a non-toxic refrigerant (Refrigerant 12). The compresser is driven by a belt from the engine through a magnetic clutch. The clutch only operates when the system is selected and the temperature in the car rises above the selected temperature. With the system operating, the compressor draws the gas from the evaporator and pressurizes it. The operating pressure from the compressor varies between

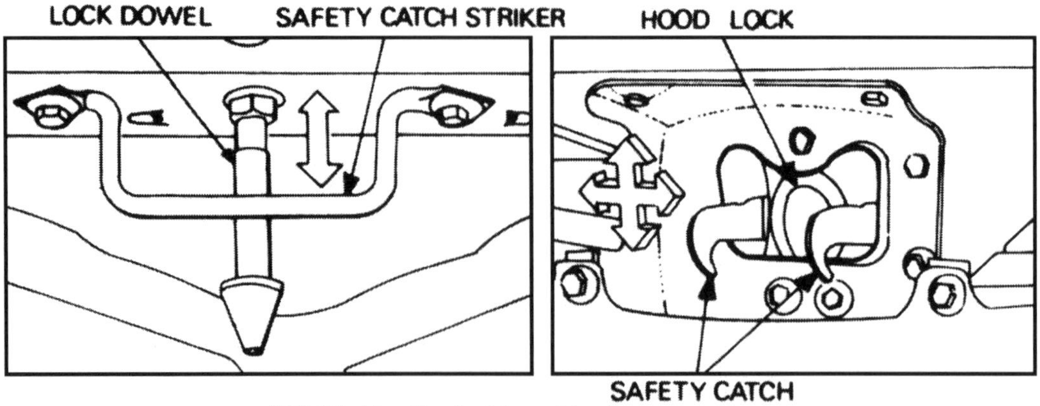

FIG 13:13 Typical hood lock mechanism

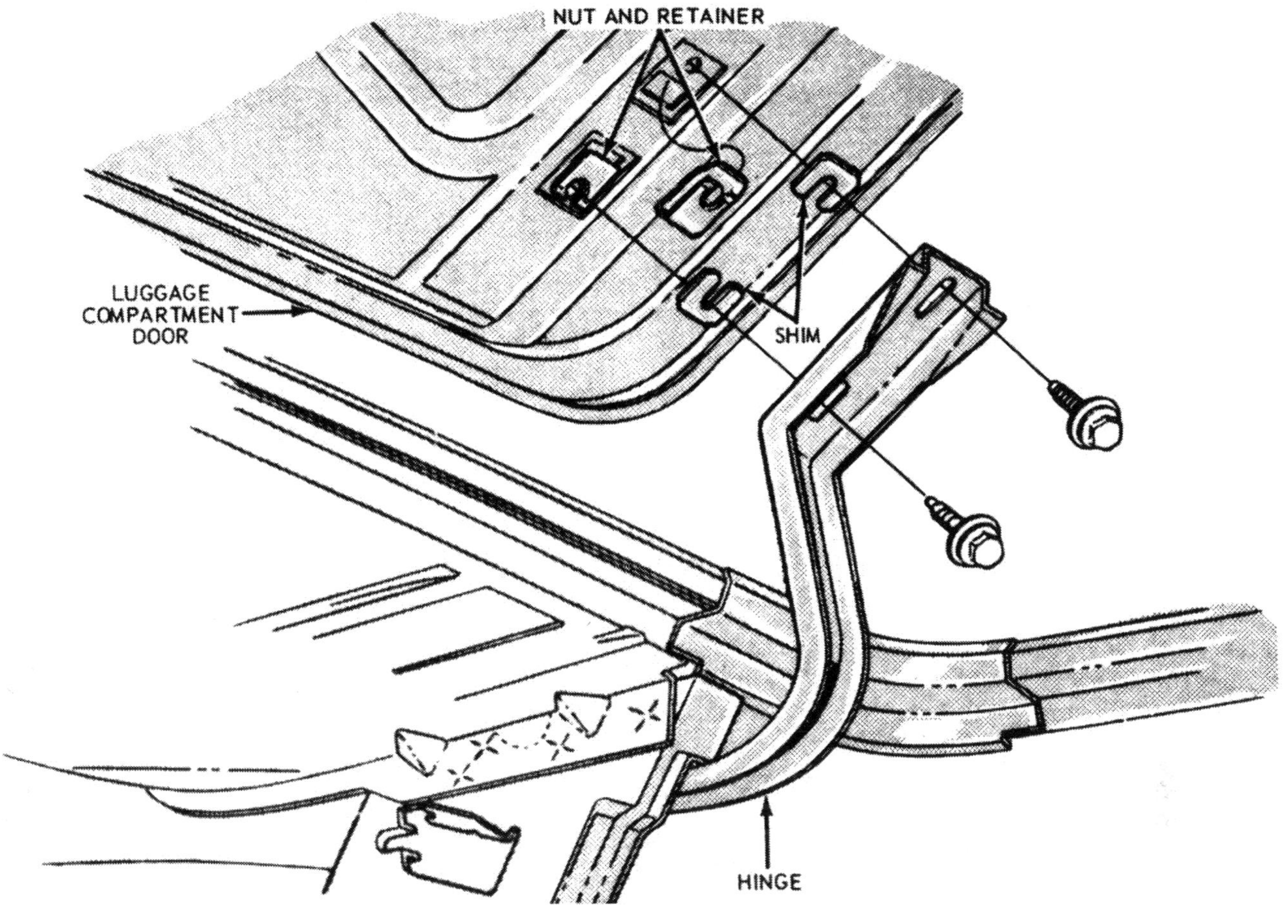

FIG 13:14 Luggage locker door attachments

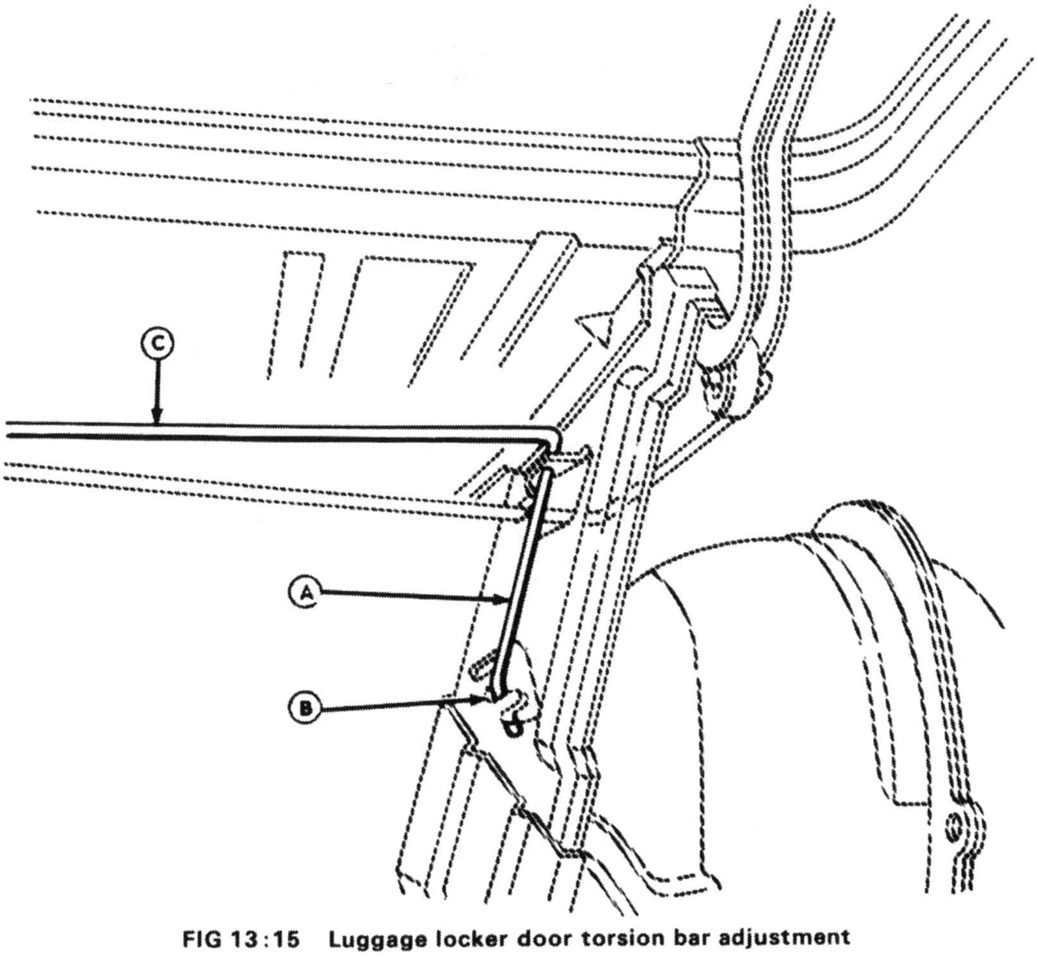

FIG 13:15 Luggage locker door torsion bar adjustment

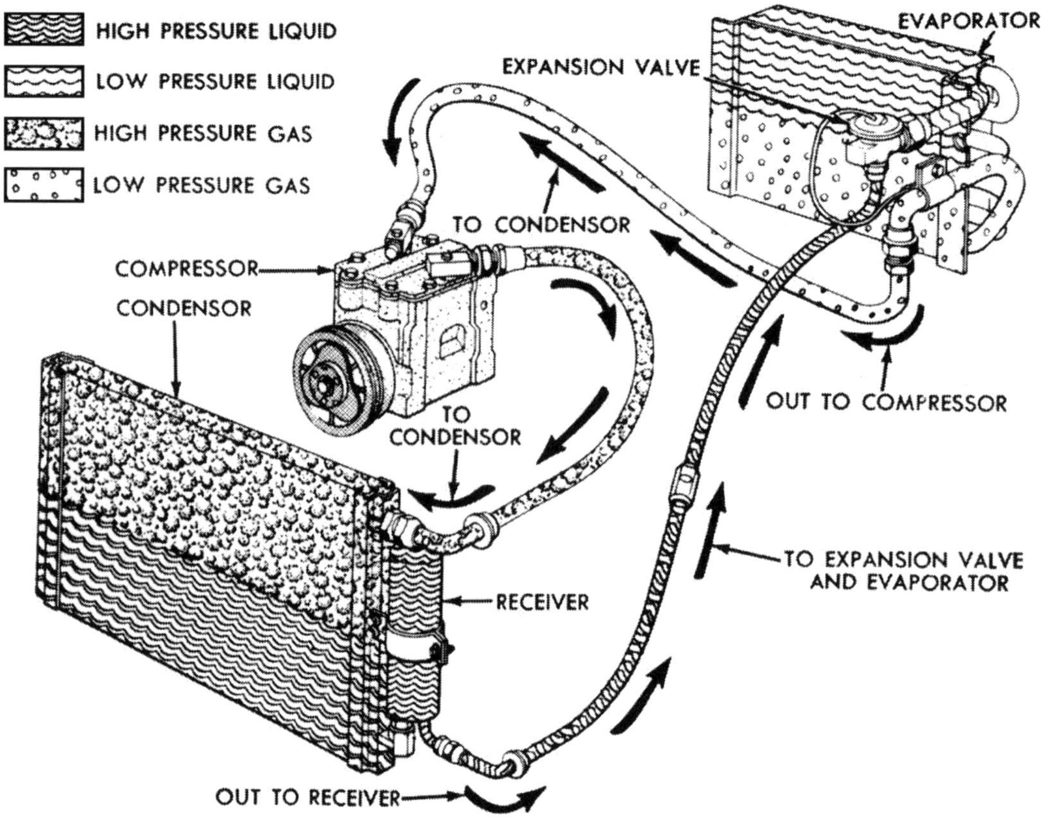

FIG 13:16 Air conditioning system

80 to 300 lb/sq inch depending on the ambient air temperature and compressor operating speed. The high pressure gas, hot because of compression, passes from the compressor to the condensor where it is cooled by the passage of air over the condensor. The refrigerant changes from gas to liquid in the condensor and passes on to the receiver. The receiver contains a water-absorbing material to ensure that the refrigerant is perfectly dry. From the receiver the liquid refrigerant, still at high pressure passes through a sight glass and hoses to the expansion valve. The expansion valve controls the amount of refrigerant passing through and reduces the pressure to between 12 and 50 lb/sq inch depending on the conditions and compressor speed. The low pressure liquid passes into the evaporator unit inside the vehicle and the heat in the air passing through it causes the liquid to change back to gas, at low pressure, and cool the air. The low pressure gas is then drawn back into the compressor. A thermostatic switch is fitted to control the compressor magnetic clutch.

Servicing of the system should only be carried out by trained personnel as though the refrigerant is non-toxic serious injury can result if the safety precautions are ignored.

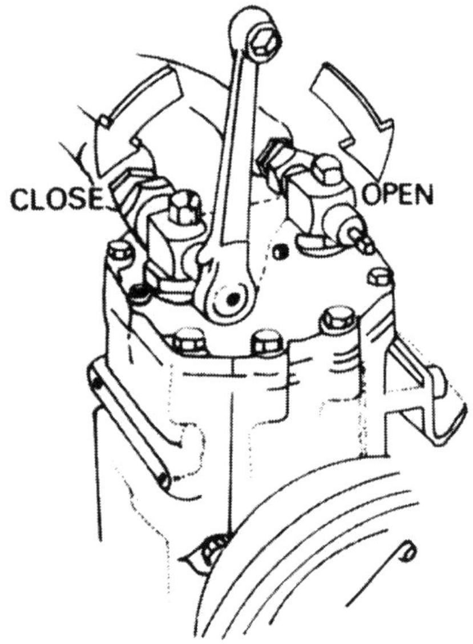

FIG 13:17 Air conditioning compressor low pressure service valve

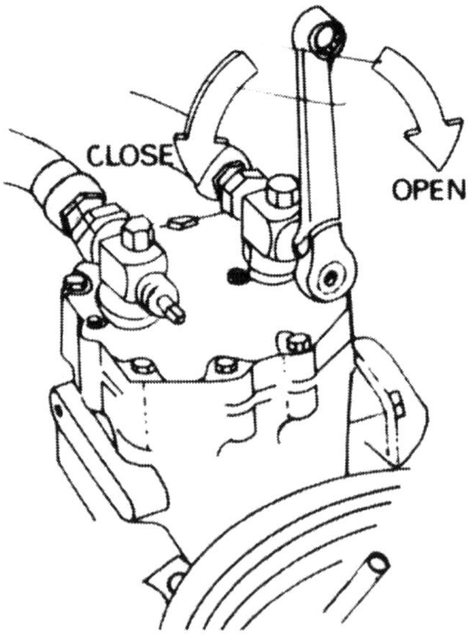

FIG 13:18 Air conditioning compressor high pressure service valve

Safety precautions:

At normal pressure and temperature, the refrigerant evaporates practically instantaneously, absorbing heat rapidly. For this reason spillage of the fluid should be avoided as it will cause 'burns' on the skin. **Contact with the eyes is more serious, so wear safety goggles when working on the system.** If the refrigerant does go into an eye, use a few drops of sterile mineral oil to wash out the eye (as the refrigerant is readily absorbed by mineral oil) and then clean the eye with a weak boric acid solution. **Seek medical aid immediately even if the irritation has ceased.**

If the system is discharged, do so slowly (to avoid flushing the oil out of the compressor) and discharge out directly into the open. If the refrigerant is discharged into a confined space, the rapid evaporation will cause the refrigerant to displace the air and reduce the amount of oxygen present by a serious amount. This could lead to asphyxiation.

Do not heat the system, such as when steam cleaning, soldering, or using a blow lamp in the area, as the rise in pressure could cause a component to burst explosively.

The refrigerant itself is non-toxic but it will produce a very poisonous gas if passed through a naked flame. The leak tester uses a small flame to heat a copper disc so avoid breathing the fumes from the tester when checking for leaks.

Sight glass:

Normally the glass should appear perfectly clear and free from bubbles. Check by running the engine at 1500 rev/min and cycling the magnetic clutch. Bubbles should appear when the clutch is off. If there are no bubbles then the system is empty.

Continuous bubbles indicate a low charge in the system. Under conditions of high temperature, bubbles or foam may appear in the sight glass.

Compressor removal:

This is the only servicing that should be carried out by the owner.

Turn both the high and low pressure valves fully clockwise. The valves are shown in **FIGS 13:17** and **13:18.** Loosen the cap on the high pressure valve, by a small amount only, and allow the refrigerant to escape from the compressor. Disconnect the service valves from the compressor. Undo the compressor attachments and remove it from the engine.

The compressor is installed in the reverse order of removal. Before opening the service valves, evacuate the compressor using a vacuum pump. After installation, the owner will probably find it most convenient to leave the service valves closed and drive the vehicle to a service agent, without using the air conditioner, so that the compressor can be evacuated.

APPENDIX

TECHNICAL DATA

Unless otherwise stated, all dimensions are given in inches

ENGINE

Type	V8, water-cooled. OHV using rocker arms and hydraulically-operated valve lifters
Capacity	289 cu in or 302 cu in
Compression ratio:	
289 2V	9.3:1
289 4V	10:1
302 2V	9.5:1
302 4V	10.5:1
Cylinder numbering sequence:	
Right bank, from front	1, 2, 3, 4
Left bank, from front	5, 6, 7, 8
Firing order	1-5-4-2-6-3-7-8
Bore	4.00 inch nominal, all models
Stroke:	
289	2.87
302	3.00
Crankshaft:	
Main bearings	Five, using renewable steel-backed bearing shells. Thrust taken at centre bearing
Connecting rod bearings	Four, two connecting rods on each crankpin
End play	.004 to .008. Wear limit .012
Journal diameter	2.2482 to 2.2490
Crankpin diameter	2.1228 to 2.1236
Journal maximum taper	.0003 per inch
Crankpin maximum taper	.0004 per inch
Journal and crankpin maximum out of round	.0004
Journal to bearing clearance:	
Desirable	.0005 to .0015
Allowable	.0005 to .0024 (except front which should be .0018 maximum)
Crankpin to bearing clearance:	
Desirable	.0008 to .0015
Allowable	.0008 to .0026
Bearing shells	Coded Red or Blue and —.002 inch for standard crankshafts
Camshaft:	
End play:	
1964, 1965 and 1966	.001 to .007. Wear limit .012
All other years	.0005 to .0055. Wear limit .0070
Journal maximum out of round	.0005
Lobe lift:	
All 1964, 1965 and later inlet	.2303
Later exhaust	.2375
Maximum lobe lift loss	.005 all models
Journal to bearing clearance	.001 to .003. Wear limit .006
Timing chain maximum deflection	.5
Cylinder head:	
Valve seat width	.060 to .080
Valve seat angle	45 deg.
Gasket surface flatness:	
In any 6 inches	.003 maximum distortion
Over full length	.007 maximum distortion
Valve guide bore diameter	.3433 to .3443 (standard)

Cylinder bores:

Out of round:		
New	.001 maximum	
Wear limit	.005 maximum	
Taper:		
New	.001 maximum	
Wear limit	.010 maximum	

Pistons:

Diameter, at right angle to pin centreline:		
Coded Red	3.9984 to 3.9990	
Coded Blue	3.9996 to 4.0002	
Piston to bore clearance	.0014 to .0022	
Ring groove width:		
Compression rings	.080 to .081	
Oil control ring	.188 to .189	
Piston pin bore diameter	.9123 to .9126	

Piston pin:

Length	3.010 to 3.040	
Diameter:		
Standard	.9119 to .9124	
+.001 oversize	.9130 to .9133	
Pin to piston clearance	.0003 to .0005. Wear limit .0008	

Piston rings:

Width:		
Compression rings	.077 to .078	
Side clearance:		
Compression rings	.002 to .004. Wear limit .006	
Oil control	Snug	
Fitted gap	.010 to .020	
Rail of oil control ring	.015 to .069	

Valves:

Seat angle	44 deg.	
Valve stem to guide clearance:	*Inlet*	*Exhaust*
1964 to 1967	.0010 to .0027 Wear limit .0045	.0020 to .0037 Wear limit .0055
1968 and 1970	.0010 to .0027 Wear limit .0045	.0015 to .0032 Wear limit .0047
Stem diameter:		
All inlet valves	.3416 to .3423	
1965 to 1967 exhaust valves	.3406 to .3413	
1968 to 1970	.3411 to .3423	
Oversize stems available	+.003, +.015, +.030	
Valve head diameter:		
1964 to 1966 Inlet	1.662 to 1.667	
1967 onwards Inlet	1.773 to 1.783	
All exhaust	1.442 to 1.457	
Valve tappet diameter	.8740 to .8745 standard	
Valve tappet to bore clearance	.0005 to .0020 inch. Wear limit .005	

Valve springs:

Free length:		
1964 to 1966	2.09	
1967 early and Non-Thermactor	1.93	
1967 late and Thermactor	1.97	
1968, 1969 and 1970	1.97	

Test data ...	Length in inches at load in lb
1964 to 1966 ...	1.780 at 71 to 79 Wear limit 1.780 at 64 1.390 at 161 to 177 Wear limit 1.390 at 145
1967 (early and non-Thermactor) ...	1.66 at 71 to 79 Wear limit 1.64 at 68 1.27 at 174 to 192 Wear limit 1.25 at 165
Later 1967 and onwards ...	1.66 at 71 to 79 Wear limit 1.66 at 68 1.23 at 171 to 189 Wear limit 1.23 at 162
Oil pump relief valve spring:	
Test length ...	1.704 inch at 11.15 to 11.75 lb
Oil pressure:	
Hot at 2000 rev/min ...	35 to 55 lb/sq inch

FUEL SYSTEM

Type ...	Single multi-barrel carburetor with sealed mechanically-operated fuel pump	
Fuel pump:		
Filter ...	Early models fitted with cartridge. Late models fitted with renewable line filter between pump and carburetor	
Delivery ...	1 pint in 20 seconds	
Delivery pressure:		
1964 to 1966 ...	4 to 6 lb/sq inch at 500 rev/min	
All later ...	4.5 to 6.5 lb/sq inch at 500 rev/min	
Minimum static suction:		
1964 and 1966 ...	6 in Hg at 500 rev/min	
All later ...	10 in Hg at 500 rev/min	
Carburetor:		
Early ...	Ford 1100 1-V	
Later ...	Autolite 1100 1-V	
1968 only ...	Autolite 4300	
Approximate idle setting ...	1½ turns open	
Idle speed ...	Given in rev/min. Headlights should be onto load alternator and if air-conditioning is fitted this should be run for 20 minutes	
	Manual-shift	*Automatic**
1964 to 1966 ...	575 to 600	475 to 500
1967:		
Standard ...	575	475
With emission control ...	625	550
1968 ...	625	550
1969 ...	650	550
1970 ...	800/500	575

**Set with a drive gear selected*

Emission control:	
Manual-shift ...	Thermactor air pump
Automatic transmission ...	Imco distributor and carburetor

Air inlet duct:
Test by immersing in water, heating water and noting temperature with thermometer

1964 to 1967:

Valve in full hot position	75°F and below
Valve should start to open	85°F
Valve in full cold position	100°F or above

1968 to 1970:

Valve in full hot position	100°F and below
Valve in heat off position...	135°F*

**Soak at temperature for 5 minutes before condemning valve*

IGNITION SYSTEM

Firing order	1-5-4-2-6-3-7-8	
Spark plug gap	.032 to .036	
Spark plug type	BF42, 18 mm	
Distributor contact gap:		
	Gap (inch)	*Dwell (deg.)*
1964 and 1965	.014 to .016	26 to 28½
1966	.017	30 to 33
1967	.017	26 to 31
1968	.021	24 to 29
1969	.021*	24 to 29
	.017	26 to 31
1970 single diaphragm	.017	26 to 31
dual diaphragm	.021	24 to 29

**Standard manual-shift transmission*

Breaker arm tension	17 to 21 oz
Condenser...	.21 to .25 microfarads
Initial timing setting:	
All models, except Thermactor equipped	
1966 and 1967	6 deg. BTDC
1966 and 1967 with Thermactor ...	0 deg. TDC
Ignition coil:	
Primary resistance	1.40 to 1.54 ohms at 75°F
Secondary resistance	7600 to 8800 ohms at 75°F
Current draw, engine stopped	4.5 amps
Current draw, idle	2.5 amps
Primary circuit resistance...	1.30 to 1.40 ohms at 75°F

COOLING SYSTEM

Antifreeze...	Ethylene Glycol, should always be used to give minimum protection to 0°F	
Thermostat:	*Starts opening*	*Fully open*
Low temperature...	155 to 162°F	182°F
High temperature...	188 to 195°F	210 to 212°F

CLUTCH

Operation	Mechanical
Adjustment	Minimum ½ inch free travel of pedal at 3000 rev/min
Springs—quantity:	
Cover assembly	9
Disc assembly	12
Clutch diameter	10 inch (10⅜ inch for 1964 and 1965)

Cover assembly identification:

	Springs	Cover	Pressure plate
1964	Bronze	Pink	White
1965	Bronze	Pink	White
1966	Bronze	Grey	Green
1967	Bronze	Purple	Blue
1968	Bronze	White	Green
1969	Bronze	White	Green
1970	Bronze	White	Green

Disc assembly identification:

	Springs	Linings
1964 to 1966	Orange	—
1967	Orange	Purple
1968 to 1970	6 Pink, 6 Orange	Blue

MANUAL SHIFT TRANSMISSION

Ratios:

	First	Second	Third	Fourth	Reverse
4-speed (all)	2.78:1	1.93:1	1.36:1	1:1	2.78:1
3-speed (1964 and 1965)	2.79:1	1.70:1	1:1	—	2.87:1
3-speed (1966 to 1970) ...	2.99:1	1.75:1	1:1	—	3.17:1

Lubrication:

	Type	Quantity (US pints)
3-speed	C3RZ-19547-B (Ford Standard transmission fluid)	$3\frac{1}{2}$
4-speed	C3RZ-19547-B	4

REAR AXLE

Lubricant:

Quantity ... ... 5 pints

Type:

Grade ... ... SAE.90 above —25°F
SAE.80 below —25°F

Make ... ... Ford Hypoid Gear Lube

SUSPENSION

Castor:

	Min.	Max.	Desired	Max. variation between wheels
1964	0 deg.	+1 deg.	$+\frac{1}{2}$ deg.	$\frac{1}{2}$ deg.
1965	$-1\frac{1}{4}$ deg.	$+\frac{3}{4}$ deg.	$-\frac{1}{4}$ deg.	$\frac{1}{2}$ deg.
1966	—1 deg.	+1 deg.	0 deg.	$\frac{1}{2}$ deg.
1967	$-1\frac{1}{2}$ deg.	$+\frac{1}{2}$ deg.	—1 deg.	$\frac{1}{2}$ deg.
1968	$-1\frac{1}{2}$ deg.	$+\frac{1}{2}$ deg.	$-\frac{1}{2}$ deg.	$\frac{1}{2}$ deg.
1969 and 1970	$-1\frac{3}{4}$ deg.	$+\frac{1}{4}$ deg.	$-\frac{3}{4}$ deg.	$\frac{1}{2}$ deg.

Camber:

	Min.	Max.	Desired	Max. variation between wheels
1964	0 deg.	+1 deg.	$+\frac{1}{2}$ deg.	$\frac{1}{2}$ deg.
1965	$-\frac{1}{4}$ deg.	$+1\frac{1}{4}$ deg.	$+\frac{1}{2}$ deg.	$\frac{1}{2}$ deg.
1966	$-\frac{1}{2}$ deg.	+1 deg.	$+\frac{1}{4}$ deg.	$\frac{1}{2}$ deg.
1967	$-\frac{1}{2}$ deg.	+1 deg.	$+\frac{1}{2}$ deg.	$\frac{1}{2}$ deg.
1968	$-\frac{1}{2}$ deg.	+1 deg.	$+\frac{1}{4}$ deg.	$\frac{1}{2}$ deg.
1969	$-\frac{1}{2}$ deg.	+1 deg.	$+\frac{1}{4}$ deg.	$\frac{1}{2}$ deg.
1970	$-\frac{1}{2}$ deg.	+1 deg.	$+\frac{1}{4}$ deg.	$\frac{1}{2}$ deg.

Toe-in:

	Min.	Max.	Desired
1964	$\frac{1}{4}$ inch	$\frac{5}{16}$ inch	—
1965	$\frac{5}{32}$ inch	$\frac{13}{32}$ inch	$\frac{9}{32}$ inch
1966	$\frac{1}{8}$ inch	$\frac{3}{8}$ inch	$\frac{1}{4}$ inch
1967	$\frac{1}{8}$ inch	$\frac{1}{4}$ inch	$\frac{3}{16}$ inch
1968	$\frac{1}{8}$ inch	$\frac{3}{8}$ inch	$\frac{1}{4}$ inch
1969	$\frac{1}{8}$ inch	$\frac{3}{8}$ inch	$\frac{3}{16}$ inch
1970	$\frac{1}{8}$ inch	$\frac{3}{8}$ inch	$\frac{3}{16}$ inch

STEERING

Type ... ... Recirculating ball. Power steering can be added as optional extra

Power steering fluid ... ... Ford Automatic transmission fluid

Steering box data:

	Gear ratio	Turns lock to lock	Lubricant* quantity
1964:			
Manual	19.9:1	4⅝	8 oz
Power	19.9:1	4⅝	8 oz
1965:			
Manual	19.9:1	4⅝	8 oz
Power	16:1	3¾	8 oz
1966:			
Manual	22:1	5	11 oz
Power	16:1	4	11 oz
1967:			
Manual	22:1	5	.87 lb
Power	16.1:1	4	.87 lb
1968 to 1970:			
Manual	22:1	5.5	.87 lb
Power	16.1:1	4	.87 lb

**Use C3AZ-19578-A*

Steering box preloads:

	Given in inch lb Worm bearing preload	Total preload
All models:		
Manual	4 to 5	9 to 10
Power	3 to 4	8 to 9

BRAKING SYSTEM

Type ... ... Hydraulically-operated service brakes, cable operated parking brake

Fluid ... ... Ford Super Heavy Duty

Adjustments ... ... All service brakes self-adjusting. Parking brake adjusted at cable

Drum brake lining identification:

	Size in inches Front	Rear	Color
1965:			
Primary	2.25 x 8.35	1.75 x 8.35	Silver
Secondary	2.25 x 10.75	1.75 x 10.75	Yellow-White
1966 Sedans:			
Primary	2.25 x 8.35	1.75 x 8.35	Silver
Secondary	2.25 x 10.75	1.75 x 10.75	Yellow-White
1966 Station wagons:			
Primary	2.50 x 8.35	2.00 x 8.35	Yellow-Black
Secondary	2.50 x 10.75	2.00 x 10.75	Black-Black
1967 to 1970 Sedans:			
Primary	2.25 x 8.43	1.75 x 8.43	Yellow-Black
Secondary	2.25 x 10.82	1.75 x 10.82	Blue-Blue
1967 to 1970 Station wagons:			
Primary	2.50 x 8.43	2.00 x 8.43	Yellow-Black
Secondary	2.50 x 10.82	2.00 x 10.75	Blue-Blue

Disc brake friction linings ... ... Bonded FoMoCo

Maximum honing on wheel and master cylinders .003

ELECTRICAL SYSTEM

Polarity	Negative earth
Voltage	12 volts
Battery	Lead/acid
Charging	Alternator
Starter motor (all models):	
No load current draw	70 amps
Brush length:	
New	.5 inch
Minimum	.25 inch
Brush tension	40 oz
Maximum commutator runout	.005 inch
Maximum voltage drop in leads	.5-volt
Normal current	Stall current and torque (battery down to 5 volts)
1964 to 1967:	
4 inch diameter—100 to 150 amps	460 amps at 9 lb ft
4½ inch diameter—250 amps	670 amps at 15.5 lb ft
1968:	
4 inch diameter—150 to 200 amps	460 amps at 9 lb ft
4½ inch diameter—150 to 200 amps	670 amps at 15.5 lb ft

CAPACITIES*

Automatic transmission	Refill
1964 and 1966	8 quarts
1967 on	9 quarts
Cooling System:	
1964	15 quarts
1965	14 quarts
1966 and 1967	15 quarts
1968 and 1969	13.6 quarts

Capacity given including heater. If no heater fitted subtract 1 quart

Engine	5 quarts (including 1 quart for filter)
Manual-Shift transmission:	
3-speed	3½ pints
4-speed	4 pints

Steering—See Steering section in Technical Data

**To convert to Imperial measure, multiply by .83*

TORQUE WRENCH SETTINGS

Given in lb ft unless otherwise stated

Engine:	
Cylinder head bolts;	
1st stage	50
2nd stage	60
Final stage	65 to 72
Inlet manifold:	
1st stage	15 to 17
Final stage	20 to 22
Exhaust manifold	15 to 20
Oil pan	9 to 11
$\frac{5}{16}$ bolts	18

Flywheel	75 to 80
Water outlet housing	12 to 15
Main bearing cap bolts	60 to 70
Oil pan drain plug	15 to 20
Oil pump attachment	23 to 28
Oil pump plate	9 to 12
Oil filter adaptor plate to block	60 to 100
Cylinder front cover	12 to 15
Water pump attachments	12 to 15
Camshaft sprocket	40 to 45
Camshaft thrust plate	6 to 9
Damper to crankshaft	70 to 90
Connecting rod bearing caps	19 to 24
Sparking plugs	15 to 20

Clutch

Flywheel housing to block	40 to 50
Cover to flywheel	12 to 20

Manual shift transmission:

	Threespeed	*Fourspeed*
Extension housing bolts	42 to 50	42 to 50
Input shaft retainer	30 to 36	19 to 25
Speedometer cable retainer	3 to 4.5	3 to 4.5
Transmission to flywheel housing	37 to 42	37 to 42

Automatic transmission:

Converter to flywheel	20 to 30
Converter housing to case	28 to 40
Rear servo cover	12 to 20
Intermediate servo cover	16 to 22
Cooler line fittings	80 to 120 lb inch
Converter drain plug	20 to 30
Transmission to engine	40 to 50
Extension housing	28 to 40
Governor body to collector	80 to 120 lb inch

Suspension:

Front shock absorber upper attachment	20 to 28
Front shock absorber bracket, upper	15 to 20
Brake backing plate	28 to 42
Suspension ball joints	60 to 90
Strut to lower arm	75 to 105
Stabilizer mounting bracket	17 to 25
Strut to underbody	60 to 80
Stabilizer link	5 to 10
Caliper:	
Upper bolt	110 to 120
Lower bolt	60 to 70
Spring shackle bars	15 to 22
Rear shock absorber, upper and lower	15 to 25
Spring to axle	30 to 45
Spring front hanger bolt	90 to 120
Spring front hanger nut	40 to 60
Wheel attachments:	
4-lug	55 to 85
5-lug	70 to 115

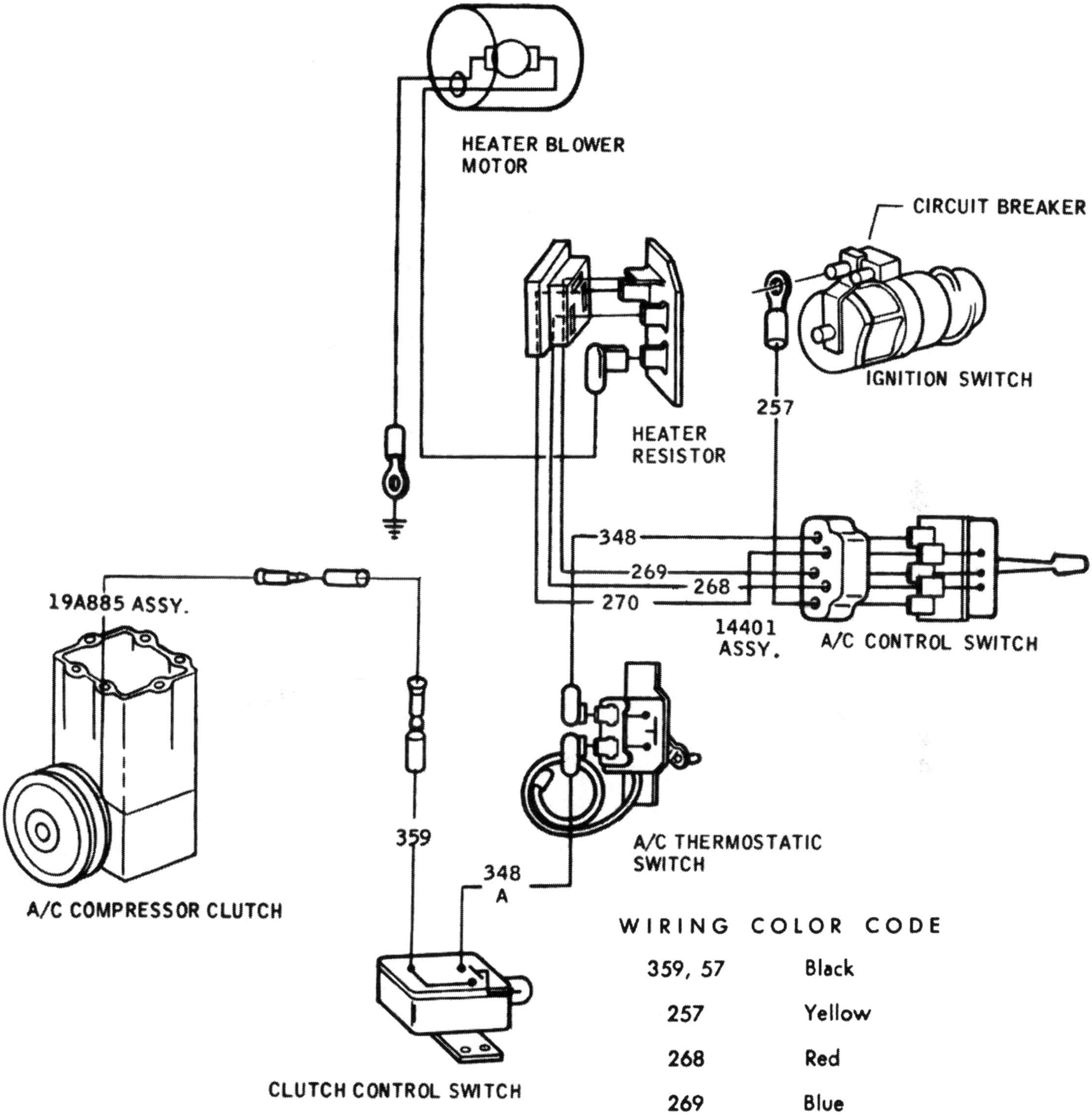

WIRING COLOR CODE

359, 57	Black
257	Yellow
268	Red
269	Blue
270	Black—Yellow Strip
348, 348A	Green

FIG 14: Air conditioner

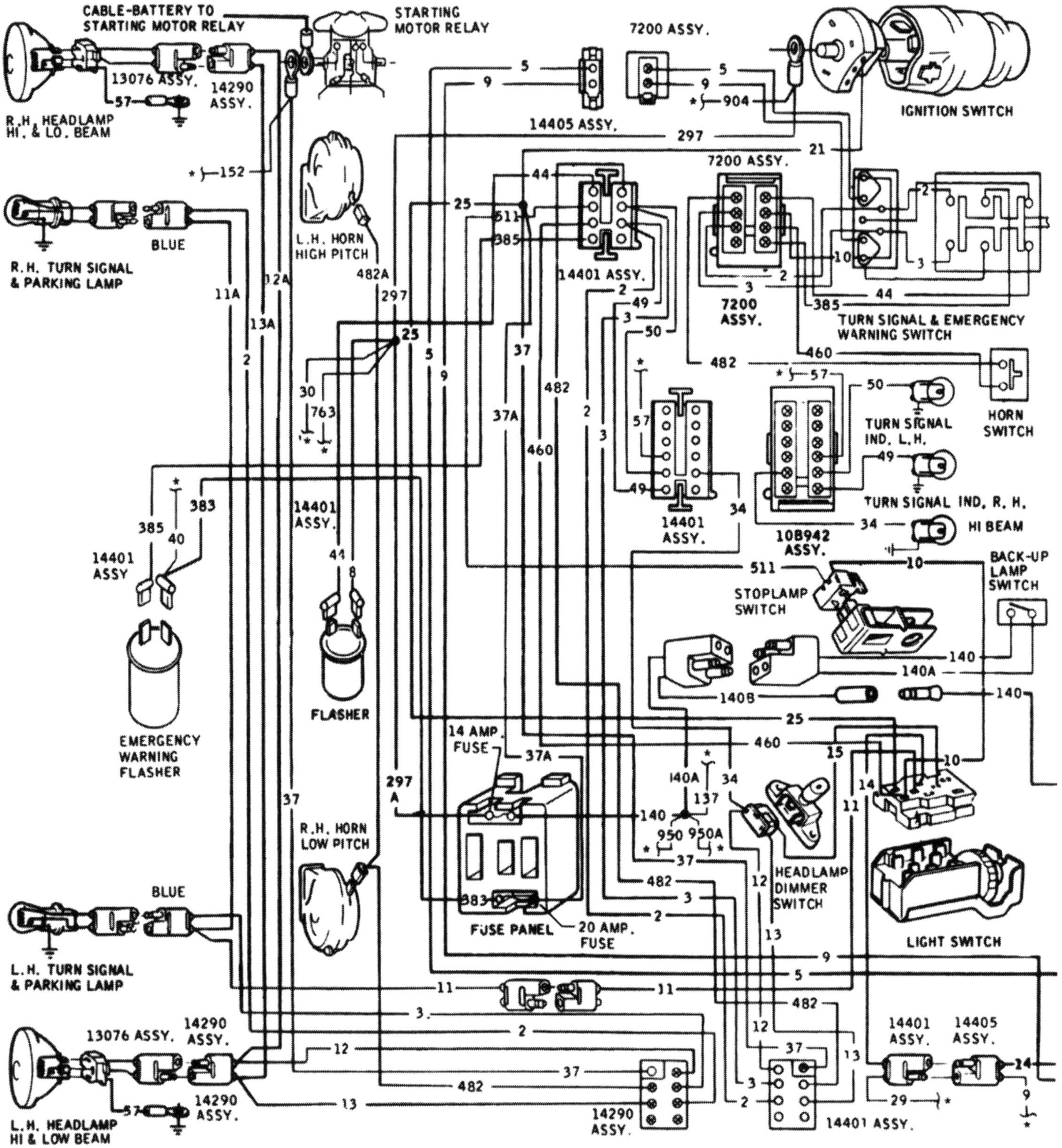

FIG 14:2 Exterior lighting

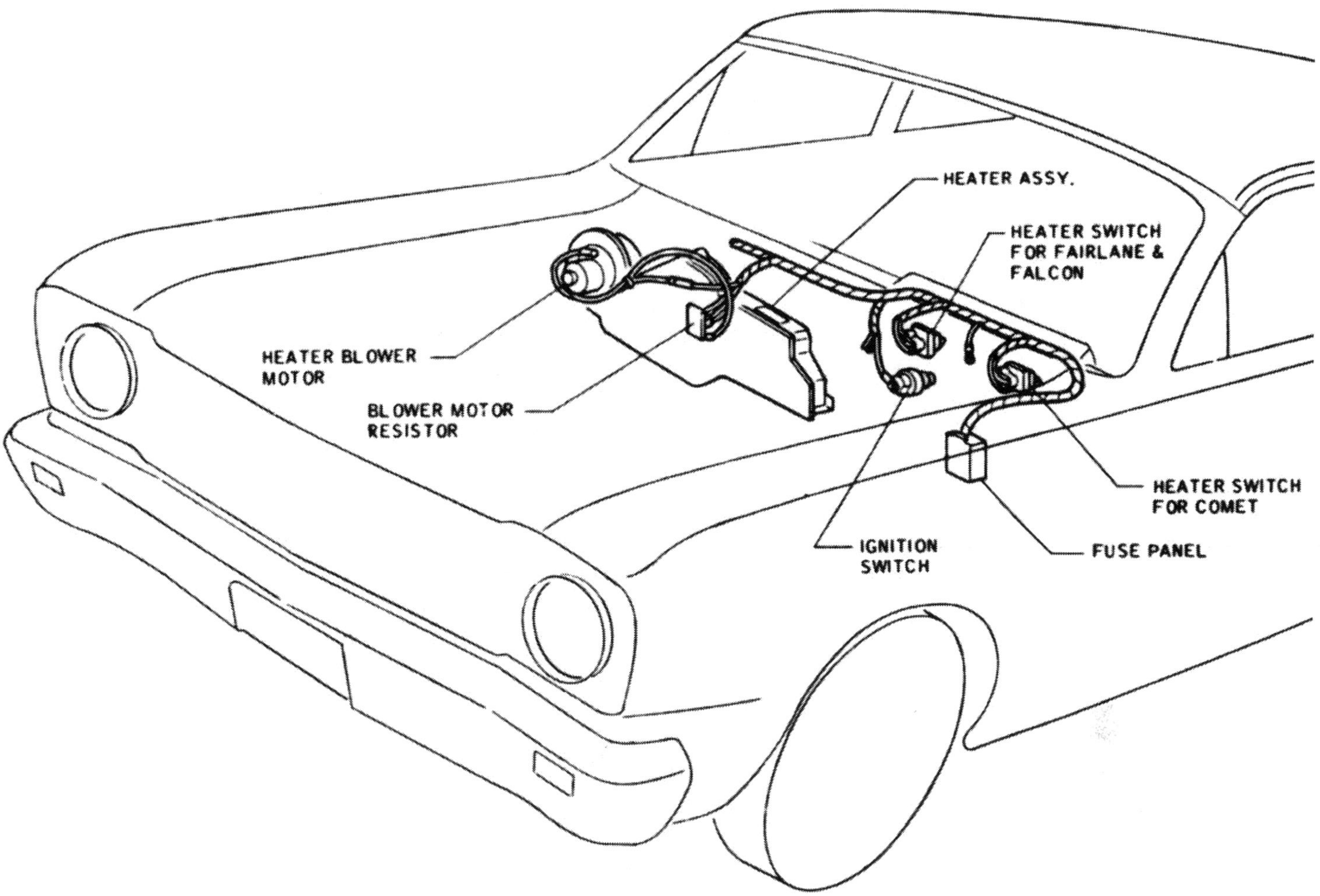

WIRING COLOR CODE

57A	57	BLACK	268	RED	297A	297	BLACK-GREEN STRIPE
	181	ORANGE	269	BLUE		913	YELLOW

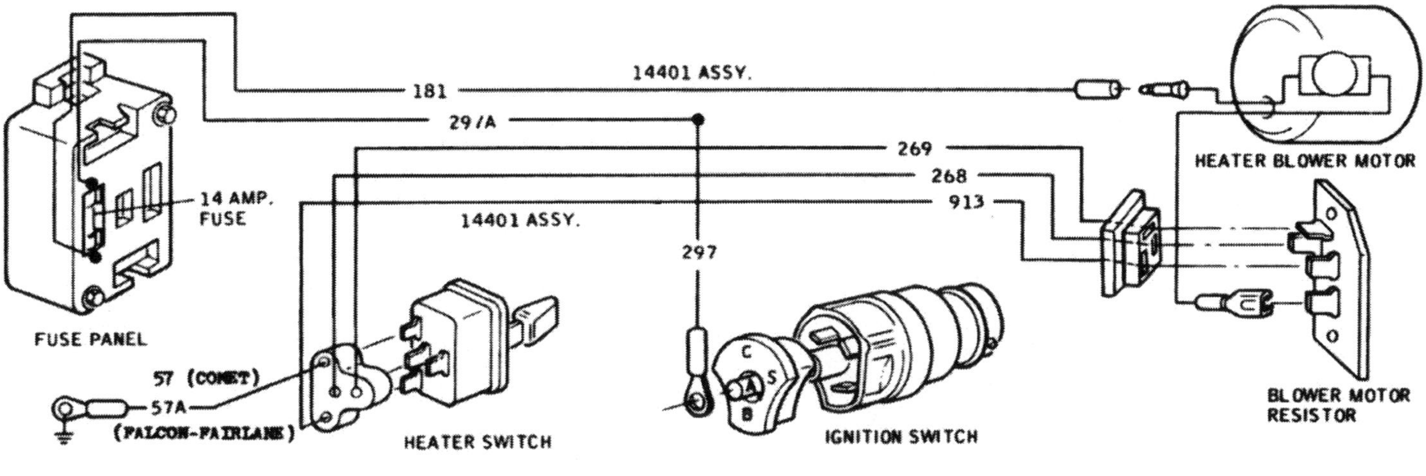

FIG 14:3 Heater

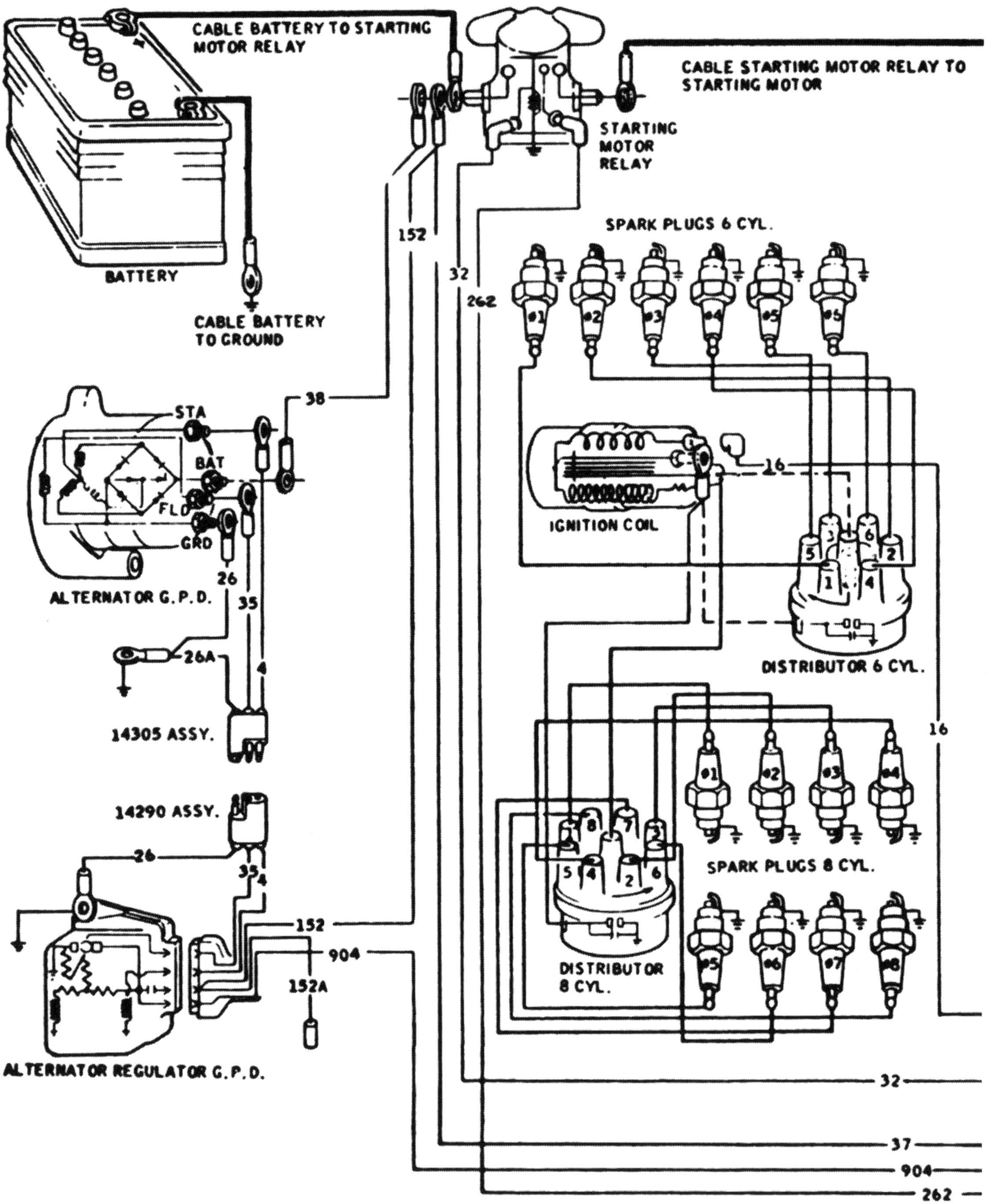

FIG 14:4 Ignition, charge, start

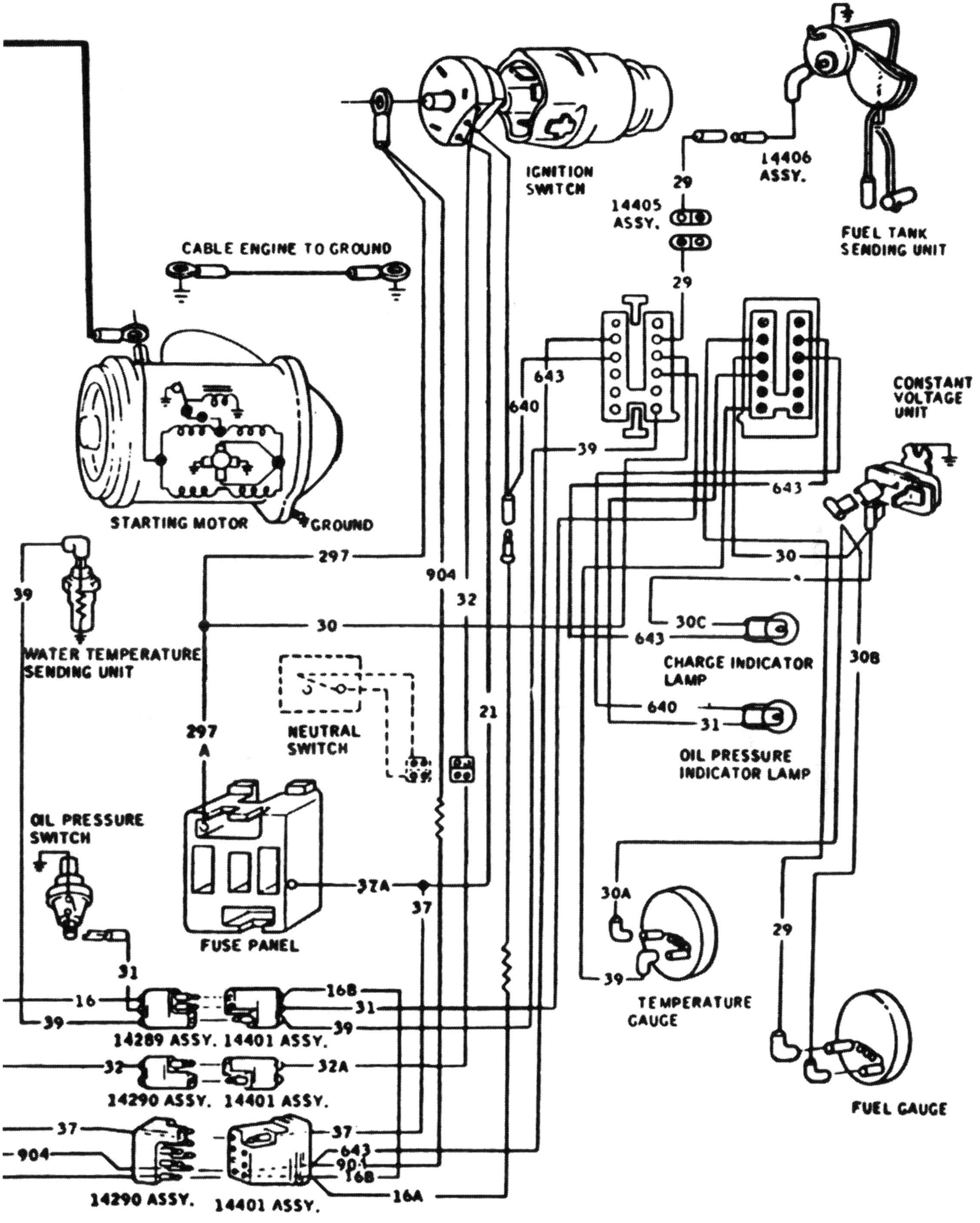

IGNITION SWITCH
14406 ASSY.
29
14405 ASSY.
FUEL TANK SENDING UNIT
CABLE ENGINE TO GROUND
643
640
39
CONSTANT VOLTAGE UNIT
STARTING MOTOR
GROUND
297
904
32
30
30C
CHARGE INDICATOR LAMP
30B
WATER TEMPERATURE SENDING UNIT
NEUTRAL SWITCH
21
297 A
640
31
OIL PRESSURE INDICATOR LAMP
OIL PRESSURE SWITCH
37A
37
30A
FUSE PANEL
TEMPERATURE GAUGE
16
16B
14289 ASSY.
14401 ASSY.
32A
14290 ASSY.
FUEL GAUGE
16A

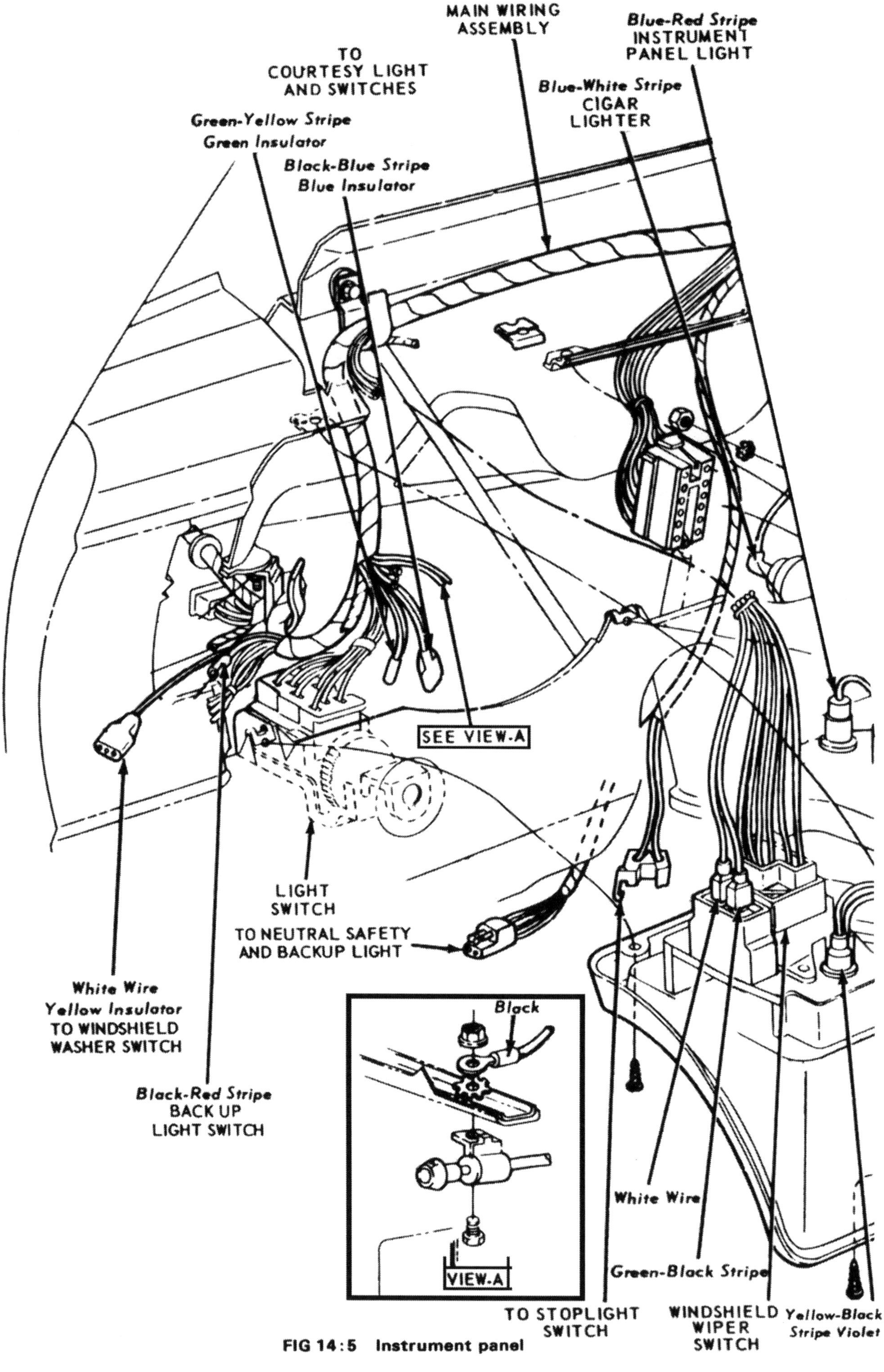

FIG 14:5 Instrument panel

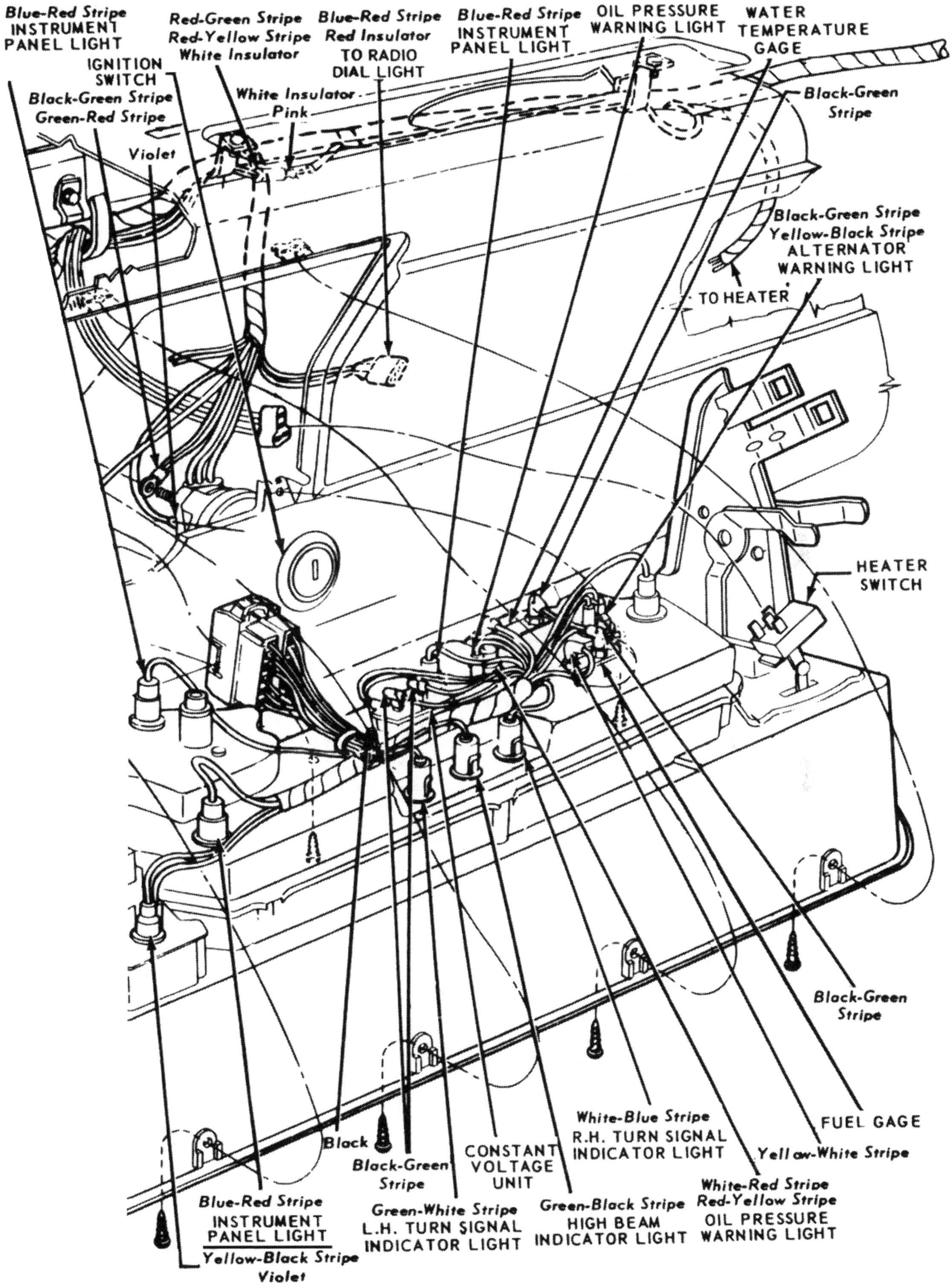
Blue-Red Stripe
INSTRUMENT
PANEL LIGHT
IGNITION
SWITCH
Black-Green Stripe
Green-Red Stripe
Violet
Red-Green Stripe
Red-Yellow Stripe
White Insulator
White Insulator
Pink
Blue-Red Stripe
Red Insulator
TO RADIO
DIAL LIGHT
Blue-Red Stripe
INSTRUMENT
PANEL LIGHT
OIL PRESSURE
WARNING LIGHT
WATER
TEMPERATURE
GAGE
Black-Green
Stripe
Black-Green Stripe
Yellow-Black Stripe
ALTERNATOR
WARNING LIGHT
TO HEATER
HEATER
SWITCH
Black-Green
Stripe
FUEL GAGE
Yellow-White Stripe
White-Blue Stripe
R.H. TURN SIGNAL
INDICATOR LIGHT
White-Red Stripe
Red-Yellow Stripe
OIL PRESSURE
WARNING LIGHT
Green-Black Stripe
HIGH BEAM
INDICATOR LIGHT
CONSTANT
VOLTAGE
UNIT
Green-White Stripe
L.H. TURN SIGNAL
INDICATOR LIGHT
Black
Black-Green
Stripe
Blue-Red Stripe
INSTRUMENT
PANEL LIGHT
Yellow-Black Stripe
Violet

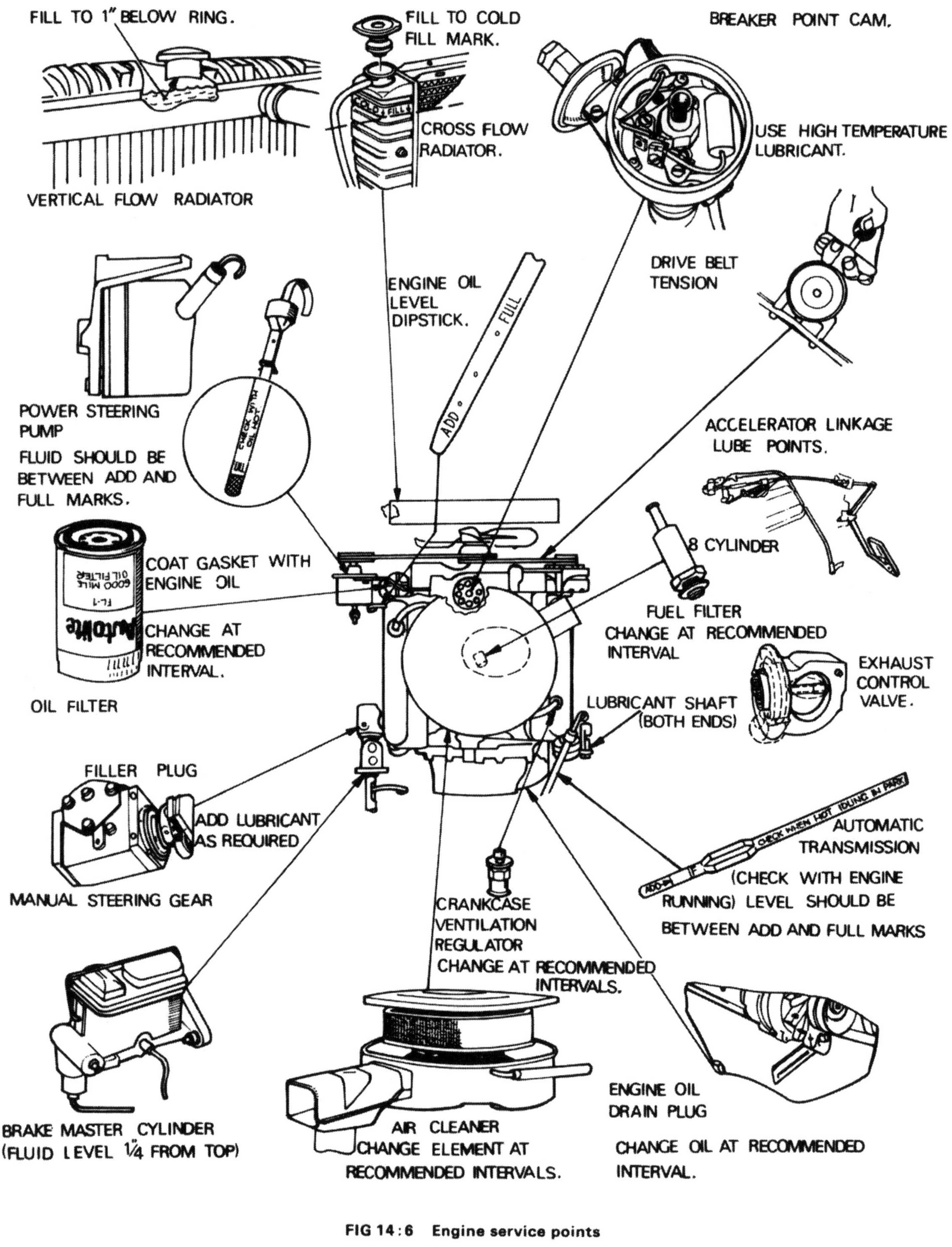

FIG 14:6 Engine service points

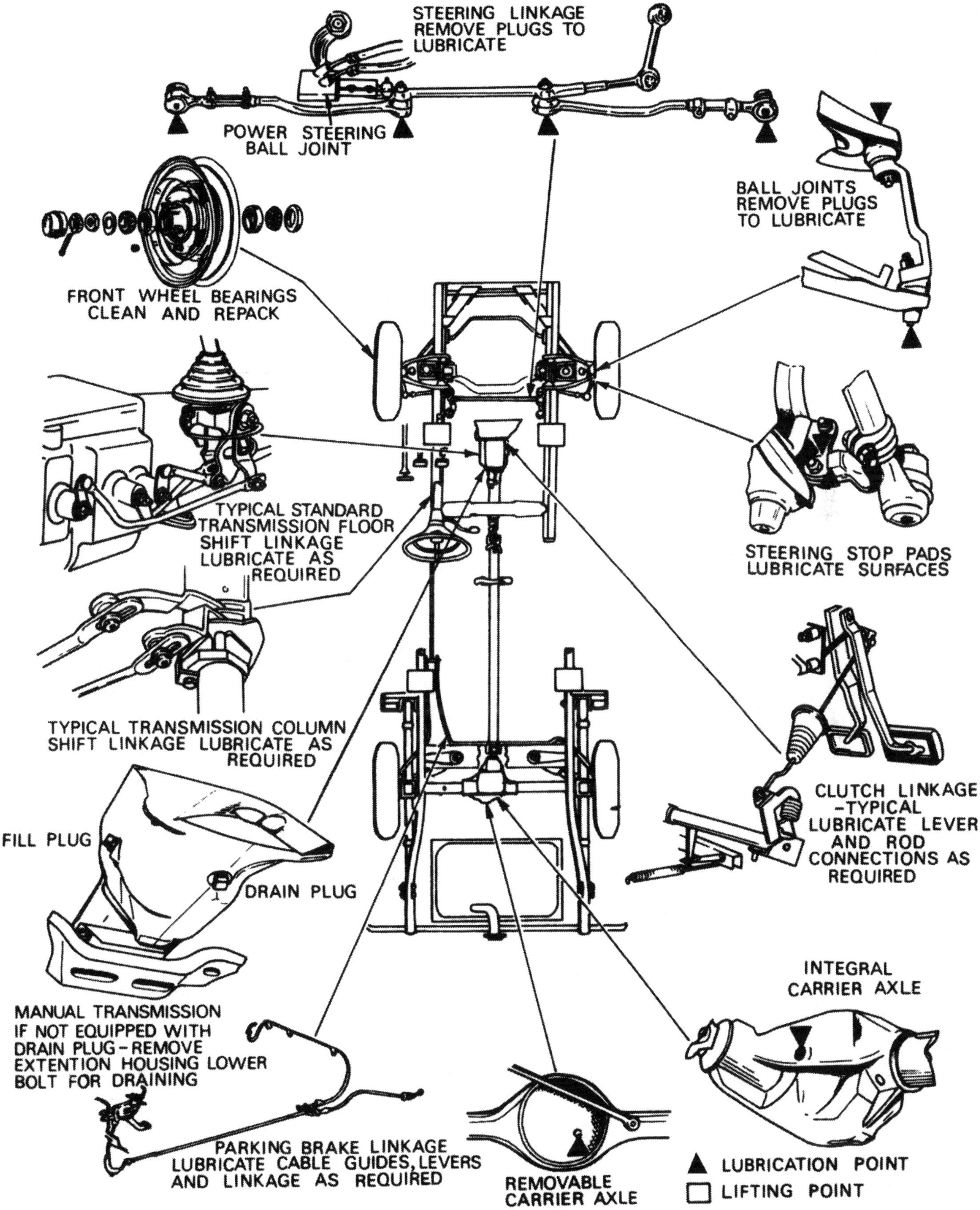

FIG 14:7 Chassis lube points

Glossary of Terms

Allen key Cranked wrench of hexagonal section for use with socket head screws.

Alternator Electrical generator producing alternating current. Rectified to direct current for battery charging.

Ambient temperature Surrounding atmospheric temperature.

Annulus Used in engineering to indicate the outer ring gear of an epicyclic gear train.

Armature The shaft carrying the windings, which rotates in the magnetic field of a generator or starter motor. That part of a solenoid or relay which is activated by the magnetic field.

Axial In line with, or pertaining to, an axis.

Backlash Play in meshing gears.

Balance lever A bar where force applied at the centre is equally divided between connections at the ends.

Banjo axle Axle casing with large diameter housing for the crownwheel and differential.

Bendix pinion A self-engaging and self-disengaging drive on a starter motor shaft.

Bevel pinion A conical shaped gearwheel, designed to mesh with a similar gear with an axis usually at 90 degree to its own.

bhp Brake horse power, measured on a dynamometer.

bmep Brake mean effective pressure. Average pressure on a piston during the working stroke.

Brake cylinder Cylinder with hydraulically operated piston(s) acting on brake shoes or pad(s).

Brake regulator Control valve fitted in hydraulic braking system which limits brake pressure to rear brakes during heavy braking to prevent rear wheel locking.

Camber Angle at which a wheel is tilted from the vertical.

Capacitor Modern term for an electrical condenser. Part of distributor assembly, connected across contact breaker points, acts as an interference suppressor.

Castellated Top face of a nut, slotted across the flats, to take a locking splitpin.

Castor Angle at which the kingpin or swivel pin is tilted when viewed from the side.

cc Cubic centimetres. Engine capacity is arrived at by multiplying the area of the bore in sq. cm by the stroke in cm by the number of cylinders.

Clevis U-shaped forked connector used with a clevis pin, usually at handbrake connections.

Collet A type of collar, usually split and located in a groove in a shaft, and held in place by a retainer. The arrangement used to retain the spring(s) on a valve stem in most cases.

Commutator Rotating segmented current distributor between armature windings and brushes in generator or motor.

Compression ratio The ratio, or quantitative relation, of the total volume (piston at bottom of stroke) to the unswept volume (piston at top of stroke) in an engine cylinder.

Condenser See 'Capacitor'.

Core plug Plug for blanking off a manufacturing hole in a casting.

Crownwheel Large bevel gear in rear axle, driven by a bevel pinion attached to the propeller shaft. Sometimes called a 'ring gear'.

'C'-spanner Like a 'C' with a handle. For use on screwed collars without flats, but with slots or holes.

Damper Modern term for shock absorber, used in vehicle suspension systems to damp out spring oscillations.

Depression The lowering of atmospheric pressure as in the inlet manifold and carburetter.

Dowel Close tolerance pin, peg, tube or bolt which accurately locates mating parts.

Drag link Rod connecting steering box drop arm (pitman arm) to nearest front wheel steering arm in certain types of steering systems.

Dry liner Thinwall tube pressed into cylinder bore.

Dry sump Lubrication system where all oil is scavenged from the sump and returned to a separate tank.

Dynamo See 'Generator'

Electrode Terminal part of an electrical component such as the points or 'Electrodes' of a sparking plug.

Electrolyte In lead-acid car batteries a solution of sulphuric acid and distilled water.

End float The axial movement between associated parts, end play.

EP Extreme pressure. In lubricants, special grades for heavily loaded bearing surfaces, such as gear teeth in a gearbox, or crown wheel and pinion in a rear axle.

Fade Of brakes. Reduced efficiency due to overheating.

Field coils Windings on the polepieces of motors and generators.

Fillets Narrow finishing strips usually applied to interior bodywork.

First motion shaft Input shaft from clutch to gearbox.

Fullflow filter Filters in which all the oil is pumped to the engine. If the element becomes clogged, a bypass valve operates to pass unfiltered oil to the engine.

FWD Front wheel drive.

Gear pump Two meshing gears in a close fitting casing. Oil is carried from the inlet round the outside of both gears in the spaces between the gear teeth and casing to the outlet. The meshing gear teeth prevent oil passing back to the inlet and the oil is forced through the outlet port.

Generator Modern term for 'Dynamo'. When rotated produces electrical current.

Grommet	A ring of protective or sealing material. Can be used to protect pipes or leads passing through bulkheads.
Grubscrew	Fully threaded headless screw with screwdriver slot. Used for locking or alignment purposes.
Gudgeon pin	Shaft which connects a piston to its connecting rod. Sometimes called 'wrist pin' or 'piston pin'.
Halfshaft	One of a pair transmitting drive from the differential.
Helical	In spiral form. The teeth of helical gears are cut at a spiral angle to the side faces of the gearwheel.
Hot spot	Hot area that assists vapourisation of fuel on its way to cylinders. Often provided by close contact between inlet and exhaust manifolds.
HT	High Tension. Applied to electrical current produced by the ignition coil for the sparking plugs.
Hydrometer	A device for checking specific gravity of liquids. Used to check specific gravity of electrolyte.
Hypoid bevel gears	A form of bevel gear used in the rear axle drive gears. The bevel pinion meshes below the centre line of the crownwheel, giving a lower propeller shaft line.
Idler	A device for passing on movement. A free running gear between driving and driven gears. A lever transmitting track rod movement to a side rod in steering gear.
Impeller	A centrifugal pumping element. Used in water pumps to stimulate flow.
Journals	Those parts of a shaft that are in contact with the bearings.
Kingpin	The main vertical pin which carries the front wheel spindle and permits steering movement. May be called 'steering pin' or 'swivel pin'.
Layshaft	The shaft which carries the laygear in the gearbox. The laygear is driven by the first motion shaft and drives the third motion shaft according to the gear selected. Called the 'countershaft' or 'second motion shaft'.
lb ft	A measure of twist or torque. A pull of 10lb at a radius of 1 ft is a torque of 10 lb ft.
lb/sq in	Pounds per square inch.
Little-end	The small, or piston end of a connecting rod. Sometimes called the 'small-end'.
LT	Low Tension. The current output from the battery.
Mandrel	Accurately manufactured bar or rod used for test or centring purposes.
Manifold	A pipe, duct, or chamber, with several branches.
Needle rollers	Bearing rollers with a length many times their diameter.
Oil bath	Reservoir which lubricates parts by immersion. In air filters, a separate oil supply for wetting a wire mesh element to hold the dust.
Oil wetted	In air filters, a wire mesh element lightly oiled to trap and hold airborne dust.
Overlap	Period during which inlet and exhaust valves are open together.
Panhard rod	Bar connected between fixed point on chassis and another on axle to control sideways movement.
Pawl	Pivoted catch which engages in the teeth of a ratchet to permit movement in one direction only.
Peg spanner	Tool with pegs, or pins, to engage in holes or slots in the part to be turned.
Pendant pedals	Pedals with levers that are pivoted at the top end.
Phillips screwdriver	The cross-point screwdriver for using with the cross-slotted heads of Phillips screws.
Pinion	A small gear in relation to another gear.
Piston-type damper	Shock absorber in which damping is controlled controlled by a piston working in a closed oil-filled cylinder.
Preloading	Preset static pressure on ball or roller bearings not due to working loads.
Radial	Radiating from a centre, like the spokes of a wheel.
Radius rod	Pivoted arm confining movement of a part to an arc of fixed radius.
Ratchet	Toothed wheel or rack which can move in one direction only, movement in the other being prevented by a pawl.
Ring gear	A gear tooth ring attached to outer periphery of flywheel. Starter pinion engages with it during starting.
Runout	Amount by which rotating part is out of true.
Semi-floating axle	Outer end of rear axle halfshaft is carried on bearing inside axle casing. Wheel hub is secured to end of shaft.
Servo	A hydraulic or pneumatic system for assisting, or, augmenting a physical effort. See 'Vacuum Servo'.
Setscrew	One which is threaded for the full length of the shank.
Shackle	A coupling link, used in the form of two parallel pins connected by side plates to secure the end of the master suspension spring and absorb the effects of deflection.
Shell bearing	Thinwalled steel shell lined with anti-friction metal. Usually semi-circular and used in pairs for main and big-end bearings.
Shock absorber	See 'Damper'.
Silentbloc	Rubber bush bonded to inner and outer metal sleeves.
Socket-head screw	Screw with hexagonal socket for an Allen key.
Solenoid	A coil of wire creating a magnetic field when electric current passes through it. Used with a soft iron core to operate contacts or a mechanical device.
Spur gear	A gear with teeth cut axially across the periphery.

Stub axle Short axle fixed at one end only.

Tachometer An instrument for accurate measurement of rotating speed. Usually indicates in revolutions per minute.

TDC Top Dead Centre. The highest point reached by a piston in a cylinder, with the crank and connecting rod in line.

Thermostat Automatic device for regulating temperature. Used in vehicle coolant systems to open a valve which restricts circulation at low temperature.

Third motion shaft Output shaft of gearbox.

Three-quarter floating axle Outer end of rear axle half-shaft flanged and bolted to wheel hub,which runs on bearing mounted on outside of axle casing. Vehicle weight is not carried by the axle shaft.

Thrust bearing or washer Used to reduce friction in rotating parts subject to axial loads.

Torque Turning or twisting effort. See 'lb ft'.

Track rod The bar(s) across the vehicle which connect the steering arms and maintain the front wheels in their correct alignment.

UJ Universal joint. A coupling between shafts which permits angular movement.

UNF Unified National Fine screw thread.

Vacuum servo Device used in brake system, using difference between atmospheric pressure and inlet manifold depression to operate a piston which acts to augment brake pressure as required. See 'Servo'.

Venturi A restriction or 'choke' in a tube, as in a carburetter, used to increase velocity to obtain a reduction in pressure.

Vernier A sliding scale for obtaining fractional readings of the graduations of an adjacent scale.

Welch plug A domed thin metal disc which is partially flattened to lock in a recess. Used to plug core holes in castings.

Wet liner Removable cylinder barrel, sealed against coolant leakage, where the coolant is in direct contact with the outer surface.

Wet sump A reservoir attached to the crankcase to hold the lubricating oil.

Made in United States
Orlando, FL
04 November 2026

90200402R00107